Discrimination Testing
in Sensory Evaluation

## A series of books on selected topics in the field of Sensory Evaluation

The first book in the Sensory Evaluation series is *Sensory Evaluation: A Practical Handbook*, published in May 2009. It focuses on the practical aspects of sensory testing, presented in a simple, 'how to' style for use by industry and academia as a step-by-step guide to carrying out a basic range of sensory tests. In-depth coverage was deliberately kept to a minimum. Subsequent books in the series cover selected topics in sensory evaluation. They are intended to give theoretical background, more complex techniques and in-depth discussion on application of sensory evaluation that were not covered in the *Practical Handbook*. However, they will seek to maintain the practical approach of the handbook and chapters will include a clear case study with sufficient detail to enable practitioners to carry out the techniques presented.

# Discrimination Testing in Sensory Evaluation

*Edited by*

*Lauren Rogers*
*Joanne Hort*
*Sarah E. Kemp*
*Tracey Hollowood*

WILEY

This edition first published 2024


*Registered Office(s)*
John Wiley & Sons, Inc., 111 River Street, Hoboken, NJ 07030, USA
John Wiley & Sons Ltd, The Atrium, Southern Gate, Chichester, West Sussex, PO19 8SQ, UK

For details of our global editorial offices, customer services, and more information about Wiley products visit us at www.wiley.com.

Wiley also publishes its books in a variety of electronic formats and by print-on-demand. Some content that appears in standard print versions of this book may not be available in other formats.



*Library of Congress Cataloging-in-Publication Data applied for:*

| | |
|---|---|
| Hardback: | 9780470671405 |
| ePDF: | 9781118635452 |
| epub: | 9781118635445 |
| oBook: | 9781118635353 |

Cover Design: Wiley
Cover Image: © Getty Images/iStockphoto

Set in 9.5/12.5pt STIXTwoText by Straive, Pondicherry, India
Printed and bound by CPI Group (UK) Ltd, Croydon, CR0 4YY

C9780470671405_020824

## Dedications

*This book is dedicated to:*

*Lawrence, Alexandra and James*
*George, Elizabeth, George and William*
*Campbell, Emma and Lara*
*Mike*

# Contents

# List of Contributors

***Cindy Beeren***
EdiBee
Altea
Spain

Whyteleafe Community Hub
Whyteleafe
UK

***Isabelle Cayeux***
Perception & Cognitive Neurosciences
Department, Science & Research,
dsm-firmenich
Geneva, Switzerland

***Caterina Dinnella***
Department of Agriculture, Food,
Environment and Forestry (DAGRI)
University of Florence
Florence
Italy

***Carlos Gómez-Corona***
Human Insights Department, Taste
Texture, and Health BU dsm-firmenich
Neuilly-sur-Seine
France

***Michael J. Hautus***
School of Psychology, Faculty of Science
The University of Auckland
Auckland
New Zealand

***Tracey Hollowood***
Formerly Managing Director of
Sensory Dimensions Ltd
UK

***Joanne Hort***
Food Experience and Sensory Testing
Laboratory (Feast)
Massey University
Palmerston North
New Zealand

***Danielle van Hout***
Aigora
Richmond, VA
USA

***Sarah E. Kemp***
Consultant and formerly Head of Global
Sensory and Consumer Guidance
Cadbury Schweppes
UK

***Min-A Kim***
Division of Food and Nutrition
Chonnam National University
Gwangju
South Korea

***Harry T. Lawless***
Cornell University
Ithaca, NY
USA

***Hye-Seong Lee***
Department of Food Science and Biotechnology, College of Engineering
Ewha Womans University
Seoul
South Korea

***Stella Lignou***
University of Reading
Reading
UK

***Linda Lopez***
S&T, Givaudan International
Kemptthal
Switzerland

***Curtis R. Luckett***
University of Tennessee
Knoxville, TN
USA

SCiFi Foods
San Leandro, CA
USA

***Michael Meyners***
Discovery & Innovation Platforms,
Procter & Gamble Service GmbH
Schwalbach am Taunus
Germany

***Karine Miot***
Sensory Department, Perfumery & Beauty
BU dsm-firmenich
London
United Kingdom

***Erminio Monteleone***
Department of Agriculture, Food, Environment and Forestry (DAGRI)
University of Florence
Florence
Italy

***Victoria Norton***
University of Reading
Reading
UK

***Lauren Rogers***
Freelance Sensory Scientist
UK

***Sara Spinelli***
Department of Agriculture, Food, Environment and Forestry (DAGRI)
University of Florence
Florence
Italy

***Bernard Thomas Carr***
Carr Consulting
Evanston, IL
USA

National Wine and Grape Industry Centre
Charles Sturt University
Wagga Wagga, New South Wales
Australia

## Preface to the Series

Sensory evaluation is a scientific discipline used to evoke, measure, analyze and interpret responses to products perceived through the senses of sight, smell, touch, taste and hearing (Anonymous 1975). It is used to reveal insights into the way in which sensory properties drive consumer acceptance and behavior and to design products that best deliver what consumers want. It is also used at a more fundamental level to provide a wider understanding of the mechanisms involved in sensory perception and consumer behavior.

Sensory evaluation emerged as a field in the 1940s. It began as simple "taste testing" typically used in the food industry for judging the quality of products such as tea, cheese, beer, and so on. From the 1950s to the 1970s, it evolved into a series of techniques to objectively and reliably measure sensory properties of products and was typically used to service quality assurance and product development. Through the 1980s and 1990s, the use of computers for data collection and statistical analysis increased the speed and sophistication of the field, so that sensory, consumer, and physicochemical data could be combined to design products that delivered to consumer needs.

Today, sensory evaluation is a sophisticated, decision-making tool that is used in partnership with marketing, research and development, and quality assessment and control throughout the product lifecycle to enable consumer-led product design and decision-making. Its application has spread from the food industry to many others, such as personal care, household care, cosmetics, flavors, fragrances, and the automotive industry. Although it is already widely used by major companies in the developed market, its use continues to grow in emerging markets, smaller companies, and new product categories, as sensory evaluation is increasingly recognized as a necessary tool for competitive advantage.

The field of sensory evaluation will continue to evolve and it is expected that faster, more flexible, and more sophisticated techniques will be developed. Social networking tools are transforming the way research is undertaken, enabling direct and real-time engagement with consumers. The use of sensory evaluation by marketing departments will continue to grow, particularly in leveraging the link between product sensory properties and emotional benefits for use in branding and advertising. Advances in other fields, such as genomics, brain imaging, and instrumental analysis, will be coupled with sensory evaluation to provide a greater understanding of perception.

Owing to the rapid growth and sophistication of the field of sensory evaluation in recent years, it is no longer possible to give anything but a brief overview of individual topics in a single general sensory science textbook. The trend is toward more specialized sensory books that focus on one specific topic, and to date, these have been produced in an ad hoc fashion by different authors/editors. Many areas remain uncovered.

We, the editors, wanted to share our passion for sensory evaluation by producing a comprehensive series of detailed books on individual topics in sensory evaluation. We are enthusiastic devotees of sensory evaluation, who are excited to act as editors to promote sensory science. Between us, we have over 120 years of industrial and academic experience in sensory science, covering food, household, and personal care products in manufacturing; food service; consultancy; and provision of sensory analysis services at local, regional, and global levels. We have published and presented widely in the field; taught workshops, short courses, and lecture series; and acted as reviewers, research supervisors, thesis advisors, project managers and examiners. We have been active in many sensory-related professional bodies, including the Institute of Food Science and Technology Sensory Science Group, of which many of us are past Chairs, the European Sensory Science Society, of which one of us is a past Chair, the Institute of Food Technologists, the British Standards Institute, and ASTM International, to name but a few. As such, we are well placed to have a broad perspective of sensory evaluation and pleased to be able to call on our network of sensory evaluation colleagues to collaborate with us.

The book series Sensory Evaluation covers the field of sensory evaluation at an advanced level and aims to:

- be a comprehensive, in-depth series on sensory evaluation
- cover traditional and cutting-edge techniques and applications in sensory evaluation using the world's foremost experts
- reach a broad audience of sensory scientists; practitioners; and students by balancing theory, methodology, and practical application
- reach industry practitioners by illustrating how sensory can be applied throughout the product life cycle, including development, manufacture, supply chain, and marketing
- cover a broad range of product applications, including food, beverages, personal care, and household products.

Our philosophy is to include cutting-edge theory and methodology, as well as illustrating the practical application of sensory evaluation. As sensory practitioners, we are always interested in how methods are actually carried out in the laboratory. Often, key details of the practicalities are omitted in journal papers and other scientific texts. We have encouraged authors to include such details in the hope that readers will be able to replicate methods themselves. The focus of sensory texts often tends to be food and beverage products assessed using olfaction and taste. We have asked authors to take a broad perspective to include non-food products and all the senses.

The book series is aimed at sensory professionals working in academia and industry, including sensory scientists, practitioners, trainers, and students; and industry-based professionals in marketing, research and development, and quality assurance/control, who need to understand sensory evaluation and how it can benefit them. The series is suitable as:

- reference texts for sensory scientists, from industry to academia
- teaching aids for senior staff with responsibility for training in an academic or industrial setting

- course books, some of which to be personally owned by students undertaking academic study or industrial training
- reference texts suitable across a broad range of industries; for example, food, beverages, personal care products, household products, flavors, fragrances.

The first book in the series, *Sensory Evaluation: A Practical Handbook*, was published in May 2009 (Kemp et al. 2009). This book focuses on the practical aspects of sensory testing, presented in a simple, "how to" style for use by industry and academia as a step-by-step guide to carry out a basic range of sensory tests. In-depth coverage was deliberately kept to a minimum. Further books in the series cover the basic methodologies used in the field of sensory evaluation: discrimination testing, descriptive analysis (Kemp et al. 2018), time-dependent measures of perception (Hort et al. 2017), and consumer research. They give theoretical background, more complex techniques, and in-depth discussion on application of sensory evaluation, while seeking to maintain the practical approach of the handbook. Chapters include clear case studies with sufficient detail to enable practitioners to carry out the techniques presented. Later books will cover a broad range of sensory topics, including applications and emerging trends.

The contributors we have selected are world-renowned scientists and leading experts in their field. Where possible, we have used originators of techniques. We have learned a lot from them as we have worked with them to shape each book. We wish to thank them for accepting our invitation to write chapters and for the time and effort they have put in to making their chapters useful and enjoyable for readers.

We would also like to thank our publisher, Wiley Blackwell, and the team for seeing the potential in this series and helping us bring it to fruition. We would also like to thank the anonymous reviewers of the series for their constructive comments.

We hope you will find the Sensory Evaluation book series both interesting and beneficial, and enjoy reading it as much as we have producing it.

*Sarah E. Kemp*
*Joanne Hort*
*Tracey Hollowood*
*Lauren Rogers*

## References

Anonymous (1975). *Minutes of Division Business Meeting. Institute of Food Technologists – Sensory Evaluation Division*. Chicago, IL: IFT.

Kemp, S.E., Hollowood, T., and Hort, J. (2009). *Sensory Evaluation: A Practical Handbook*. Oxford: Wiley-Blackwell.

Kemp, S.E., Hollowood, T., and Hort, J. (2018). *Descriptive Analysis in Sensory Evaluation*. Oxford, UK: Wiley-Blackwell.

Hort, J., Kemp, S.E., and Hollowood, T. (2017). *Time-Dependent Measures in Sensory Evaluation*. Oxford, UK: Wiley-Blackwell.

# Preface

Discrimination tests are one of the key sets of methods in a sensory scientist's toolbox. The aim of this book is to provide a comprehensive and up-to-date overview of the area, the tests, and their applications.

Discrimination tests are covered in classic general sensory science texts, including Meilgaard et al. (2016), Lawless and Heymann (2010), and Stone et al. (2020). These have limited space to give to the topic, which makes it difficult to strike a balance between theory and practical application. To the editors' knowledge, there are two previous publications devoted to discrimination tests. Bi (2015, updated from the first edition in 2008) produced a very detailed book that thoroughly covers the statistical aspects of discrimination testing, as well as information about consumer emotions, time intensity, and shelf-life. Rogers (2017) edited a practical handbook for sensory scientists with chapters and case studies dedicated to the most popular sensory discrimination tests.

The editors saw a need for a book devoted to discrimination testing that would provide in-depth theoretical and practical coverage of traditional and recently developed approaches, as well as cover several of the many applications of sensory science such as food, home and personal care, and ingredient testing. In addition, the book also considers discrimination testing with different subjects and those with differing discrimination abilities, as well as replicated discrimination tests and the use of discrimination tests in determining sensory quality. The scope of this book includes history, theory, techniques and applications of discrimination tests. The book also includes case studies to bring the applications to life, making it an essential reference for sensory scientists.

The book is structured in two sections. Section 1 is an introduction covering general topics in discrimination testing, including history, general considerations, determining discrimination abilities, similarity, and approaches to replicated tests. This section also delves into Thurstonian modeling and the use of sureness judgments to quantify the size of the sensory differences. Section 2 covers some applications of discrimination testing such as the use of children to conduct the tests, the use of attribute-specific tests with many samples, and conducting discrimination tests in the flavors and fragrances industries.

Each chapter includes theory, psychological aspects, methodology, statistical analysis, applications, practical considerations, case studies and examples, future developments, and a reference list. The aim is to give a balance between theory and practice, with enough theory for readers to fully understand the background and underlying mechanisms of the technique, and in many instances enough detail to enable the reader to carry out the methodology.

We thank all authors for giving their time and effort to their chapter despite their busy schedules, and for their patience with the process.

We hope you find this book as interesting and beneficial to read as we did to produce.

*Lauren Rogers*
*Tracey Hollowood*
*Joanne Hort*
*Sarah E. Kemp*

## References

Bi, J. (2015). *Sensory Discrimination Tests and Measurements: Sensometrics in Sensory Evaluation*. Wiley.

Bi, J. (2008). *Sensory Discrimination Tests and Measurements: Statistical Principles, Procedures and Tables*. Wiley.

Meilgaard, M., Civille, G., and Carr, B. (2016). *Sensory Evaluation Techniques*. Boca Raton, Florida: Taylor & Francis.

Lawless, H.T. and Heymann, H. (2010). *Sensory Evaluation of Food. Principles and Practices. Sensory Evaluation of Food – Principles and Practices*. New York: Springer.

Rogers, L. (2017). *Discrimination Testing in Sensory Science: A Practical Handbook*. Woodhead Publishing.

Stone, H., Bleibaum, R., and Thomas, H.A. (2020). *Sensory Evaluation Practices*. Academic press.

## Editor Biographies

***Tracey Hollowood BSc (Hons), PhD,*** is the former director of Sensory and Consumer Research for Sensory Dimensions Ltd, she has over 25 years of academic and industrial experience and has been published extensively. She established the UK's first postgraduate certificate in Sensory Science and managed Nottingham University's Sensory Science Centre. She was a previous chair of the Institute of Food Science and Technology (IFST) Midland Branch and the Sensory Science Group (SSG).

***Joanne Hort, BEd (Hons), PhD, CSci, FIFST, MNZIFST, RSensSci,*** is the Fonterra-Riddet Chair of Consumer and Sensory Science at Massey University in New Zealand (NZ) following on from her various academic roles, latterly SABMiller Chair of Sensory Science at the University of Nottingham. Initially, Professor Hort studied food technology and began her career in teaching. However, she returned to university to receive her doctorate concerning the modelling of the sensory attributes of cheese from analytical and instrumental measures in 1998 and then took up a post as lecturer at Sheffield Hallam University. She was appointed as a lecturer in Sensory Science at the University of Nottingham in 2002. There she established the University of Nottingham Sensory Science Centre, obtaining her Chair in 2013. Her multidisciplinary approach combining analytical, brain imaging and sensory techniques provided rich insight into multisensory interactions, individual variation, temporal changes in flavor perception and the emotional response to sensory properties. In 2017, she took the position of inaugural Fonterra Riddet Chair in Consumer and Sensory Science at Massey University where she has established the Food Experience and Sensory Testing (Feast) lab, complete with digital immersive space for research on emotional response in ecologically valid environments. She has over 120 publications and sits on the editorial board for Food Quality and Preference. She is a fellow of the Institute of Food Science and Technology (IFST), member of the NZIFST and NZ Royal Society. She is vice president of the Pangborn Sensory Science Trust, a founder member and past chair of the European Sensory Science Society, and a founder member and past chair of the UK IFST Sensory Science Group.

***Sarah E. Kemp, BSc (Hons), PhD, CSci, FIFST, RSensSci,*** is a chartered sensory and consumer science professional with more than 35 years of experience in academia and industry. Dr Kemp gained a BSc in Food Technology in 1986 and a PhD in Taste Chemistry in 1989 from the Food Science and Technology Department at the University of Reading, UK. In

1990, she did a postdoctoral research fellowship in sensory science at the Monell Chemical Senses Center in Philadelphia, USA. Dr Kemp has held many positions in industry, including manager of Sensory Psychology (US) and director of European Consumer and Marketing Research (France) in the Fragrance Division at Givaudan; product area leader and sensory science leader in Foods Consumer Science at Unilever Research, Colworth, UK; head of Global Sensory and Consumer Guidance at Cadbury Schweppes, UK; and director of Sensory and Consumer Services at Reading Scientific Services Limited, UK. Dr Kemp has also set up and run her own consultancy service and catering company. She has written numerous scientific articles in the field of sensory evaluation, has provided sensory training courses, including lecturing on the European Masters Course in Food Science, and has worked on bodies developing standards in sensory evaluation, including ASTM and the British Standards Institution, where she is chair of Committee AW/012 Sensory Analysis. She is a fellow of the Institute of Food Science and Technology and a founder member, past chair, and examiner for the IFST's Sensory Science Group, as well as being a member of other professional sensory societies and a past Governor of East Kent College, UK.

***Lauren Rogers*** is a freelance sensory scientist living in the beautiful Staffordshire Peak District. Lauren fell in love with sensory science in the early 1990s when she worked for Dalgety Plc, a food ingredients company. Later, she worked for GlaxoSmithKline, mainly on beverages such as Ribena and Horlicks. Lauren became a freelancer in 2009 and, as well as working for several different clients on many different product types and many different qualitative and quantitative methods, Lauren lectures on sensory science at the University of Nottingham and Le Cordon Bleu. Lauren is particularly interested in the sensory, consumer, and emotional attributes of brands; sensory shelf-life; panel performance; and sensory claim substantiation. She has experience in many food products including ingredients, snack products, drinks, and pet-food. She also has worked on many home and personal care products such as shampoos, conditioners, toothpastes, and creams. Lauren is a fellow of the Institute of Food Science and Technology (IFST) and an active member of the IFST's Sensory Science Group, maintaining their sensory science qualifications. She is also a member of the Market Research Society, the Sensometrics Society, the Society of Sensory Professionals and the ASTM E-18 (Sensory) Committee. Lauren is a member of the British Standards Society working on various international sensory science standards. Lauren has previously published two books, *Sensory Panel Management* and *Discrimination Testing in Sensory Science.*

# Section 1

# Introduction

# 1

# Introduction

*Sarah E. Kemp[1], Tracey Hollowood[2], Joanne Hort[3] and Lauren Rogers[4]*

[1] *Consultant and formerly Head of Global Sensory and Consumer Guidance, Cadbury Schweppes, UK*
[2] *Formerly Managing Director of Sensory Dimensions Ltd, UK*
[3] *Food Experience and Sensory Testing Laboratory (Feast), Massey University, Palmerston North, New Zealand*
[4] *Freelance Sensory Scientist, UK*

## 1.1 Introduction

Sensory discrimination tests are methodologies used in fields such as foods and beverages, home and personal care, cosmetics, and consumer goods to assess whether two samples are perceptibly different. The two samples being tested generally come from two production batches. These different batches might use, for example, different ingredients, different manufacturing processes, or different packaging. Discrimination tests are also used to evaluate and quantify individuals' abilities to perceive and differentiate various sensory attributes or a holistic difference between products. Participants are typically presented with pairs of products and are tasked with identifying differences or similarities in sensory characteristics, such as appearance, aroma, taste, flavor or texture. These tests play a crucial role in quality control, product development, and market research, for example, providing valuable insights into consumer preferences and aiding in the improvement of product formulations based on perceived sensory attributes.

### 1.1.1 Discrimination Testing as a Tool

Imagine that you are the new product development manager for a major snack manufacturer and have been tasked with an innovative global wellness project to further reduce the levels of salt and fat in the company's flagship product. Several designs of experiment studies are currently being discussed by the team and the next step is to discuss the project timings for the assessment of the sensory aspects of the products arising from these experiments. You have a meeting with the sensory manager to discuss the plans and the sensory methods available to help meet the project's action standards. There are several classes of sensory methods available to you, including descriptive, temporal and consumer methods, however, you agree

*Discrimination Testing in Sensory Evaluation*, First Edition. Edited by Lauren Rogers, Joanne Hort, Sarah E. Kemp and Tracey Hollowood.

with the global sensory manager that discrimination tests[1] are the best choice for this stage of the project as the differences between products should be slight: there is no point in conducting a discrimination test on products that are obviously different. The sensory manager has explained that if you need to know *how* products are different, you will need a different method (for example, quantitative descriptive analysis or free choice profiling). The sensory discrimination tests will allow you to determine if consumers will notice the difference between the chemically and physically different products from the experimental design studies and the current gold standard products.

To use a sensory discrimination test, three things are usually required: people to assess the samples, the choice of the test and some samples to assess. Obviously, a sensory scientist is also required to choose and administer the test, as well as to analyze and report the results. The person who takes part in the test is often someone who has been screened for their sensory acuity (ISO 2023) and trained in the test method, although they might also be naïve consumers. By training in the method first, the participant is focused on the items in the test and not the execution of the test itself. This person assesses the items presented to them using the training they received, answering the question posed by the test; for example, selecting the odd sample in a discrimination test known as a triangle test (Sinkinson 2017) or rating the difference between samples on a scale (Whelan 2017). The results from all participants in the test are then analyzed to make a final decision based on the objectives and action standard(s).

There are many different sensory discrimination tests available to the sensory scientist to help meet the types of objectives they might encounter. The choice of test will depend on several factors, with the primary considerations being the type of product being tested and the objectives of the test. There is a tendency for businesses to habitually use one discrimination test for all objectives, but not all objectives are equal: for an interesting discussion about the choice of sensory test related to the objective(s) of the research, see Lawless and Heymann (2010, pp. 8–10).

When the focus of the discrimination test is to answer a product-related question, the intended use of the results and the test's position within the project stages are important factors in the choice of test (Meilgaard et al. 2016).

## 1.2 Developments of Discrimination Testing

Sensory discrimination testing began early in the 20th century and was originally referred to as difference testing (Meiselman et al. 2022), however, the initial "discrimination tests" were conducted by Weber and Fechner in the early 19th century to examine the relationship between physical stimuli and sensory experience (David 1963). This field of study, now referred to as psychophysics, was characterized by the researchers' belief that sensations could not be measured directly. Consequently, they employed a variety of indirect methods and experiments to gauge perception. Weber's experiments in 1834 explored the

1 Ideally, sensory discrimination tests should be referred to as sensory discrimination "methods", however, the word "test" has been used so frequently that it is in common usage, and so the word 'test' has been used throughout this chapter.

"just noticeable difference" between different weights using blindfolded subjects (Holden et al. 2011). He found that the detectable difference was proportional to the initial weight. This method has since become known as "the method of constant stimuli" and bears resemblance to paired comparison methods such as the 2-alternative forced choice test commonly used in sensory science today.

Fechner expanded on Weber's findings, formulating Weber's law and introducing two additional methods: "the method of limits" and "the method of adjustment or average error." Although the latter method is challenging to implement in food research, as it requires subjects to adjust levels of ingredients such as salt, the method of limits – which involves changing the stimulus incrementally and asking subjects if they perceive any sensation – is akin to same-different tests and the ascending forced-choice method of limits employed to determine thresholds today. Thus, the technique of comparing pairs of stimuli began with Weber and was further refined by Fechner (Fechner 1860; Lawless and Heymann 2010).

Louis Thurstone (1927) also made extensive use of paired comparison tests in his psychophysical work to describe the discrimination process, which has been of great interest in the field of sensory discrimination testing. Researchers employ Thurstonian modelling and signal detection theory (SDT) to study differences in people's sensitivity to specific stimuli.

Perhaps the earliest example of a sensory discrimination test in the literature with a focus on food is Fisher's famous article on tea tasting, which would become the cornerstone reference for hypothesis testing. The test originated in the 1920s in Cambridge, when a group of friends, including Ronald Fisher, were debating the proper way to pour tea and whether the milk should be added to the cup before or after the tea. One member of the group claimed that she could discern whether the milk had been added first or last. To test this claim, Fisher designed an experiment to determine whether she could accurately identify the order in which the milk was added (Fisher 1935). The original documentation provides an interesting account of the experiment's design, statistical analysis, and results. Although the method employed in the tea testing was not given a name, Gridgeman (1959) referred to it as a "double-tetrad sorting design." Today, we would likely classify it as an octad or under the M + N tests (Lockhart 1951).

Of course, food discrimination testing has been in use for much longer than the past nine decades. Sensory tests, though informally, have been employed to assess food edibility and check drinking water quality for centuries. Additionally, nonfood testing has also been conducted throughout history, such as evaluating the suitability of housing, e.g. caves and weapons, e.g. assessing the sharpness of flint tools (Meilgaard et al. 2016).

Ranking is possibly the earliest sensory discrimination method, with references to its use for assessing preferences for eggs and sweetcorn in very early studies (as reported by Morse 1942). This type of grading, although not a formal sensory test, was one of the first methods used to assess the quality of a product. Experts would assess the quality based on a grading system, and these grading experts were possibly the first example of panel training and maintenance in the literature. However, there was no mention of screening for sensory acuity, and experts checked their assessments against those of their colleagues or standard samples. Some of these grading methods are still used for tea, wine, and coffee (Kilcast 2010).

The first *named* discrimination testing method was published in 1936 by Sylvia Cover, who wanted to determine if the cooking temperature of roasts made the meat more or less tender (Cover 1936). The new method was referred to as "the paired-eating method," which was based on the name "paired-feeding method," which was used in animal husbandry. The paired comparison method, which Cover used, had been in use for many years in other disciplines, including psychology, statistics, mathematics, and economics, however, Cover was the first to apply it in the context of sensory science to test food (David 1963).

Cover's paper on the assessment of meat tenderness is an interesting read, and the method she developed is now known as "the paired comparison." This discrimination test is actually referred to as the first published method in sensory science. Cover presented the judges with two carefully selected samples, known as "paired bites," in her pioneering discrimination testing method for meat tenderness assessment. To ensure the reliability of the testing, the cuts were taken from the left and right sides of the same animal, and the bite-sized pieces were from the same muscle type. Despite the research having excellent sampling control, the use of three-digit codes or balanced designs was yet to appear. Cover did not disclose how many judges were employed, nor whether they were screened or trained, although she conducted 261 paired comparisons over a period of two years. According to Cover, the advantages of using the paired-eating method were its simplicity and the ability to directly compare samples. She used the binomial method to determine the statistical significance of her results but did not mention this as an advantage.

In 1939, Cover made several improvements to her method based on suggestions from colleagues, aiming to enhance its applicability to various aspects of food research (Cover 1940). Among the significant modifications was the alteration of the sample presentation order, which Cover determined by employing a deck of playing cards, effectively mitigating any potential bias introduced by her technician. To minimize the occurrence of expectation errors, Cover removed meat identification information from the sheet provided to the judges. In addition to the binomial method she had previously employed, Cover employed the Chi-square method to analyze the data. Addressing the issue of judge selection, training, and the number of judges, Cover stated in her publication that no method had yet been devised to identify individuals who would serve as superior judges. Consequently, Cover recommended the involvement of more than two judges for each experiment, and for the majority of tests, she employed a panel of six judges. Each judge evaluated between eight and 11 pairs of the same two samples. Cover could not have anticipated that the analysis of replicated discrimination tests would remain a topic of contention even in contemporary times.

If the paired comparison method is considered to be the first sensory discrimination test in the field of food research, the second test is believed to be the triangle test, which was independently developed by two groups of researchers. The Seagram Quality Research Laboratory developed the test in 1941–1942 but did not publish the details until later (Peryam and Swartz 1950), while the Carlsberg Breweries Research Laboratory developed the test in 1946 (Bengtsson and Helm 1946; Helm and Trolle 1946). The 1946 papers refer to earlier publications that also discuss methodologies and statistics, although it is unclear if the triangle test was mentioned. Both groups of authors refer to the "triangular test" as having been in use for several years, suggesting its origin a few years earlier.

Carlsberg Breweries published two papers on the triangle test (Bengtsson and Helm 1946; Helm and Trolle 1946), which are noteworthy in a historical sense beyond discrimination testing. These papers contain interesting references to consumer testing, or "mass testing" as it was referred to, and include photographs and a consumer questionnaire with just two questions (six if personal information is included). The publications also describe the essential elements to consider before conducting a sensory study, which are now commonly taken for granted. It is challenging to imagine the difficulties faced by food scientists at that time when evaluating new products or changes in ingredients without such tools and advice.

This early version of the triangle test involved giving each judge three bottles of beer labelled with a number or letter, where two of the bottles contained the same beer and the third contained a different beer. The appearance of the bottles and beers was identical except for the identifying label. The tasters poured the beer into three glasses and were asked to identify "which two samples are identical?" The presentation design of the six possible A and B combinations was randomized across four replicate tests, and the judges worked independently in a room kept at 20°C. Additionally, judges were asked for their preference after the test and were informed whether their choice was correct or incorrect, both of which are no longer recommended. The statistical analysis was conducted using a table of values created by Bengtsson, who adapted the Chi-square analysis for use with triangle tests. The judges were allowed to specify that they could not detect a difference, resulting in three possible answers for each test: the number correct, the number incorrect, and those who could not detect any difference.

Another difference between the early and modern versions of the test is that each judge participated in each test around 24 times, and there was no provision for replicated testing in the analysis. However, the authors aimed to select "expert" tasters and needed this type of data to determine each individual's tasting ability.

The description of the triangle test as outlined by Peryam and Swartz in 1950 is more similar to the method employed today. It entailed presenting three samples to the judge, with two of them being identical and the third being different, and the judge was asked to identify the odd sample. However, there is one major difference to the test in use today, in that the control sample would always be presented twice, resulting in only three presentation designs in total. This is markedly different from the Bengtsson, Helm, and Trolle's description of the test, who used all six presentation designs as we do today. The triangle test was also frequently used by the authors for determining preference, which is now avoided.

In its early days, the triangle test was referred to by various names, such as the Helm technique, the trio comparison, the triad, the triangular test, the odd sample method, the oddity (which understandably fell out of favor quickly), and the three-glass test. The ISO standard was still referred to as the "triangular test" until the name triangle became officially adopted in the title in 2004. In 1948, Roessler and colleagues at the University of California developed the first table of critical values for the triangle test, which must have been a significant benefit to researchers at the time, given the absence of computers (Roessler et al. 1948).

The Seagram Quality Research Laboratory paper (Peryam and Swartz 1950) also describes two additional discrimination tests: the duo-trio and the dual-standard. These tests were likely developed in 1941–1942 alongside the triangle test. The duo-trio involves presenting judges with three samples, one of which is labelled as the control and the other two coded: one sample is the test sample and the other a repeat of the control. The judge was asked to identify the sample that was different from the control. In the dual standard, four samples are presented in pairs, with the judge becoming familiar with the differences between the first pair before being presented with coded versions and asked to identify which sample is like each standard.

The difference-preference test was developed in the 1940s (Dove 1947). This test was part of the "Subjective-Objective Approach," which aimed to emphasize the importance of subjective assessments in food testing. The test involved paired comparisons with an added preference question using a 10-point scale. The author also listed requirements for the laboratory, sample preparation, and judges, including selecting judges based on vocabulary, experience, and ability to detect small differences. The paper also describes taste tests with animals instead of humans on products such as lettuce and cabbage.

In 1932, Arthur Fox discovered the taste anomaly with phenylthiocarbamide (PTC) in his famous "dust flying" experiment (Fox 1932). Harris and Kalmus (1949) discussed various methods to assess whether people were tasters or nontasters of PTC in an attempt to resolve conflicting findings. They proposed a test based on the method of limits, where the subject is presented with eight small glasses, four containing PTC and four not, and is asked to sort them into two groups. This test has been referred to as the Harris-Kalmus test, the Harris-Kalmus threshold procedure, the double tetrad, or the octad (Lawless and Heymann 2010). The method is similar to Fisher's tea-tasting experiment mentioned earlier in the chapter.

In the 1940s and 1950s, there was significant progress in sensory testing, especially in sensory discrimination tests, due to the focus on nutrition during the war years and interest in the development of new food products. In 1950, the US Bureau of Human Nutrition and Home Economics held a conference to gather information and make recommendations for future food testing, attended mainly by academics and research associations (Howgate 2015). The conference proceedings provide valuable insights into the difficulties of testing food using sensory methods, including discussions between David Peryam, Claude Hills, and Sylvia Cover (Dawson and Harris 1951). Lockhart (1951) proposed many potential sensory discrimination tests and suggested a useful table for test layouts, including M + N tests, which encompass over 40 tests for sensory discrimination testing (Bi 2015). In 1949, Wood recommended the tetrad test, which at the time involved presenting four samples and asking the subject to identify the samples as either A or B. Neither Wood, Lockhart, nor Peryam referred to the test as the tetrad, it was only later renamed by Renner and Romer (1973). Munson and Gardner (1950) created the ABX task, which involved presenting three samples and asking the subject to match the third sample to one of the first two. While the ABX task is not commonly used in sensory science, being mainly used in psycholinguistics and acoustic testing, it has potential for projects involving masking techniques (Greenaway 2017).

Statistical methods for analyzing discrimination tests were widely discussed in publications (Fisher 1935; Friedman 1937; Kendall 1938, 1954; Roessler et al. 1948; Lockhart 1951; Bradley 1953; Bliss 1960; Gridgeman 1960). Distinctions between trained panellists and untrained consumers were also debated (Platt 1937; Morse 1942; Bliss et al. 1943; Dove 1947), with Kiehl and Rhodes (1956) calling for more rigorous consumer testing. Kiehl notes the use of small numbers of people in "difference-preference" methods and emphasizes the importance of not assuming expert preferences represent those of the general public. This area has been discussed more recently: see Ares and Varela (2017) and the subsequent comments for a very interesting discussion.

Interest in experimental design for sensory testing was also prominent, with publications discussing issues such as the number of samples assessed in a session, panel member selection, number of panellists required, sample presentation order, and the use of forms versus oral recordings (Platt 1937; Cartwright et al. 1952; Mason and Koch 1953; Hopkins 1954; Gridgeman 1955). Pfaffmann et al. (1954) introduced the "single stimulus" method, assessing one sample at a time: later called the A-not A test. Davis and Hanson (1954) introduced the adapted triangle test and four-sample test, which addressed not only whether there was a difference but also the intensity and direction of the difference.

The n-alternative forced choice tests (e.g. n-AFC: 2-AFC, 3-AFC, 4-AFC, etc.) aimed to minimize response bias, i.e. differences in subjects' criteria for determining differences. Blackwell (1953) suggested this type of approach instead of the method of limits for vision thresholds with paired samples, and Jones (1956) suggested using any number of blank samples with the stimulus material, e.g. A-BBB (4-AFC). Byer and Abrams (1953) used the 2-AFC method (referred to as the "2-sample test") and compared it to the triangle test method for assessing beer. Although Dawson and Dochterman (1951) have been attributed with creating the 2-AFC, the experiment they conducted at a palatability testing conference asked participants "which of the two samples had an added flavoring ingredient," which appears to be a mix of 2-AFC and same-different instructions.

The multiple comparison test was created by Mahoney et al. (1957a, b) to compare several samples to a control sample. Initially, Carlin et al. (1955), Kramer and Ditman (1956), and Hogue and Briant (1957) were believed to have created similar methods, but they were actually quality-grading approaches. Sather (1958) clarified this and distinguished between the "real" multiple comparison test and the quality-grading-type approaches. The multiple comparison test has also been called the degree of difference (DOD) test and the difference from control (DFC) test, and it can be used to test products with variability between batches (Whelan 2017). Pecore et al. (2006) and Young et al. (2008) adapted the method to measure control and test product variability.

In 1958, Peryam introduced two new tests: the multiple standards test and the multiple pairs test (Peryam 1958). The multiple standards test presents the trial sample along with several standards, and the subject must identify the sample that is the "most different" from the set. Peryam suggested that this approach could be beneficial for inherently variable products, but it has not been widely used. The multiple pairs test involves presenting the subject with an equal number of coded standards and trial samples, and their task is to sort them into two groups. Peryam notes that the number of samples can vary, referring to the tetrad test. The second edition of the ASTM Sensory Testing Methods manual includes an interesting case study on the use of the multiple standards test for salt reduction in vegetable soup (Chambers and Wolf 1968).

After a decade, a new sensory discrimination test called "two-out-of-five" was introduced by Amoore et al. (1968), although the test layout was proposed by Lockhart in 1951. The motivation behind this test was to study individuals with specific anosmias (inability to perceive specific odors) while minimizing replication requirements. The authors dismissed the paired comparison and triangle tests as they required excessive replication to determine an individual's threshold for a particular odorant. They also found the Harris-Kalmus test too complicated for subjects. Ultimately, they settled for a two-out-of-five test, which they deemed the most efficient in terms of equipment and sample usage. In this test, subjects were presented with a series of dilutions and asked to select the two samples containing an odor. Later versions of the test did not include any blanks and assessors were informed that there were two samples of one type and three of another, and they were asked to identify the group of two samples.

Over the next few years, the pace of publications related to assessor choice, experimental design, and statistics increased significantly, however, no notable new discrimination tests were reported. In 1960, Gridgeman discusses the role of statisticians in sensory discrimination testing, emphasizing the use of mathematical models, experimental design, and the consideration of the "no difference" option. Several influential texts in sensory science include works by Amerine et al. (1965), Larmond (1967), Green and Swets (discussing SDT and psychophysics, 1966), Chambers and Wolf (1968), Taylor (1968), Pridmore (1968), Watts (1968), Macmillan et al. (1977), and Moskowitz et al. (1980).

The four-interval same-different test, widely used in speech research, was renamed the "dual-pair" test for food research by Rousseau and O'Mahony (2001). This test involves presenting four samples, where one pair is identical and one pair is different (e.g. AA and AB or BB and BA). The authors conducted investigations on the dual-pair method and concluded that, among the methods tested, the same-different test was statistically more powerful.

There has also been interest in the tetrad test in more recent years (Sanderson 2017; Hautus et al. 2022; Bi 2023). Ennis (2012a) developed a guideline to assist sensory scientists considering a switch from the triangle test to the tetrad test. The triangle test requires fewer samples to assess, has lower memory requirements, and has less potential for carryover or adaptation effects, while the tetrad test requires fewer panellists as it is more powerful in detecting differences. The tetrad test also requires less product preparation due to fewer panellists being needed. However, both tests have their disadvantages: the triangle test requires more panellists compared to some other tests, which can increase the resource and logistical requirements. The tetrad test involves assessing more samples, which can increase the workload and time required. It also has higher memory requirements and a higher potential for carryover or adaptation effects, making it less suitable for highly flavored products. Clear instructions should be given to panellists: research by Ennis (2012b) suggests that instructing assessors to sort samples into two groups enhances the test's statistical power compared to asking them to select a specific pair.

Several newer sensory discrimination methods have emerged in the 21st century and are worth exploring. These include the self-defined 2-AFC test introduced by McClure and Lawless in 2010, the 2-AFC with reminder (2-AFCR) developed by van Hout et al. in 2011, the multiple dual pair method proposed by Rousseau and Ennis in 2002, and the dual reference duo-trio test presented by Bi et al. in 2015. These tests offer different approaches and are considered noteworthy in sensory research. Furthermore, there are interesting case studies that employ various M + N options (Bi et al. 2014) for products like supplement tablets, powdered drinks, and nonstick pan coatings. One such study was conducted by Kuesten et al. in 2017, highlighting the application of these methods in real-world scenarios.

The ASTM Sensory Testing Manual (Wolf 2020) includes an excellent chapter on discrimination tests with several case studies. These case studies provide valuable insights into the practical use of sensory discrimination tests and how they can be applied in different business scenarios. The references for each of the ASTM discrimination test standards are also included in the chapter, making it an excellent resource.

## 1.3 Discrimination as a Technique in Sensory Science

### 1.3.1 Factors Influencing Sensory Discrimination

Sensory physiology plays a crucial role in sensory discrimination testing as it involves understanding the functioning of our sensory organs and how they transmit information to our brain. Each sensory modality, such as taste, smell, touch, vision, and hearing, has its own physiological mechanisms that influence discrimination capabilities. For example, in taste discrimination, the taste buds on the tongue can detect different taste qualities like sweet, sour, salty, bitter, and umami. Variations in the number and sensitivity of taste receptors among individuals can impact their ability to discriminate between subtle taste differences. These differences can arise, for example, from age, gender, taster status, and acuity. Similarly, in olfactory discrimination, the receptors in the nose detect various odor molecules, and factors like olfactory receptor genes and nasal airflow patterns can influence an individual's odor discrimination abilities. For more information about all of these areas, please see Chapter 4.

Cognitive processes also play a significant role in sensory discrimination tests, as they encompass the mental activities involved in perceiving, interpreting, and making decisions about sensory information. Training, attention, memory, and decision-making processes influence an individual's ability to discriminate between sensory stimuli (Lawless and Heymann 2010). Attentional focus determines which sensory cues are prioritized and processed, affecting the ability to detect subtle differences. Memory plays a role in comparing current sensory experiences with past experiences, aiding in discrimination and recognition tasks. Decision-making processes involve evaluating and integrating sensory information to make judgments about differences or similarities. Factors like cognitive load, fatigue, and individual cognitive abilities can influence sensory discrimination performance (Lee et al. 2007).

Environmental factors can impact sensory discrimination outcomes by influencing the context and conditions under which testing occurs. Factors such as lighting, temperature, noise, and distractions in the testing environment can affect an individual's sensory perception and discrimination abilities (Meilgaard et al. 2016 and Chapter 4). Additionally, the presence of other stimuli or individuals in the testing environment may introduce biases or distractions that can affect discrimination performance (Kemp et al. 2009). Controlling and standardizing environmental factors during testing is essential to ensure reliable and valid discrimination results. Moreover, cultural influences, familiarity with specific sensory attributes, and individual experiences related to the stimuli being tested can also influence sensory discrimination outcomes and should be considered in the test design (Oleszkiewicz et al. 2023).

### 1.3.2 Action Standards and Sensory Discrimination Testing

There are many sources of variation associated with discrimination testing, for example: the products and samples, the processing and production methods, the packaging, the time taken from production to testing, as well as the panellists and the test environment. It is important to remember that any two products will be found to be different if the sample size (that is the number of people taking part in the test) is large enough (Rousseau 2015). When initiating any investigation, irrespective of its focus on testing personal care products, home-care products, foods, beverages, textiles, clothing, fragrances, or sportswear, several fundamental questions are important to consider. First, it is imperative to define the aims and objectives of the investigation, outlining the specific goals that need to be achieved through the study. This can be achieved by asking several questions to the client regarding the test objectives and how the results will be used. These questions might include:

- What are the objectives?
- What are you trying to find out?
- How many samples?
- What sample preparation is needed?
- Any previous discrimination testing results that might be useful?
- How much of each product is available?
- When are the results required?
- Any novel ingredients?
- Do the samples look different?
- Any potential issues with carryover?
- At what stage is the project?
- Is this a low, medium, or high business-risk product?
- What will the results be used for?

One useful approach to understanding the aims and objectives of the discrimination test is to ask the client to imagine the test has been completed and ask them to describe the outcome, as well as their next steps based on the test findings. This technique allows the client to visualize the various outcomes and share their thoughts on how to proceed based on that information. It is a valuable approach to gain insights and understand the client's expectations regarding the test outcomes and their potential actions.

Second, the underlying business need and rationale for conducting the test must be clearly documented and agreed. It is important to understand how the investigation fits into the larger business strategies and requirements, as this helps put the test significance into perspective. The business aspects must be understood prior to the experiment being designed. Last, consideration should be given to the potential consequences and implications that will follow from the test outcome. Identifying the actionable outcomes and understanding their implications for decision-making is crucial in ensuring the investigation's relevance and impact. The use of action standards (AS) is a useful technique in the test design process, as they offer significant guidance to the test organizer. An AS serves to define the overarching objective of the experiment while also specifying the course of action or next steps based on the test results. For instance, when evaluating a new supplier for a flavoring agent, the AS could have been framed as follows:

**AS1:** "If the sensory test confirms a flavour difference between the new supplier's product and our existing supplier's product for product X, we will not proceed with the new supplier."

This AS explicitly outlines the objective: to determine whether there is a difference in flavor between the products and to reject the new supplier in case such a difference is identified. Consequently, a straightforward discrimination test could have been selected. However, for some products, a "settling in" or "conditioning" time is required prior to testing to ensure that the product is in an optimal state for testing, and in that case, the AS could have been formulated as:

**AS2:** "If the sensory test confirms a flavour difference between the new supplier's product and our existing supplier's product for product X, we will test again after Y days/weeks when the product is at its optimum."

Again, a straightforward discrimination test, conducted twice, could have been selected based on this AS, but an alternative testing approach might have been adopted with a different AS:

**AS3:** "If the sensory test confirms a flavour difference between the new supplier's product and our existing supplier's product for product X at week Y, we will investigate the nature of the flavour difference and if there are any other changes to product X as a result of the supplier change. Our existing supplier will stop producing at the end of December so it is critical we find a new supplier."

In this scenario, the sensory scientist might have consulted the client to gauge the likelihood of a significant difference between the new and existing suppliers' ingredients. If a substantial difference was anticipated, the scientist could have chosen to take a descriptive analysis approach to understand the nature of the differences (Ford 2017). Such insights are crucial in understanding the impact of this change on consumers' perceptions of product X. If the flavor changes jeopardized key drivers of preference, further discussions with the new supplier might have been required to achieve a match.

Another AS could have been as follows:

**AS4:** "If the sensory test confirms that there is a textural difference between the new supplier and our existing supplier for product X, we will need to understand what the textural differences are and if there are any other changes to product X as a result of the supplier change. We also need to understand how this difference is related to the natural variance in our product."

In this case, the sensory scientist might have chosen to conduct batch-to-batch variability tests to assess where the new ingredient batch fits. The DFC method could have served as a suitable starting point to gather data for this evaluation, as multiple batches can be included in a single test.

An example of a deficient AS is shown below (AS5):

**AS5:** "Is there a flavour difference between the new supplier and our existing supplier for product X?"

This AS5 is inadequate since it fails to document the subsequent steps after identifying the presence or absence of a difference. In other words, it fails to include any "action"! As a result, selecting an appropriate test becomes challenging. When reporting the results, the client would likely question the subsequent course of action, and the sensory scientist might not have the necessary information to guide them effectively.

These examples demonstrate the critical importance of clearly defining the objective and outlining the intended outcomes of the sensory study through ASs, as these influence the selection of an appropriate sensory test.

Once the scope of the investigation has been confirmed, and the objectives and AS(s) documented, the choice of which tests might be appropriate can then be evaluated. It can be tempting to default to familiar standard tests, like a triangle or tetrad test, simply because they have been used before and the methodology is readily available. One useful approach is to compile a list of potential tests and carefully evaluate their advantages and limitations. This thoughtful examination will aid in selecting the most fitting test and may also raise additional questions that warrant consideration.

### 1.3.3 Advantages and Disadvantages

Sensory discrimination testing offers several advantages that contribute to the improvement of products and consumer satisfaction. One main benefit is its ability to provide precise information about the sensory differences and similarities between products. By comparing and contrasting different samples, these tests help identify subtle differences or similarities in appearance, aroma, taste, flavor, or texture, or give the sensory scientist important information about holistic differences and similarities between products. This precision is crucial for industries such as foods and beverages, and home and personal care, where slight variations can significantly impact consumer preferences and market success. Additionally, sensory discrimination testing is a useful technique in quality control, ensuring that products consistently meet the required standards, helping protect brand reputation and improve customer loyalty (Lawless and Heymann 2010). Discrimination tests in their simplest forms are easy to use compared to other sensory methods. They require little training, are easy to conduct, and the analysis is straightforward, often simply involving reading a table.

However, like any method, sensory discrimination testing has its share of disadvantages. Setting up, designing, interpreting, and analyzing sensory data requires skilled and trained people to minimize variability and maintain the reliability of results. An incorrect test can be easily chosen to answer a specific question, and erroneous conclusions made based on the results collected from a suitably selected test. Another challenge, particularly for more complex discrimination tests, is the time and resources required for conducting these tests, making them potentially expensive and time-consuming for companies. Discrimination tests typically require a greater number of assessors and a larger amount of sample than other sensory methods. Sample variation may obscure differences, and so sample consistency is important. Where differences in one sensory attribute are being investigated, it can be difficult to manipulate or assess this attribute without changing or being influenced by other sensory properties. The task for assessors to perform in discrimination tests is relatively straightforward, but judgments are typically made by assessing several samples simultaneously, which can increase carryover effects, fatigue, and memory load compared to other sensory methods, and therefore test protocols need to be carefully considered. Striking a balance between the benefits and limitations is essential to leverage the insights gained from sensory discrimination testing effectively. Despite these challenges, the valuable information obtained can significantly enhance product development and market success in various industries.

## 1.4 Applications

Sensory discrimination testing plays a crucial role in various industries and fields where the evaluation of sensory attributes is important. Understanding the differences and similarities between sensory stimuli is essential for product development, quality control, and consumer research. By employing rigorous sensory discrimination tests, companies can make informed decisions about product formulations, optimize sensory characteristics, and ensure consistency and quality.

The new global wellness project scenario described above is a typical business project that requires sensory discrimination tests to achieve the end goal: a healthier, but still delicious, snack product. However, there are many other applications of sensory discrimination tests. For example, a company might be investigating the possibility of a new factory location, a new supplier for a washing powder ingredient, a new improved flavor for a soft drink, or a more sustainable packaging option for pet food.

### 1.4.1 Product Design and Development

In product development, sensory discrimination testing provides valuable insights into the differences or similarities of sensory attributes of different formulations or prototypes (Lawless and Heymann 2010; Meilgaard et al. 2016). By systematically comparing samples, researchers can identify the key sensory differences and determine which formulation contributes to the desired sensory experience. Discrimination tests can detect taints from packaging materials and ingredients, and are also used in shelf-life studies to determine when perceptible changes occur. This information enables companies to refine product formulations and processes, optimize flavors, packaging, textures, or aromas, and create products that meet consumer preferences.

### 1.4.2 Quality Assurance/Quality Control

In quality control, sensory discrimination testing serves as a tool for monitoring the consistency and uniformity of products over time. By conducting regular sensory evaluations, manufacturers can identify any variations or deviations in sensory attributes and take corrective actions to maintain the desired sensory profiles. Sensory discrimination tests also help detect any sensory defects or off-notes that might affect product quality or consumer acceptance. Discrimination testing plays a valuable role in quality assurance and control, particularly when establishing new product standards or selecting equivalent products in similarity testing mode. It is crucial to use discrimination tests appropriately to ensure reliable and accurate results (Munoz 2002, 2013; Meilgaard et al. 2016).

### 1.4.3 Consumer Research

In consumer research, sensory discrimination testing allows researchers to investigate consumer preferences and perceptions, for example, in shelf life testing and sensory claim substantiation. By presenting consumers with different product variations or prototypes, researchers can understand which sensory attributes drive consumer preference and purchase behavior. This information guides companies in developing products that align with consumer expectations and preferences, ultimately enhancing consumer satisfaction and loyalty.

### 1.4.4 Panel Selection and Training

Discrimination tests are also employed in the screening and recruitment of trained panellists. The triangle test, for example, can be used to assess the sensory acuity of potential panellists. Threshold tests, using ranking or alternative forced choice tests, are used to determine taste sensitivity during panel screening, and discrimination tests aid in panel monitoring and retraining (O'Mahony and Rousseau 2002). These tests ensure panellists maintain their sensitivity and improve panel performance over time.

### 1.4.5 Research

Discrimination tests are also used in basic and medical research, for example, in eye examination tests such as measuring visual acuity, checking for refractive errors (like nearsightedness, farsightedness, and astigmatism), evaluating eye coordination and focusing ability, and examining the overall health of the eye, including the retina and optic nerve. They have also been used in the determination of the impact of semantic memory, optimism and mindset, for example, on odor discrimination (Sabiniewicz et al. 2021) and the investigation of eating behaviors (Zhong et al. 2018; Garcia-Burgos et al. 2022). Investigators have also used discrimination testing to understand individual differences (see Chapter 4) and human sensitivity (for example, see Yang et al. 2022 and Norton et al. 2022). Discrimination tests have also been used extensively in COVID-19 research (see, for example, Hannum et al. 2020 and Hunter et al. 2023).

## 1.5 Overview of Book

This book covers many aspects and considerations of discrimination testing in sensory science. This chapter, Chapter 1, has given an overview of discrimination testing including the background and history associated with its development. It includes a definition and classification of sensory discrimination tests, including their importance and applications. It also includes a short history of the development of sensory discrimination testing, from the early tests used by Weber and Fechner in the early 19th century (David 1963), and the development of the methodological approaches (Meiselman et al. 2022) through to the tetrads, hexads, and octads (Bi 2023).

Chapter 2 covers the general considerations in setting up and running discrimination tests by starting with a discussion about the goals of discrimination tests and leading on to general considerations for conducting the tests, such as blind testing and independent judgments. The statistical analysis of discrimination testing data is also included in this chapter, including a detailed discussion on test power, sample size, and equivalence testing. A particularly interesting part of this chapter is the discussion on psychological factors and how a panellist might approach a test as a "test of their abilities." This part of the chapter also includes an outline of the different cognitive strategies associated with the different tasks. For example, in the triangle test, the panellist might use a "comparison of distances" strategy for all three pairs of products presented in the test, whilst, for the 3-AFC (ASTM 2019), the panellist has the easier "skimming" task when selecting the sample that is the strongest.

Chapter 3 gives an easily accessible overview of the many different tests that are useful in meeting the wide range of applications facing the sensory scientist in both industry and academia by presenting a valuable table that compares and contrasts 16 of the main discrimination tests. This chapter also offers insights into selected tests, providing background information on the main tests, the test layouts, and presentation orders. Each test is accompanied by an example questionnaire for practical use, along with references for conducting data analysis. To illustrate how these discrimination tests are applied in real-world scenarios, the chapter also includes examples from relevant literature.

Chapter 4 gives an overview of factors affecting individual differences in oral and smell responsiveness, such as age, gender, thermal taster status, training, familiarity, environmental conditions, and psychological traits. The authors provide a fascinating and unique perspective on discrimination testing. They offer an overview of the diverse factors that impact oral and smell responsiveness, as well as the conditions and variables that can influence an individual's ability to discriminate. These factors encompass involvement, knowledge, expertise, and environment. The chapter covers a range of elements affecting discrimination ability, including a panellist's age, training, familiarity (whether using trained or naive individuals for discrimination tests), environmental conditions, thermal taster status, oral tactile acuity (and various screening methods as discussed), psychological traits (such as food neophobia), and individual differences in odor sensitivity (including anosmia).

Chapter 5 covers the important area of similarity or equivalence testing and includes useful information for people working in quality control and manufacturing, as well as reformulations in product development and sensory claim substantiation. The chapter includes information about the evolution of similarity testing, different analysis approaches, and examples for parity and equivalence claims, as well as recipe changes.

Thurstonian modelling and SDT have proved useful tools to analyze sensory discrimination tests (Thurstone 1927a, b; Ennis et al. 2014) and Chapter 6 introduces and builds upon both these topics, using the A-Not A (ISO 2017) and 2AFC tests (ISO 2005) to enable the sensory scientist to understand several intricacies such as response bias, sensitivity, and different panellist strategies to the tasks they have to undertake. This will help the sensory scientist learn more, not just about how our senses work and how our brains decide what we perceive, but how to select the right test to match the objectives.

Chapter 7 adds to the information in Chapter 6 by use of the R-Index (Lee and van Hout 2009) analysis method incorporating confidence ratings for the A-Not A, same-different and ranking tests (ISO 2006). When planning a study involving a calculation of the R-index, considerations such as participant familiarity with the reference, the number of products, and sample sizes require careful thought. This chapter offers comprehensive guidance for analyzing A-Not A and ranking data through the R-index, and explores various objectives for which R-index analysis proves valuable. These include comparing multiple products, hedonic measures, emotions, and determining individuals' sensitivity. The chapter finishes with a case study of the application of R-Index to preference ranking data.

Chapter 8 will prove to be a very useful chapter for all sensory scientists who conduct replicated sensory discrimination tests, often performed due to a lack of available assessors (Meyners et al. 2023). An important consideration when using replicated discrimination tests is that observations may lose their independence, leading to potential complexities in determining sample size and statistical power. Consequently, standard analyses based on the binomial law may become less suitable. The chapter covers key considerations, as well as advantages (such as determining the overall discrimination rate and panel performance) and limitations (for example, fatigue and learning effects) of replicated discrimination tests. The chapter also provides an overview of the available methods for data analysis along with a discussion of their pros and cons. One of the most useful aspects of this chapter is the glossary of terms and mathematical notation.

The topic of sensory quality management is presented in Chapter 9 where the authors explain how discrimination tests and signal detection analysis can be used to quantify the size of the difference between products. A new classification of discrimination tests is introduced, emphasizing performance distinctions among the different tests and explaining how they are conducted, highlighting performance differences between different test methods. A seven-step strategy for developing ASs is shared in the chapter and demonstrated through a case study, and the chapter also includes detailed explanations of the data analysis and interpretation.

Chapter 10 includes discrimination testing in two contrasting areas: flavors and fragrances. The chapter starts with general considerations for discrimination testing for both product types, and includes several case studies, for example, sugar reduction in beverages and ingredient changes in shampoo. What is especially interesting about this chapter is the section on social responsibility, particularly regarding diversity in sensory panels (see also Gómez-Corona 2020). The authors share their experiences of using blind and visually impaired, as well as more vulnerable (e.g. low education or social deprivation) panellists on their panels. This chapter also looks to the future and includes social and environmental challenges, machine learning and artificial intelligence.

Chapter 11 describes the use of children in discrimination testing to help with the development of children's foods and beverages. Extending the human sensitivity information from Chapter 4, the author discusses the differences between children and adults in terms of children's sensitivities across the five main senses, and how these change as we age and impact our food preferences and choices. The author gives three reasons why children should be used for discrimination testing: appropriate sensitivity, understanding children's perceptions, and when appearance is key. Obviously, there are many difficulties with conducting discrimination tests with children but the author discusses each in turn with helpful suggestions to aid experimentation. The chapter includes the environmental, training, and ethical considerations of working with this challenging consumer group.

Chapter 12 has its focus on attribute-specific discrimination tests, especially when the testing program necessitates assessments involving more than two samples, a common challenge frequently encountered by sensory scientists. The chapter includes an introduction to attribute-specific tests and includes the formulae for several analysis methods, in both Microsoft Excel, Google Sheets, and R. The chapter also discusses four options to be used in multisample attribute-specific testing: Mosteller's extension of Thurstone, Bradley-Terry Models, Elo and mElo ratings, and Friedman-style rank analysis.

Finally, the last chapter includes a summary of the book and outlines future directions for discrimination testing. From the inception of discrimination tests to the present day, significant developments in methods have been implemented that provide sensory scientists with a wide range of options for many sensory applications. The editors and authors aimed to ensure that this book serves as a valuable resource, bringing sensory scientists up to date with this wide range of potential tests and approaches applicable across various scenarios.

## References


Amerine, M.A., Pangborn, R.M., and Roessler, E.B. (1965). Principles of sensory evaluation of food. *Food Sci. Technol. Monogr.* 338–339.

Amoore, J.E., Venstrom, D., and Davis, A.R. (1968). Measurement of specific anosmia. *Percept. Mot. Skills* 26: 143–164.

Ares, G. and Varela, P. (2017). Trained vs. consumer panels for analytical testing: fuelling a long lasting debate in the field. *Food Qual. Prefer.* 61: 79–86.

ASTM International (2019). *Practice for Defining and Calculating Individual and Group Sensory Thresholds from Forced-Choice Data Sets of Intermediate Size (E1432).* ASTM International. https://www.astm.org/e1432-19.html.

Bengtsson, K. and Helm, E. (1946). Principles of taste testing. *Wallerstein Lab. Commun.* 9: 171.

Bi, J. (2015). *Sensory Discrimination Tests and Measurements.* Wiley Blackwell.

Bi, J. (2023). The methods of tetrads, hexads, and octads: a type of more powerful sensory discrimination methods. *Food Qual. Prefer.* 108: 104879.

Bi, J., Lee, H.-S., and O'Mahony, M. (2014). Estimation of Thurstonian models for various forced-choice sensory discrimination methods as a form of the 'M + N' test. *J. Sens. Stud.* 29: 325–338.

Blackwell, H.R. (1953). *Psychophysical Thresholds: Experimental Studies of Methods of Measurement (Rep No. 36).* Ann Arbor: University of Michigan, Bulletin of the Engineering Research Institute.

Bliss, C.J. (1960). Some statistical aspects of preference and related tests. *J. R. Stat. Soc. Ser. C: Appl. Stat.* 9 (1): 8–19.

Bliss, C.J., Anderson, E.O., and Marland, R.E. (1943). *Technique for Testing Consumer Preferences, with Special Reference to the Constituents of Ice Cream.* Storrs Agricultural Experiment Station.

Bradley, R.A. (1953). Some statistical methods in taste testing and quality evaluation. *Biometrics* 9 (1): 22–38.

Byer, A.J. and Abrams, D. (1953). A comparison of the triangular and two sample taste-test methods. *Food Technol.* 7: 185–187.

Carlin, A.F., Kempthorne, O., and Gordon, J. (1955). Some aspects of numerical scoring in subjective evaluation of foods. *J. Food Sci.* 21: 273–281. https://doi.org/10.1111/j.1365-2621.1956.tb16921.x.

Cartwright, L.C., Snell, C.T., and Kelley, P.H. (1952). Organoleptic panel testing as research tool. *Anal. Chem.* 24 (3): 503–506.

Chambers, E. and Wolf, M.B. (1968). *Sensory Testing Methods*. West Conshohocken, PA: ASTM International.

Cover, S. (1936). A new subjective method of testing tenderness in meat – the paired-eating method. *J. Food Sci.* 1: 287–295. https://doi.org/10.1111/j.1365-2621.1936.tb17790.x.

Cover, S. (1940). Some modifications of the paired-eating method in meat cookery research. *J. Food Sci.* 5: 379–394. https://doi.org/10.1111/j.1365-2621.1940.tb17199.x.

David, H.A. (1963). *The Method of Paired Comparisons*, vol. 12, 120. London.

Davis, J. and Hanson, H.L. (1954). Sensory test methods. I. The triangle intensity (T-I) and related test systems for sensory analysis. *Food Technol.* 8: 335–339.

Dawson, E.H. and Dochterman, E.F. (1951). A comparison of sensory methods of measuring differences in food qualities. *Food Technol.* 5: 79–81.

Dawson, E. H. and Harris, B. L. (1951). Sensory methods of measuring differences in food quality. *Conference Proceedings*, Bureau of Human Nutrition and Home Economics. https://www.google.co.uk/books/edition/Sensory_Methods_for_Measuring_Difference/0WHDI0i9gcQC?hl=en&gbpv=1&dq=Sensory+methods+of+measuring+differences+in+food+quality&pg=PA1&printsec=frontcover

Dove, W.F. (1947). Food acceptability – its determination and evaluation. *Food Technol.* 1: 39–50.

Ennis, J.M. (2012a). Guiding the switch from triangle testing to tetrad testing. *J. Sens. Stud.* 27: 4.

Ennis, J.M. (2012b). Proper task instructions and the two-out-of-five test. *IFPress* 15 (4): 3,4.

Ennis, J.M., Rousseau, B., and Ennis, D.M. (2014). Sensory difference tests as measurement instruments: a review of recent advances. *J. Sens. Stud.* 29: 89–102.

Fechner, G.T. (1860). *Elemente der Psychophysik*. LeipzigL Breitkopf and Hartel.

Fisher, R.A. (1935). *The Design of Experiments*. Edinburgh: Oliver and Boyd.

Ford, R.A. (2017). Chapter 3 – Deciding which test to use in discrimination testing. In: *Discrimination Testing in Sensory Science* (ed. L. Rogers), 67–83. Woodhead Publishing. https://doi.org/10.1016/B978-0-08-101009-9.00003-4.

Fox, A.L. (1932). The relationship between chemical constitution and taste. *Proc. Natl. Acad. Sci. USA* 18 (1): 115–120.

Friedman, M. (1937). The use of ranks to avoid the assumption of normality implicit in the analysis of variance. *J. Am. Stat. Assoc.* 32 (200): 675–701.

Garcia-Burgos, D., Andres, F., Trier, S. et al. (2022). Discrimination of sweet-fat ingredients in people with weight-and eating-related problems using a signal detection theory. *J. Sens. Stud.* 37 (6): e12783.

Gómez-Corona, C. (2020). Sensory and consumer research for good: a review on social responsibility. *Curr. Opin. Food Sci.* 33: 115–123.

Green, D.M. and Swets, J.A. (1966). *Signal Detection Theory and Psychophysics*. New York: Wiley.

Greenaway, R. (2017). Chapter 13 – ABX task. In: *Discrimination Testing in Sensory Science* (ed. L. Rogers), 267–288. Woodhead Publishing. https://doi.org/10.1016/B978-0-08-101009-9.00013-7.

Gridgeman, N.T. (1955). Taste comparisons-2 samples or 3. *Food Technol.* 9 (3): 148–150.

Gridgeman, N.T. (1959). The lady tasting tea, and allied topics. *J. Am. Stat. Assoc.* 54, 1959 (288): 776–783.

Gridgeman, N.T. (1960). Statistics and taste testing. *J. R. Stat. Soc. Ser. C (Appl. Stat.)* 9 (2): 103–112.

Hannum, M.E., Ramirez, V.A., Lipson, S.J. et al. (2020). Objective sensory testing methods reveal a higher prevalence of olfactory loss in COVID-19 – positive patients compared to subjective methods: a systematic review and meta-analysis. *Chem. Senses* 45 (9): 865–874.

Harris, H. and Kalmus, H. (1949). The measurement of taste sensitivity to phenylthiourea (PTC). *Ann. Eugenics* 15: 24–31. https://onlinelibrary.wiley.com/doi/10.1111/j.1469-1809.1949.tb02419.x.

Hautus, M.J., van Hout, D., Lee, H.S. et al. (2022). The observed variance of d′ estimates compared across the 2-AFCR, triangle, and tetrad tasks. *Food Qual. Prefer. 100*: 104578.

Helm, E. and Trolle, B. (1946). Selection of a taste panel. *Wallerstein Lab. Commun.* 9: 181.

Hogue, D.V. and Briant, A.M. (1957). Determining flavor difference in crops treated with pesticides. A comparison of a triangle and a multiple comparison method. *J. Food Sci.* 22: 351–357. https://ift.onlinelibrary.wiley.com/doi/10.1111/j.1365-2621.1957.tb17022.x.

Holden, J.K. et al. (2011). An undergraduate laboratory exercise to study Weber's law. *J. Undergrad. Neurosci. Educ.* 9 (2): A71–A74.

Hopkins, J.W. (1954). Observations on sensitivity and repeatability of triad taste difference tests. *Biometrics* 10 (4): 521–530.

Howgate, P. (2015). A history of the development of sensory methods for the evaluation of freshness of fish. *J. Aquat. Food Prod. Technol.* 24: 516–532.

Hunter, S.R., Hannum, M.E., Pellegrino, R. et al. (2023). Proof-of-concept: SCENTinel 1.1 rapidly discriminates COVID-19-related olfactory disorders. *Chem. Senses* 48: bjad002.

ISO 8586 (2017). *Sensory Analysis – Methodology—"A"-"Not A" Test.* Geneva: International Organisation for Standardisation.

ISO 8586 (2023). *Sensory Analysis— Selection and training of sensory assessors*, 28. International Organisation for Standardisation: Geneva.

ISO 8588 (2005). *Sensory Analysis – Methodology – Paired Comparison Test.* Geneva: International Organisation for Standardisation.

ISO 8588 (2006). *Sensory Analysis – Methodology – Ranking.* Geneva: International Organisation for Standardisation.

Jones, F.N. (1956). A forced-choice method of limits. *Am. J. Psychol.* 69: 672–673.

Kemp, S.E., Hollowood, T., and Hort, J. (2009). *Sensory Evaluation, A Practical Handbook.* Wiley Blackwell.

Kendall, M. (1938). A new measure of rank correlation. *Biometrika* 30 (1/2): 81–93. https://doi.org/10.2307/2332226.

Kendall, M. (1954). Some recent developments in ranking methods. Institute of statistics monographs series no. 112.

Kiehl, E. and Rhodes, V. (1956). New techniques in consumer preference research. *J. Farm Econ.* 38 (5): 1335–1345. Retrieved from http://www.jstor.org/stable/1234552.

Kilcast, D. (ed.) (2010). *Sensory Analysis for Food and Beverage Quality Control: A Practical Guide.* Cambridge: Woodhead Publishing Limited.

Kramer, A. and Ditman, L.P. (1956). A simplified variables taste panel method for detecting flavor changes in vegetables treated with pesticides. *Food Technol.* 10: 155–159.

Kuesten, C., Hou, A., Wu, J., and Bi, J. (2017). Case studies of industrial applications of the "M + N" classification method with larger M and N for visual and manual inspections. *J. Sens. Stud.* 32: e12245. https://doi.org/10.1111/joss.12245.

Larmond, E. (1967). *Methods for Sensory Evaluation of Food.* Canada Department of Agriculture.

Lawless, H.T. and Heymann, H. (2010). *Sensory Evaluation of Food. Principles and Practices. Sensory Evaluation of Food – Principles and Practices.* New York: Springer.

Lee, H.S. and van Hout, D. (2009). Quantification of sensory and food quality: the R-index analysis. *J. Food Sci.* 74 (6): 57–64. https://doi.org/10.1111/j.1750-3841.2009.01204.x.

Lee, H.S., Van Hout, D., and Hautus, M.J. (2007). Comparison of performance in the A – not A, 2-AFC, and same–different tests for the flavor discrimination of margarines: the effect of cognitive decision strategies. *Food Qual. Prefer.* 18 (6): 920–928.

Lockhart, E.E. (1951). Binomial systems and organoleptic analysis. *Food Technol.* 5: 428–431.

Macmillan, N.A., Kaplan, H.L., and Creelman, C.D. (1977). The psychophysics of categorical perception. *Psychol. Rev.* 84: 452–471.

Mahoney, C.H. (1975). Evaluating flavour differences in canned foods. Part II. Fundamentals of the simplified procedure. *Food Technol.* 11 (9): 37–42.

Mason, D.D. and Koch, E.J. (1953). Some problems in the design and statistical analysis of taste tests. *Biometrics* 9 (1): 39–46.

McClure, S. and Lawless, H.T. (2010). Comparison of the triangle and a self-defined two alternative forced choice test. *Food Qual. Prefer.* 21: 547–552.

Meilgaard, M., Civille, G., and Carr, B. (2016). *Sensory Evaluation Techniques.* Boca Raton, FL: Taylor & Francis.

Meiselman, H.L., Jaeger, S.R., Carr, B.T., and Churchill, A. (2022). Approaching 100 years of sensory and consumer science: developments and ongoing issues. *Food Qual. Prefer. 100*: 104614.

Meyners, M., Carr, B.T., and Kunert, J. (2023). Equivalence and non-inferiority tests using replicated discrimination and preference data. *J. Sens. Stud.* 38: e12882.

Morse, R.L.D. (1942). *Egg Grading and Consumers' Preferences with Special Reference to Iowa Egg Marketing.* Iowa State University.

Moskowitz, H., Jacobs, B., and Firtle, N. (1980). Discrimination testing and product decisions. *J. Mark. Res.* 17 (1): 84–90. https://doi.org/10.2307/3151121.

Munoz, A.M. (2002). Sensory evaluation in quality control: an overview, new developments and future opportunities. *Food Qual. Prefer. 13* (6): 329–339.

Munoz, A.M. (ed.) (2013). *Sensory Evaluation in Quality Control.* Springer Science & Business Media.

Munson, W.A. and Gardner, M.B. (1950). Standardizing auditory tests. *The Journal of the Acoustical Society of America 22* (5_Supplement): 675–675.

Norton, V., Lignou, S., Faka, M., and Methven, L. (2022). Individual variation in mouthfeel sensitivity: Investigating influences of whey protein content, consumer age, food format and fat addition. *Food Qual. Prefer.* 101: 104638.

Oleszkiewicz, A., Schmidt, P., Smith, B. et al. (2023). Effects of blindness and anosmia on auditory discrimination of temperature and carbonation of liquids. *Food Qual. Prefer. 107*: 104852.

O'Mahony, M. and Rousseau, B. (2002). Discrimination testing: a few ideas, old and new. *Food Qual. Prefer.* 14: 157–164.

Pecore, S., Stoer, N., Hooge, S. et al. (2006). Degree of difference testing: a new approach incorporating control lot variability. *Food Qual. Prefer.* 17 (7–8): 552–555.

Peryam, D.R. (1958). Sensory difference tests. *Food Technol.* 12 (5): 231–236.

Peryam, D.R. and Swartz, V.W. (1950). Measurement of sensory differences. *Food Technol.* 4 (1): 390–395.

Pfaffmann, C., Schlosberg, H., and Cornsweet, J. (1954). Variables affecting difference tests. In: *Food Acceptance Testing Methodology* (ed. D.R. Peryam, F.J. Pilgrim, and M.S. Peterson), 4–20. Chicago: Quartermaster Food and Container Institute.

Platt, W. (1937). Some fundamental assumptions pertaining to the judgement of food flavours. *Food Sci.* 2 (3): 237–249.

Pridmore, W.A. (1968). The organization of a sensory testing panel. *J. R. Stat. Soc. Ser. D (The Stat.)* 18 (1): 1–10.

Renner, E. and Romer, G. (1973). The Tetrade-Test as a capable sensory method – I. Method and application. *Z. Lebensm. Unters-Forsch* 151: 326.

Roessler, E.B., Warren, J., Guymon, J.F. et al. (1948). Significance in triangular taste tests. *Food Res.* 13 (6): 503–505.

Rousseau, B. (2015). Sensory discrimination testing and consumer relevance. *Food Qual. Prefer.* 43: 122–125.

Rousseau, B. and Ennis, D.M. (2002). The multiple dual-pair method. *Percept. Psychophys.* 64 (6): 1008–1014.

Rousseau, B. and O'Mahony, M. (2001). Investigation of the dual-pair method as a possible alternative to the triangle and same-different tests. *J. Sens. Stud.* 16: 161–178. https://doi.org/10.1111/j.1745-459X.2001.tb00294.x.

Sabiniewicz, A., Wrage, F., and Hummel, T. (2021). The impact of cognitive and noncognitive factors on odor discrimination performance. *J. Sens. Stud.* 36 (5): e12683.

Sanderson, T. (2017). Chapter 9 – Tetrad test. In: *Discrimination Testing in Sensory Science* (ed. L. Rogers), 183–195. Woodhead Publishing. https://doi.org/10.1016/B978-0-08-101009-9.00009-5.

Sather, L (1958). Laboratory flavour panels. Oregon Agricultural Experiment Station miscellaneous paper no. 56

Sinkinson, C. (2017). Chapter 7 – Triangle test. In: *Discrimination Testing in Sensory Science* (ed. L. Rogers), 153–170. Woodhead Publishing. https://doi.org/10.1016/B978-0-08-101009-9.00007-1.

Taylor, J. (1968). The influence of statistical principles on the design and conduct of sensory tests of foodstuffs. *J. R. Stat. Soc. Ser. D (The Stat.)* 18 (1): 19–23.

Thurstone, L.L. (1927). The method of paired comparisons for social values. *J. Abnorm. Soc. Psychol.* 21: 384–400.

Thurstone, L.L. (1927a). A law of comparative judgment. *Psychol. Rev.* 34: 273–286.

Thurstone, L.L. (1927b). Psychophysical analysis. *Am. J. Psychol.* 38: 368–389.

Van Hout, D., Hautus, M.J., and Lee, H.-S. (2011). Investigation of test performance over repeated sessions using signal detection theory: comparison of three non-attribute-specified difference tests 2-AFCR, A-not A and 2-AFC. *J. Sens. Stud.* 26: 311–321. https://doi.org/10.1111/j.1745-459X.2011.00346.x.

Watts, G. (1968). Flavour test on beer. *J. R. Stat. Soc. Ser. D (The Stat.)* 18 (2): 149–156.
Whelan, V.J. (2017). Chapter 11 – Difference From control (DFC) test. In: *Discrimination Testing in Sensory Science* (ed. L. Rogers), 209–236. Woodhead Publishing. https://doi.org/10.1016/B978-0-08-101009-9.00011-3.
Wolf, M.B. (ed.) (2020). *Sensory Testing Methods*, 3e. West Conshohocken, PA: ASTM International.
Wood, E.C. (1949). Organoleptic tests in the food industry. III.-Some statistical considerations in organoleptic tests. *J. Chem. Technol. Biotechnol.* 68: 128–131. https://doi.org/10.1002/jctb.5000680411.
Yang, T., Zhang, L., Xu, G. et al. (2022). Investigating taste sensitivity, chemesthetic sensation and their relationship with emotion perception in Chinese young and older adults. *Food Qual. Prefer.* 96: 104406.
Young, T.P., Stoer, S., Hulting, N. et al. (2008). Incorporating test and control product variability in degree of difference tests. *Food Qual. Prefer.* 19: 734–736.
Zhong, L., Hadde, E.K., Zhou, Z. et al. (2018). Sensory discrimination of the viscosity of thickened liquids for dysphagia management. *J. Sens. Stud.* 33 (6): e12464.

# 2

# General Considerations in Discrimination Testing

*Harry T. Lawless*

*Cornell University, Ithaca, NY, USA*

## 2.1 Introduction: Goals of Discrimination Tests

Discrimination tests are one form of sensory tests aimed at answering a simple question: are two versions of a product perceivably different (or not)? One can consider almost all sensory tests as a search for differences. Even descriptive analysis with a trained panel first looks at differences between means on various attribute scales. But discrimination tests are the simplest to conduct and analyze, and only amass evidence to answer that one question. Having done so, we still do not know in what ways two products are different, nor do we know whether consumers would find the difference noteworthy or actionable, agreeable or disagreeable. Simply finding evidence for a difference, or for sensory equivalence, is often the starting point, before all other sensory testing is even relevant (Ennis 1993).

Simple difference tests are used in a variety of scenarios for processed foods and other consumer products. In product research, often small changes may be made in ingredients, for example, or when a new supplier is contracted. Processing changes may accompany more efficient production methods, new equipment, or any simple changes in cooking time, pasteurization, temperature, pressure, and the like. Cost reductions may involve both ingredients and process changes. It is becoming increasingly common to offer different versions of a product with enhanced nutritional profiles such as reduced sugar, sodium or fat, and/or increased fiber, vitamins or mineral content. Quality control is primarily a search for product differences that exceed some allowable specification limits. Finally, shelf-life studies may involve a search for sensory changes as well (Hough 2010). If changes are sensorially detectable further actions may be needed or consumer testing as a follow-up.

In taste, smell, and flavor research, one of the primary sensory specifications of the potency of a flavor substance is its sensory threshold. Threshold is a matter of determining when something is detectable, that is, at what minimum concentration can we perceive it. This is often rephrased and operationalized as the point at which an item containing the target flavor substance is discriminable from a product or carrier without the substance. A threshold procedure is often conducted as a series of simple difference tests (ASTM 2019a, b).

*Discrimination Testing in Sensory Evaluation*, First Edition. Edited by Lauren Rogers, Joanne Hort, Sarah E. Kemp and Tracey Hollowood.

So thresholds may be viewed as a special case of discrimination testing, one specifically aimed at finding some minimum perceivable level. The statistics may be different for threshold testing, for the goal is to find the level at which 50% of the population can perceive the difference, rather than simply testing for a statistically significant result. But the test procedures are often equivalent.

Perhaps the most common scenario for simple discrimination tests is the in-house sensory panel. This is a pool of volunteer employees, from whom a sample of from about 40 to 80 participants will be recruited for any given test. Often they are pre-screened to ensure a basic level of sensory function. Thus, people who have a poor sense of taste or smell are generally excluded. Product familiarity and product usage is a plus, but is not generally required. Such a panel is conceived of as an analytical instrument. Its job is to act as a sensory detection mechanism. Tests are carefully controlled and usually performed in a sensory testing facility, with minimal distractions and attention to standard practices. Even though the panel is not a representative sample of the target consumer population, its results are often extrapolated in various ways. For example, if no significant difference is the test result, the conclusion is often that "most consumers won't notice." The underlying logic is this: If I have failed to see evidence for a perceivable difference with my carefully selected panel under controlled conditions (i.e., the best most sensitive way of conducting the test) then the consumer will not be likely to detect this difference under the more variable and noisy situations encountered in real life. Of course, this logic is not airtight. Consumers have the opportunity to gain extended experience with the new product, and they may be intimately familiar with the characteristics of a standard well-known product they have used for years. Furthermore, frequent or heavy users of a product are likely to be unforgiving and even alienated if you change their well-known and favorite version of a brand they have consumed for a long time. So there are risks involved. Of course, if a difference is detected, consumer testing may be the next step to see if the difference actually matters.

Most of these tests involve frequency counts of "correct" responses, statistical tests on frequency (such as chi-square) or proportions are commonly used and appropriate. The chi-square test is mathematically equivalent to the test on differences of proportions based on the $z$-score approximation to the binomial (Lawless and Heymann 2010). Most sensory texts include tables for the minimum number of correct choices required to establish significant differences. The tests are straightforward to execute and can be rapidly analyzed simply by knowing the proportion correct and the panel size ("N"). Other tests are amenable to parametric statistics such as ratings on a degree of difference line scale, where a $t$-test on means can be done comparing the control pair's mean score to the test pair's mean. Almost all the common tests now have psychometric models for converting the proportion correct into a metric for sensory difference, based upon signal detection theory and Thurstonian modeling. This provides an opportunity for enriching the analysis by taking it a step beyond simple statistical significance. Finally, a statistically significant result is no indication that the difference is perceivable to everyone or even a majority of consumers, and the proportion of people actually discerning the difference could be quite small indeed. A correction for guessing can be applied using Abbott's formula (Filipello 1956; Ferdinandus et al. 1970). This is used to separate the proportion of correct guessers from the most likely proportion of actual discriminators, as will be discussed below (Lawless 2010).

Further general information on sensory testing and discrimination tests can be found in the texts by Lawless and Heymann (2010), Meilgaard et al. (2015), and Stone and Sidel (1993). Concise works on sensory evaluation include the book by Kemp et al. (2009) and the ASTM manual by Wolf (2020).

## 2.2 Types of Tests

### 2.2.1 Classification by Task

Simple sensory difference tests fall into several broad categories. First, there are general tests for any overall difference. These tests are appropriate when the sensory results cannot be predicted ahead of time or the results are complex or multi-attribute in nature. Within the overall difference tests, there are two large groups: one group based on sorting a set of products, and another group based on matching items to a previously inspected (viewed, tasted, smelled, or touched) reference item. An example of the sorting task is the triangle test, where two products are from the same lot and another is from a different version of the product, and the participant must choose which one is most different from the other two (i.e., the "odd" sample). In the matching category, an example is the duo-trio test, in which the participant must choose from two items which one matches a previewed reference item. A separate category of choice tests consists of those based on finding differences in a specified attribute. These are known as n-AFC tests for n-alternative forced-choice. The "n" stands for the number of alternatives. So a 2-AFC is a paired test. All the aforementioned tests are based on a choice of products from a set, and are considered criterion-free. In one sense, these kinds of tests are truly objective. You must prove to the experimenter that you can tell the difference by choosing the correct item.

Usually there is no opportunity for the respondent to say, "I can't tell." If the person feels they cannot discriminate, they are still forced to pick from the alternatives, i.e., to guess. Allowing a non-response or an "uncertain" category in a paired test is known as 2-AC rather than 2-AFC, the "F" is dropped because the choice is no longer forced, that is, the assessor can simply opt out. The uncertain response has been used in psychophysics (Watson et al. 1973) and there are statistical models for dealing with the uncertain response option in a paired test (Braun et al. 2004) similar to the approaches to dealing with "no preference" in paired preference testing. Although most sensory professionals avoid the uncertain response and force a choice, there is occasional interest in the 2-AC method (e.g., Chacon and Sepulveda 2011). Common sensory wisdom dictates against allowing the "I can't tell" response for two reasons. First, it is seen as an easy way out for a person to avoid making a decision under conditions of uncertainty, which is uncomfortable for some people. Second, it is a longstanding maxim of signal detection theory that people often have more information or more evidence than they think. That is, even though they feel unsure, they may still discriminate. The counter-argument questions why persons should be counted in a test if they cannot discriminate after all (Chacon and Sepulveda 2011). The rejoinder is that to eliminate them would lower the sample size and thus the power of the test. As noted above, perhaps they are just being "lazy."

Another class of tests is those that are not criterion-free, and the participant must choose from a set of responses, rather than pointing to an item or sorting items. Response-choice

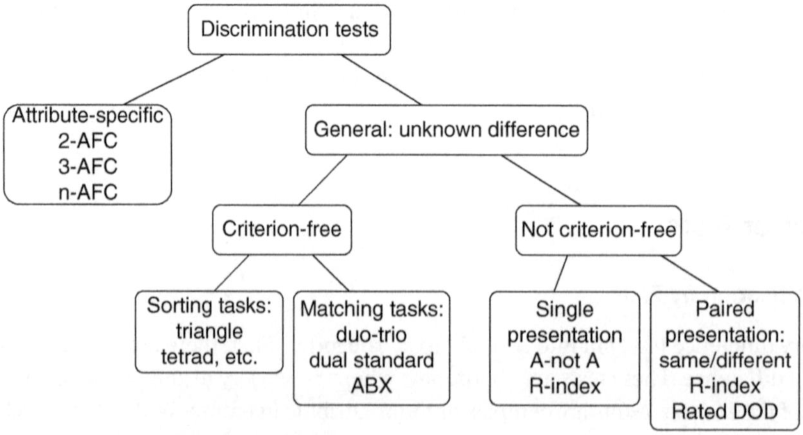


**Figure 2.1** A classification scheme for discrimination tests.

tests include the A-not A test and the same/different test. With these tests, the participant decides where to set his or her criterion for what is an example of "A" or "not A" (as each item is presented singly) or how much of a difference justifies a decision of sameness versus difference (as pairs are presented). These tests can be extended to include rating scales for sureness (certainty) and in the case of same/different a rating scale for the degree of difference. Because these tests are not criterion-free, it is imperative to include control items (i.e., both A and not A items must be shown) or control pairs (i.e., an identical control pair in the same/different test) in order to establish a baseline of response. Then the frequency of test item or test pair responses can be compared against the control baseline as discussed below.

The classification groupings can be seen in Figure 2.1. Table 2.1 lists the actual tests and some of their important characteristics such as the expected chance performance level.

### 2.2.2 Examples of Tests, Common and Uncommon

Historically, the first two discrimination tests to be described in the published literature were the triangle and duo-trio procedures. The triangle test appeared in a publication from the Carlsberg breweries by Helm and Trolle (1946). In the triangle, the difference is not specified, and the participants must choose which of three products is most different from the other two. The test setup is of course three samples, two of which are physically identical (duplicates). The triangle test is considered by some to be a special case of Torgerson's method of triads, in which three different products are given and the person must choose which two are most similar and which two are most different. The duo-trio procedure seems to have evolved as a quality control test in the distilled spirits industry (Peryam and Swartz 1950). In the duo trio, a reference sample is given for inspection, and then two test samples, one of which is a duplicate of the reference. In the original wording, the participants must specify which sample is different from the reference, although it is common in current practice to phrase it in terms of choosing which item matches the reference (Lawless and Heymann 2010). The wording change is probably inconsequential.

**Table 2.1** Discrimination Tests – Classification.

| Class of test | Test | Samples: inspection phase | Samples: test phase | Task/instructions | Chance Probability |
|---|---|---|---|---|---|
| **Item choice:** | | | | | |
| Matching | Duo-trio | Ref – A | A, B (see Note 1) | Match sample to reference | 1/2 |
| | 2-AFCR-M | Ref given second | A, B | Match samples to item in middle of series | 1/2 |
| | ABX | Ref-A, Ref-B | A (or B) | Match sample to reference | 1/2 |
| | Dual standard | Ref-A, Ref-B | A, B | Match both items to ref. | 1/2 |
| | Pick-2 | Ref-A | A, A′, B, B′ | Match 2 samples to reference | 1/6 |
| Forced-choice | Paired comparison | (None) | A, B | Choose sample with most of specified attribute | 1/2 |
| | 3-AFC | (None) | A, A′, B | (Same) | 1/3 |
| | *n*-AFC | (None) | $A_1 - A_{n-1}$, B | (Same) | 1/*n* |
| Sorting | Triangle | (None) | A, A′, B (or A, B, B′) | Choose the most different sample | 1/3 |
| | 2 out of 5 | (None) | A, A′, B, B′, B″ | Sort into two groups | 1/10 |
| | 4/8 "Harris-Kalmus" | (None) | $A_1 - A_4$, $B_1 - B_4$ | Sort into two groups | 1/70 |
| | Tetrad/ non-directional | (None) | A, A′, B, B′ | Sort into two groups | 1/3 |
| | Dual pair | (None) | A, B and A, A′ | Choose A, B (different pair) | 1/2 |
| **Response choice:** | Same–different | (None) | Pairs: A, A′ or A, B | Choose response: "same" or "different" | N/A (see Note 2) |
| | A, not-A | Ref-A | A or B | Choose response: "A" or "not-A" | N/A (see Note 2) |
| | R-index | Both | See Note 3 | Choose response on sureness scale | N/A (see Note 3) |

Note 1: Throughout the table, the letters A and B refer to different versions of the product, e.g., control item (A) and test item (B).
Note 2: For the tests involving a response choice, a criterion may be set by each individual, and therefore the chance probability may not be equal to 1/2. The baseline is set by the false alarm rate (see text).
Note 3: A sureness (certainty) scale can be used with both A-not-A tests and same/different.

In the landmark paper by Peryam and Swartz in 1950, another test was also described, called the dual standard test. This was similar to the duo trio but with two references. In the dual standard, both versions of the product are given in the inspection phase, and then two blind coded test samples to attempt to match to the appropriate references, if possible. The triangle test was also described in that paper. A little-known detail in the 1950 paper was that for all three tests, a warm-up sample was recommended. The respondent was told to ignore it and it was given primarily to adjust the state of the oral cavity to the taste at hand. This part of the procedure has largely been forgotten, but is worthy of reconsideration. The paper also described the general statistical test for significance in choice tests, which was the simple binomial-based test on proportions, using the normal distribution approximation, as discussed below.

So in this early publication, two of the major categories of overall difference tests were described, namely sorting tests (triangle) and matching tests (duo trio and dual standard). Another example of matching, similar to the dual standard is referred to as the ABX test (Huang and Lawless 1988) in which the two versions of the product are presented in the inspection phase but only one of the products to match in the blind-coded test phase. The triangle test is the simplest example of a sorting task (make two piles) and is sometimes called an oddity task (Macmillan and Creelman 1991). Another example of sorting is the tetrad test (Masuoka et al. 1995) in which four samples are given at one time. They consist of two duplicate pairs, and the judge must sort them into two groups of two. A more elaborate test was the two-out-of-five test invented by Amoore et al. (1968) in order to study odor thresholds and specific anosmia (smell blindness to a specific compound or to a related family of compounds). In this task, the participant is provided with five sniff bottles, two of which contain the test compound at a specific concentration, and three of which contain the diluent only. The task once again is to make two groups by choosing the two with the target odor. In the threshold testing, the concentration would be raised in semi-logarithmic steps to see at what point the individual could correctly discriminate. An even more challenging version of this was the four-out-of-eight test attributed to Harris and Kalmus (1949) in testing for taste blindness to PTC (phenylthiocarbamide, phenylthiourea) and related compounds. The potential fatigue aspects of having four bitter samples, even near threshold was recognized by later workers and so the task was modified to a three-out-of-six task by Lawless (1980).

As noted above, in all of these tasks, an answer is forced. If they feel uncertain, they must guess. This forced aspect of the test allows the specification of the chance performance level, and thus the statistical test against a mathematical statement of the null hypothesis as discussed below. The forced nature of the test renders them bias-free or criterion-free. That is, the respondent cannot set some arbitrary criterion for response versus non-response. The nature of the response is specified and required.

Another category of forced-choice tests are the n-AFC tests in which an attribute is specified, and the participant must choose which of two or more items has the strongest amount of that characteristic. The simplest example is the paired comparison test for differences. A common phrasing might be "which of these two samples is sweeter?" One of the samples is expected to have more of the specified attribute due to an ingredient change, for example, and so there is a correct answer, making the statistical test one-tailed. As the test is a forced-choice, another term for the paired comparison test is a two-alternative forced-choice test or 2-AFC. More than two examples are often given to lower the chance probability level and demand a higher level of discrimination. Usually the additional samples are the ones that are lower in content of the critical ingredient (e.g., sugar), and thus expected to be lower in the resulting sensory attribute (e.g., sweetness). However, in some cases, the duplicate items may be more concentrated, and the person will be asked to choose the item with the least of that attribute (e.g., Byer and Abrams 1953). This is not common. If we call the item to be chosen the "target" and the items with less of the substance the control, we can have a 3-AFC test with one target and two controls, or an n-AFC test with one target and n – 1 control samples.

The final major category of tests consists of those that are not criterion-free. In these cases, the respondent must set some mental boundary for how much evidence is needed to choose one of two (or more) possible responses. The simplest case is the A-not A test. In this task, participants are first familiarized with a standard or control product ("A"). Then they are presented with items one at a time, sometimes the control and sometimes the test product, and they must specify whether the item is a good example of the previously inspected item (answer "A") or not (answer "not A"). In the simplest version of this test, each participant would receive one of each version in the test phase. This design will allow a McNemar test to be conducted for a statistical difference, and controls for the response bias or criterion by allowing for estimation of a false alarm rate. A false alarm is calling something "not A" when it was in fact an example of the control product. Thus, it is critical to include both the control and test items in the test phase as part of the design. The McNemar test compares the frequency of those who respond "not A" to the test item and also "A" to the control item, to the frequency of those with the opposite response.

The same issue is inherent in the same/different test. In this task, samples are presented in pairs of either identical items (control/control or test/test) or different items (control/test or test/control). The participant must choose a response, either "same" or "different." In this case they must set a mental criterion for how much of a perceived difference is necessary in order to respond "different." Some people might be biased to call almost everything "different" while others might require a clear and obvious difference in order to make that choice. So, as in the A-not A test, it is critical to have a control or baseline for comparison, in this case a control pair of identical items. Once again the McNemar test can be used. It is used to compare the frequency of those who respond "different" to the different pair and "same" to the identical pair, vs the two groups that change their response, but with the opposite response pattern.

Both of these tests can be modified to allow for additional responses, rather than a binary choice. Usually the response alternatives take the form of a sureness rating or certainty rating. For example, in the A-not A, the responses may be offered as follows: "A, sure," "A, not sure," "not A, not sure," or "not A, sure." This approach is sometimes called the R-index method (O'Mahony 1979), and allows a calculation of response differences to the two items that gives a value analogous to the percent correct choice in a 2-AFC task. This method is discussed in the chapters that follow on signal detection and Thurstonian models. The response scale itself does not take the bias out of the situation, but the presentation once again of both versions of the product. A similar sureness scale can be used for the same/different procedure (Delwiche and O'Mahony 1997; Kamerud and Larson 2010).

Rather than ask for a sureness rating, other ratings may be used. One case is just to ask for a rating of overall difference (Aust et al. 1985; Pecore et al. 2006). Ratings for similarity or dissimilarity were routinely collected in multidimensional scaling studies. The response scale may be labeled "exact match" at one end and "extremely different" at the other, or some similar wordings. The respondent chooses from either response categories, or can make a mark on a line scale (visual analog scale), for example. Once again, it is critical to include pairs in which the samples are identical, in order to get a baseline for responses when there is really no difference. If each respondent receives one test pair and one baseline pair, the same experimental design is present as for the binary same–different situation. However, because there are now more response options, the Stuart–Maxwell test is appropriate for categorical responses, or the paired (dependent samples) *t*-test if the data are considered more continuous.

A little-used modification of the same/different procedure gives four items in two pairs, one pair of which is identical and one contains the different samples. So the task is to specify which of the pairs contains the different items. This is called the dual pair method or 4-IAX procedure in psychology. This returns the same/different test to a true forced-choice and thus criterion-free situation. It has generated some interest in the sensory testing community (Rousseau and O'Mahony 2000, 2001; Rousseau and Ennis 2001, 2002), but at this point there appears to be little or no consistent advantage over other methods.

## 2.3 General Test Considerations

Two overriding principles are common to all sensory evaluation procedures, namely the principles of blind testing and independent judgments. By blind testing, we mean that the participants should not know the identity of the products, especially regarding which item is likely to be a control product versus a modified or test product. Accordingly, it is common practice to label products with random three-digit codes. Of course, the participant must be aware of the general category of the product, in order to form a frame of reference and adopt a sensible sampling or tasting method, unless the test specifically instructs them how to taste the product. But beyond the general type of product, there should be no other information. Independent judgments are necessary in order to have data that are not influenced by other people's opinions or decisions. Statistically we assume an independent random sampling of events. Independence of judgments is often facilitated by conducting the test in booths in a sensory testing facility, in order to isolate judges from the responses or opinions of others. These principles are so evident to sensory evaluation professionals that they are rarely even mentioned, just assumed.

The facility for discrimination tests is usually a sensory testing laboratory with booths connected to a test kitchen or preparation area. Communication to and from the test area is often achieved by a signaling system, and products are served through a small sliding door or other apparatus that prohibits the respondent from viewing the preparation area. The test and prep areas should be free from odors, and generally there is positive pressure from the test area to the prep area. Temperature and sometimes humidity are often controlled. The primary goal is to produce a situation that is free from distractions and unwanted influences, one that allows the panelist to concentrate on the task at hand. Descriptions of sensory testing facilities and further details can be found in most sensory evaluation textbooks (e.g., Lawless and Heymann 2010). Specific guidelines have also been written by various professional groups and committees (e.g., Eggert and Zook 1986).

The panelists chosen for a standing discrimination panel in a product research facility are generally selected to have normal sensory function. For a food test, for example, they should have a normal sense of taste and smell. For fabric handfeel, they should have no skin ailments or conditions that would suggest an abnormal sense of touch (e.g., diabetic neuropathy). For taste and smell, simple screening tests can be constructed, and recommendations can be found in sensory evaluation texts. Sometimes these take the form of discrimination tests using spiked or specially constructed products. For example, for a panel dealing with fruit juices, it would be useful to ascertain whether the candidates can discriminate various levels of sugar or acid. The screening tests are most useful when they use versions of the products that the panel will actually deal with in future evaluations. Threshold testing is not recommended, as most product tests do not involve sensations near threshold. One exception would be tests for a taint or off-flavor, where it might be advisable to insure that the testers were sensitive to the substance of interest. Obviously, people judging color and appearance should not be colorblind or have some color vision anomaly. Simple standardized tests for color perception are readily available from ophthalmology suppliers, such as the famous Ishihara plates. Various on-line versions can be found but these assume you have an accurate color monitor. Note that non-employees can also be used for discrimination panels, but they would go through a similar screening procedure.

These principles assume that you are using the panel as an analytical instrument, and your primary goal is to ensure a sensitive test that does not miss differences that could potentially be important. This is the "safety net" philosophy. It states that if my screened and selected panel cannot tell the difference under highly controlled conditions, then consumers are not likely to do so out in the real world either. One must be careful to communicate this logic to management, lest they assume that the results are completely projectable and representative of what consumers would do. In other cases, finding a result that is truly representative of consumer perception is the goal. O'Mahony and colleagues (O'Mahony 1995; O'Mahony and Rousseau 2002) distinguish between these two kinds of orientations as sensory evaluation Type I and Type II (an unfortunate parallel with statistical terminology). In Type II, the goal is for the data to represent the perceptions of consumers. In that case, no screening would be done and the discrimination panel would be selected primarily for their usage of the product or product category. Sensory testing programs should be careful in choosing that route, because recent research confirms that up to 25% of consumers cannot meaningfully distinguish even the most obvious sensory differences (Stone and Sidel 1993; Chapman and Lawless 2005). If a random sample of consumers is used for discrimination testing, all parties involved must realize that there is an important limitation on the sensitivity of the test. A non-significant statistical result is almost useless, because the result may have been obtained from a group of non-discriminating persons. Of course, if a difference is found, that result is still actionable.

Other important methodological considerations are whether to allow re-tasting and whether to replicate the test using the same panelists. Re-tasting is something of a judgment call and may depend heavily upon whether the product is prone to fatigue the senses or otherwise have deleterious sequential effects on people's abilities to distinguish sensory differences (Rousseau and O'Mahony 2000). Generally, it is advantageous to retaste products, as the first impression can be misleading, especially for the first product if there is no warmup sample. Thustonian modeling considerations suggest that re-tasting can enhance discrimination. Whatever the decision about allowing re-tasting, it should be made thoughtfully and usually as a committee decision. If re-tasting is not allowed, the instructions on the ballot or questionnaire should clearly specify that participants cannot go back and try a sample again.

This brings up the question of how the product is to be sampled or handled by the panelist. Different products may dictate different strategies. In doing coffee triangle tests, for example, I found it helpful to try all three coffees quickly, just to acclimatize my palate. Then each (black) coffee would be sampled from a spoon, while "slurping" to aerate the sample and maximize the perception of the volatiles. These were "tricks" that were not specified by the instructions on the ballot, but ones that were learned from talking to more experienced tasters. If the panelists' sensitivity is to be maximized, it may be worthwhile to have some training in appropriate tasting methods to show them what seems to work for most people. Instructions for tasting or product handling should be clearly stated on the ballot, including how much of the sample they are required to ingest, if it is a food test. A sample ballot for a triangle test is shown in Figure 2.2.

**Welcome to the sensory laboratory discrimination panel.**
**Today's test will be a triangle test on**

**BRAISED TRAKE**

You will be served three samples on separate plates marked with three-digit codes. Please make sure the codes on the plates match the codes listed below.

Taste the samples from left to right as indicated on this sheet.
You (may)(may not) go back and re-taste the samples.

You must eat at least half of each sample.
You may spit out the sample in the spit cup that is provided.
Please eat a bite of cracker and rinse your mouth with water before tasting each sample.

CIRCLE THE NUMBER OF THE SAMPLE THAT IS MOST DIFFERENT FROM THE OTHER TWO.

IF YOU HAVE ANY QUESTIONS, PLEASE ASK THE SERVER NOW.

837 456 925

When you are finished, please signal the server and place all your items back on the tray.

THANK YOU FOR YOUR PARTICIPATION.

**Figure 2.2** A sample ballot for a triangle test.

Replication is generally desirable. If a person takes the time to leave their job site and duties to come to the sensory lab or test facility, it makes sense to ask them to do more than a single test. This is simply an efficient use of the person's time. Of course, if the product is very fatiguing or has lingering tastes or other sensations such as hot pepper burn, then the opportunity for replication may be limited or even counterproductive. For less experienced assessors, replication may provide a kind of practice effect (Dacremont and Sauvegeot 1997). Statistical approaches to handling replicated data vary, and are discussed in a subsequent chapter (Brockoff and Schlich 1998). A simple approach is to consider each replicate as a separate test. In a duplicate test one can use the binomial analysis, but change the chance probability level and only count those that chose correctly in both parts (e.g., 1/9 rather than 1/3 in the triangle). At the other end of the spectrum, there are complex beta-binomial models (Ennis and Bi 1998; Bi 2006a: Liggett and Delwiche 2005). These allow combination of each replicate into an overall count, but after taking into consideration the non-randomness patterns of individual assessors. In general, the sensitivity of the test and statistical power can be enhanced by replication (Brockoff 2003).

The considerations discussed here are only a brief overview of sensory testing principles and the reader is referred to other sources for more detailed information. As always, the selection of the test method, panelists and other test considerations should be consistent with the goals of the individual project. One must not use a discrimination test when the goal is to find out about consumer acceptance of a new product that is clearly different. As a final note, remember that panelists are humans and not really testing instruments and as such, they require some form of motivation in order to take the test seriously and do their best. It is common to provide token incentives such as food treats or raffle tickets to panelists serving in a routine testing program. In my experience with coffee, the sensory group would publish our individual monthly "batting average" on the wall of the testing room, which I suppose is potentially motivational if you are a competitive person.

## 2.4 Basic Statistical Analysis

### 2.4.1 Null and Alternative Hypotheses

The forced-choice nature of most discrimination tests lends itself to a straightforward statistical analysis. Because responses can be classified as correct or incorrect, there are two outcomes and thus the binomial distribution is the appropriate model. Due to the association with a clear chance-responding probability level, the null and alternative hypotheses can be set up as specific mathematical statements. This is a common pitfall for students, who are prone to phrase the null as "there is no difference," probably due to their first exposure to $t$-tests on means. That statement of the null is incorrect, although it persists in some textbooks (Stone and Sidel 1993). The correct null is a statement that the population proportion of correct choices is equal to the chance probability level, and the alternative hypothesis is that the population proportion correct is greater than the chance probability level. Thus for the triangle test, the null and alternative appear as follows:

$$H_o{:}p = 1/3 \quad \text{and} \quad H_a{:}p > 1/3 \tag{2.1}$$

where $p$ is the expected population proportion making correct choices. Do not confuse this with "$p$" as used to symbolize the chance probability level or with the probability associated with a specific experimental outcome (i.e. the "$p$-value" for an obtained statistic's value).

### 2.4.2 Exact Binomial, z-Score Approximation, and Chi-Square

The calculated statistics and $p$-values for any choice test result can be found in three related approaches. The first, and most accurate, is to calculate the actual binomial distribution $p$-value associated with the outcome (count of number correct), and all results that are greater than that count. This amounts to calculating the area of the tail of a binomial distribution from the obtained count to $N$, i.e. what is the total probability of seeing an event this extreme and all events that would be more extreme, under a true null. The binomial probability of an event where the correct choice count is $x$ out of $N$ total events (think of $N$ as the number of judges for now) is given by

$$p(x) = \binom{N}{x} p^x (1-p)^{N-x} \tag{2.2}$$

where the expression $\binom{N}{x} = (N!)/(x!(N-x)!)$ is simply the combinatorial rule for $N$ items combined $X$ at a time. Because we have to sum for all values from $x$ to $N$ to get the total probability in the tail, we can use this:

$$p(x_{total}) = \sum_{y=x}^{N} \binom{N}{y} p^y (1-p)^{N-y} \tag{2.3}$$

This calculation becomes tiresome if done by hand, especially for large $N$ and/or large values of $x$. For that reason, the $z$-score approximation to the binomial can be used to simplify matters. This amounts to a test of the observed proportion versus the chance-expected proportion as follows:

$$z = \frac{\left[P_{observed} - p\right] - 1/2N}{\sqrt{p(1-p)/N}} = \frac{\left[x - Np\right] - 0.5}{\sqrt{p(1-p)N}} \tag{2.4}$$

where $P_{observed} = x/N$ and $p$ is the chance probability. Some statisticians prefer that the experimentally obtained proportion (i.e., $x/N$ and $1-x/N$) be used instead of the chance probability in the denominator, but this often makes only a small difference. Note that we have a continuity correction in the numerator due to the fact that the normal distribution is continuous but the binomial distribution is counting heads. You can think of this as correcting for the maximum degree of inaccuracy – the most we could be off is by 1/2 of a person. As the test is one-tailed, the minimum value for $z$ is 1.645 in order to find a $p$-value less than 0.05. This fact produces the following rearrangements for the minimum number of correct answers, $X_{min}$ out of $N$ observations:

$$X_{min} \geq \frac{2N+3}{6} + 0.775\sqrt{N} \quad \text{for methods with a chance probability of 1/3.} \tag{2.5}$$

And

$$X_{min} \geq \frac{N+1}{2} + 0.8225\sqrt{N} \quad \text{for methods with a chance probability of 1/2.} \tag{2.6}$$

These latter two equations are used to construct tables of minimum numbers of correct judgments required for declaring statistical significance. But it is a rather simple matter to take any proportion you observe and place it into Eq. (2.4) to see if the resulting $z$-score exceeds 1.645.

A mathematically equivalent formula preferred by some is to use the chi-square distribution instead of the normal approximation. This takes the form:

$$\chi^2 = \frac{\left(|x - np| - 0.5\right)^2}{np} + \frac{\left(|(n-x) - n(1-p)| - 0.5\right)^2}{n(1-p)} \tag{2.7}$$

where $x$ is the observed count correct, $n - x$ is the count of incorrect choices, $np$ is the expected count of correct choices and $n(1 - p)$ is the expected count of incorrect choices. Thus we have the familiar chi-square formula for the sum of observed counts minus expected (quantity squared), each divided by the expected count. This is equivalent to the $z$-score approximation, as chi is the square of $z$, a proof of which can be found in Lawless and Heymann (2010, pp. 503–504). Note that the equivalence only holds if both equations either use or do not use the continuity correction.

### 2.4.3 McNemar and Stuart Maxwell Tests

Recall that for the A-not A tests and same/different tests there is no specific chance probability associated with a correct response. So the statistical approach must be different than the simple binomial model. One way to design the method is to have each participant only make one judgment. That is, give them one sample only in the A-not A or one pair only in the same/different test. That way a simple chi-square test on frequency counts can be done as the judgments are a priori independent. However, this is not a very efficient design and ignores the fact that judges are not like coins being tossed, but rather they have biases and their responses tend to be correlated within a person over multiple trials. Taking advantage of this fact, we can design a balanced test where each person assesses not only the test item (or test pair in same/different) but also the control item (or identical pair in same/different). Thus the data may be cast as shown in Figure 2.3.

Note that the letters, A–D, refer to the frequency counts in each of the four possible combinations. Given our paired design, then, each person contributes to only one cell, and thus the counts are once again independent. The McNemar statistic is distributed as chi-square with one degree of freedom and thus a critical value of 3.84. It is simply calculated as follows:

$$\chi^2 = \frac{\left(|\mathrm{B} - \mathrm{C}| - 1\right)^2}{\mathrm{B} + \mathrm{C}} \tag{2.8}$$

Note that this equation ignores the people who gave the same response on the two trials (people falling into counts A and D), which is perhaps appropriate as they are transmitting

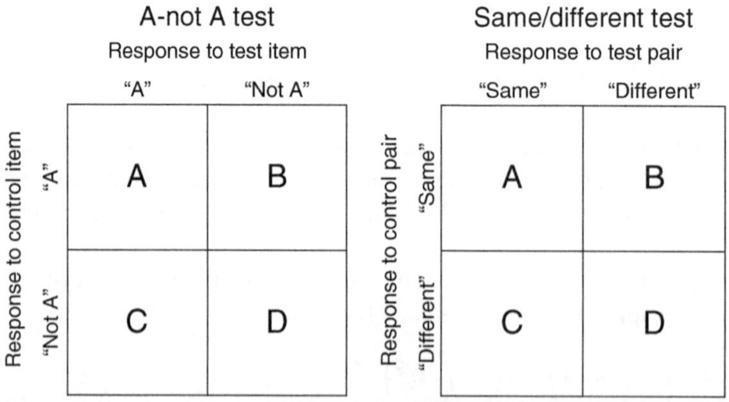


**Figure 2.3** The data tabulation for a McNemar test is based on a 2 × 2 response matrix, with comparison of the frequency counts in the diagonal cells (B and C) in which responses were different for the two products or two pairs.

no information. However, if these counts are very large compared to those in cells B and C, the sensory professional should reconsider the value of any statistically significant result with McNemar. Also note that the effective sample size, B + C, will be less than the total, *N*. So there may be some loss of test power with this approach. The test is mathematically equivalent to a nonparametric sign test on two proportions.

For rating scale data with more than two responses, the extension of the McNemar test is the Stuart–Maxwell test (if there are two samples) or the Cochran–Mantel–Haenszel test (CMH for short) for three or more samples. Discussions of these tests can be found in Fritz (2009) and Bi (2006a) as well as Gacula et al. (2009). The Stuart test has simple mathematical solutions for three categories, but requires matrix calculations for four or more categories and thus is better handled by statistical programs. Further information on analyses of A-not A tests can be found in Bi and Ennis (2001) and Christensen et al. (2011).

## 2.5 Test Power and Sample Size; Equivalence and Similarity

### 2.5.1 A Quick Look at Sample Size and Test Power

Because discrimination tests can be used to gather evidence for product similarity and equivalence, the sensory professional must be concerned with the possibility of Type II error. Avoiding Type II error can be critical in cases where the change goes undetected by the sensory test, but is noticeable to some loyal consumers who may be alienated by an unexpected difference in their favorite product. Conversely, a product improvement that goes undetected may result in lost opportunities. One way to guard against Type II error is to make sure the test has a sufficient panel size and sufficient test power to detect a difference if one exists. This is a complex topic that will be dealt with in a subsequent chapter, but the basic ideas will be introduced here.

The test power is the ability of the test to detect a difference that truly exists, and mathematically is defined as the probability given by one minus beta, where beta is the long term probability of committing a Type II error. Power is a function of several interacting factors: the panel size (*N*), the alpha level, the beta level, the standard deviations or error in the measurements, and the size of the effect one is trying to detect. The factors behave in systematic and largely predictable ways. If *N* is high, the test is more likely to find a difference. If alpha floats up, it is easier to reject the null and thus alpha and beta are inversely related, all other things being equal. If the variance or standard deviations are reduced, the test becomes more sensitive and powerful. Finally, it is easier to detect a large difference than a small one, so the effect size one must be sure to detect is a critical factor in determining the test power. The last item depends upon your product knowledge, and what the consumer implications are likely to be for any difference you miss.

Starting with the panel size is usually a basic consideration in sensory testing. How many people do I need in the test? The general relationship for *N* as a function of the other factors discussed above is given by the following equation:

$$N = \left[ \frac{Z_\alpha \sqrt{p_o q_o} + Z_\beta \sqrt{p_a q_a}}{p_o - p_a} \right]^2 \tag{2.9}$$

where *Z*'s are the *Z*-scores associated with the alpha and beta risk, $p_o$ and $p_a$ are the probabilities associated with the null and alternative hypotheses, and $q = 1 - p$. The denominator is often the critical influence on sample size, as small changes in the difference between the null and alternative probabilities can have a big effect.

A worked example. Suppose we let alpha equal 0.05 ($z = 1.645$, one tailed) and beta equal 0.10 ($z = 1.645$, two tailed) for a triangle test, and we wish to detect a difference where 2/3 get the test correct. The value of 2/3 is equivalent to a proportion of true detection of 50%, after the correction for guessing. This might be considered a rather large proportion of people detecting the difference. Putting these values into Eq. (2.9), and rounding to the nearest whole number, we get

$$N = \left[ \frac{1.645\sqrt{0.33(0.67)} + 1.645\sqrt{0.67(0.33)}}{0.33 - 0.67} \right]^2 = 22$$

To detect a smaller difference, where $p_a$ is only 0.5 (giving 25% detection) the required panel size now goes to 92. The denominator is the major influence. Detecting a small difference is more difficult than detecting an obvious one.

The correction for guessing is given by Abbott's formula, as

$$P_{observed} = P_{corrected} + P_{chance}\left(1 - P_{corrected}\right) \tag{2.10}$$

where $P_{observed}$ is the required proportion to obtain as a result, in order to claim there is a $P_{corrected}$ proportion of actual discrimination. There are various other rearrangements of this formula (Finney 1949). Working from an observed result, the $P_{corrected}$ is given as follows:

$$P_{corrected} = \frac{P_{observed} - P_{chance}}{1 - P_{chance}} \tag{2.11}$$

Rearranging Eq. (2.9), and solving for $Z_\beta$, the power can be found from taking the cumulative normal distribution function of the expression equal to $Z$, and subtracting it from one, since power is equal to one minus beta. The equation for test power in a discrimination test then becomes

$$Power = 1 - \Phi\left[\frac{Z_\alpha \sqrt{p_o q_o / N} - (p_o - p_a)}{\sqrt{p_a q_a / N}}\right] \tag{2.12}$$

where $\Phi$ represents the cumulative normal distribution function that coverts a $z$-score into a proportion.

### 2.5.2 Equivalence and Similarity

Using a discrimination test to gather evidence for product equivalence is a complex undertaking. To the non-sensory researcher, it might seem to makes sense to accept a non-significant result as evidence for "no difference." However, a failure to reject the null hypothesis can occur for many reasons. For example, you may have tested too few people to detect a small effect. There may have been errors in serving or preparation. Perhaps the panel was not properly vetted, that is, screened and oriented to the task. There are many ways to do a sloppy experiment and get a null result. So a failure to reject the null hypothesis is often ambiguous, and thus it is difficult to draw any conclusions.

These opportunities for failure raise the question of whether a result of "no significant difference" is ever evidence for product similarity. The answer is yes, but the conditions, record and sensitivity (i.e., power) of the test must be known. For decades, workers in sensory evaluation have used a non-significant discrimination test as evidence for product similarity. Common sense informs us of when this is a trustworthy result. The first concern is the known discriminative ability of the panel as an instrument for detecting differences. If the panel has a known track record of finding significant differences, we can have some assurance that for any subsequent test done in the same manner with the same pool of testers, that they will continue to detect such differences. Second, was the test conducted properly? This involves the methodological considerations discussed previously. Do we have a sufficient panel size, controlled conditions, experienced workers in the prep and serving areas, a specified protocol for preparing, handling and discarding products, blind testing and independent judgments, and of course an appropriate statistical analysis. Finally, a test result that is close to the boundary for statistical significance is more problematic than one that is near the chance performance level. The sensory professional should know and appreciate the ability of the panel to detect differences, as well as how the discrimination panel results relate to consumer perception and possibly preferences.

A somewhat more sophisticated approach is to make use of such nonsignificant results, but only under conditions of known test power and appropriate sample size, using the kind of calculations shown above. An important consideration is the effect size one wishes to be able to detect. Another way to phrase this is, "how much of a difference can we tolerate?" No two products are ever exactly the same. Zero difference is not an actionable goal. So a reasonable limit on the degree-of-difference must be set by management, with careful forethought and if possible, consideration of the consumer impact of a given size of a change.

Specifying the degree of difference can be done in several ways, as discussed below. Once the degree of difference is specified, the denominator of Eq. (2.9) is known, and we can determine our sample size and test power (Eq. 2.12).

Direct tests of similarity are also possible. In order to do so, one must specify the upper limit on the degree of difference, and then do a statistical test to see if the obtained result is significantly below this upper limit. This logic turns the statistical test on its head. Rather than looking down to see if the result is significantly above chance, the researcher is now looking up at a ceiling, to see if the 95% one-tailed confidence interval remains below that upper limit. As noted above, the tolerable degree of difference must be specified by management. There are two ways to specify the degree of difference. One is to use Abbott's formula to calculate an allowable proportion of discriminators, and then translate this into a percent correct that serves as the upper bound. For example, 50% discriminators in a triangle test would require 66.7% correct after the correction for those who luckily guessed and chose the correct item as the odd sample. This approach, however, does not take into account the inherent variability or difficulty of the test, only its chance performance level. Thus a 3-AFC test and a triangle will yield the same proportion of discriminators, but choosing the odd sample is a much more demanding task than simply picking the strongest tasting of three. An alternative measure is the Thurstonian model's sensory difference measure, called $d'$ (d-prime) in signal detection, and known also as a delta value in much of the literature (e.g., Bi 2006a). The derivation of this measure is too complex to fully discuss here, but is covered in a subsequent chapter. The important fact to remember is that all of the common sensory tests now have known psychometric functions that allow the conversion of a proportion correct into a Thurstonian delta value and vice versa. These models take into account the inherent variability of the tests as well as their associated psychological or cognitive strategy.

Returning to the tests for similarity, the idea was introduced by three sensory scientists in the early 1990s, who took similar approaches and all used the concept of proportions of true discriminators as a way to set the upper bound. Around 1990, Carr (in Meilgaard et al. 2006) published tables for the maximum allowable proportions correct, given different proportions of discriminators, alpha and beta levels, for specific test methods. Schlich (1993) published tables with useful crossover points. Given levels for alpha, beta and proportions of discriminators, the tables showed a required panel size, and then a crossover point. If the number of correct choices was below this point, one could conclude significant similarity, and if the count was below, one could declare a significant difference. MacRae (1995) used a similar approach to Carr's, but wrote in terms of 95% confidence intervals. Bi (2011) showed how to use the Thurstonian delta value as a limit, but converted it back to the expected percent correct. He then suggested using the simple binomial expansion to prove that a result was below that level by the 95% confidence interval, an approach very similar to the first three authors mentioned above. Specifically, you first refer to the psychometric function tables to find the proportion correct, $P_c$ that will be produced by your maximal allowable delta value, and then solve for the maximal observed number correct, $X_c$. This will leave 5% or less in the tail of the binomial distribution based on $N$ and $P_c$. Using binomial expansions has also been discussed by Ennis and Ennis (2009) who showed how to set up a rejectable null hypothesis of non-equivalence, but mostly within the realm of preference testing for parity claims (a two-tailed situation, rather than one-tailed as in discrimination testing).

Whatever the strategy that is taken for gathering evidence of product similarity, the sensory specialist should be prepared to use a much larger panel size than that which is normally employed for simple tests of difference. There are several reasons for this. First, the power of the test is related to the sample size, meaning the panel size or number of observations. Second, if the panel size is not large enough, the confidence intervals will be too large to fit under the allowable ceiling. Remember that a large "*N*" will reduce the standard error, and thus the confidence interval. Finally, if the ceiling given by the maximum allowable difference gets lower, that ceiling moves towards the chance probability level, producing a very narrow "window" within which the result and its upper confidence interval must fit. Requiring a result below chance is not practical. For example, Carr's tables show blank values where this would happen. Historically, it was common to do discrimination testing with panels of 25–50 testers, but many research groups are now using panels twice that size when conclusions about product equivalence have important consequences.

## 2.6 Psychological Factors

Any discrimination test can be approached by a panelist as a test of their abilities. So it is important to realize that most people will strive to get the "right answer." They will try to figure out what is going on. For example, in a duo-trio test that is repeated, they may assume that each product serves as the reference once. This may or not be the experimental design, but is an example of how a panelist can outsmart himself. This motivation to make the correct response results in a limitation on how the tests can be conducted. A triangle test must be a test of overall difference. You cannot ask the panelist to pick out the sample that is most different in "attribute *X*." They may or may not pay attention to that attribute. They simply want to give the right answer. Furthermore, most product changes produce multiple sensory effects, and often these effects are correlated or redundant. For example, you cannot change the sweetness of a fruit beverage without also altering the sourness, due to mixture interactions in taste. As one increases, the other will decrease. At least two sensory effects are present if you change the sweetener level (and probably more).

Different tasks require different cognitive strategies. This leads to major differences in the overall sensitivity of a test to differences or power (Ennis 1993; Ennis and Jesionka 2011). In a triangle test, the task requires one to make three paired comparisons, and decide which two items are the duplicates based on the smallest sensory difference (Frijters 1979). The odd sample is by default the remaining item. This is sometimes called a "comparison of distances" strategy. In the 3-AFC test, on the other hand, one merely has to choose the strongest of three items in the specified attribute. This is less demanding and less inherently variable than making three paired comparisons. This strategy is sometimes referred to as "skimming." Thus for a given amount of sensory difference, the triangle test will yield a smaller percent correct than the 3-AFC test. This formed a widely known paradox originally attributed to Byer and Abrams (1953), that some people could make the wrong choice under triangle instructions, but choose correctly under 3-AFC. Frijters (1979) showed that this paradox was not real, and resolved it by showing that the Thurstonian model's sensory distances were the same for both outcomes. The practical importance of this finding is that AFC tests are generally more sensitive to differences and yield a higher percent correct than their counterparts in the overall difference category. Thus they also require fewer panelists to get the same level of statistical significance. These power issues will be discussed in the chapter on Thurstonian models later in this book.

Other psychological considerations involve memory load and sequential effects such as adaptation and carry-over, sensations that linger and can't be effectively removed. Remembering the reference sample in a procedure like a duo-trio test may impose some difficulties for some panelists. For this reason, some researchers have suggested that the reference sample should come in the middle of the series, that is, between the two test items, (Rousseau and Ennis 2002) or be repeated at the end of the trial (Lee and Kim 2008; Kim et al. 2010). Other modifications involve repeating or refreshing the reference periodically (Hautus et al. 2011; Van Hout et al. 2011). Thorough experimental evaluations of these modifications are lacking, and so the advantages or potential advantages are unclear.

With regard to sequential effects, O'Mahony and colleagues have done extensive studies of different orders of presentation (O'Mahony and Odbert 1985; O'Mahony and Goldstein 1986). Most sensory professionals stick to the basic idea that orders of sample presentation should be randomized or counterbalanced across panelists and/or trials. However, based on considerations in taste adaptation, O'Mahony has pointed out that not all sequences are created equal, due to sensory adaptation. Adaptation is a kind of fatigue, decrement or adjustment in response of the sensory system after stimulation changes. So for example, upon leaving a dark cinema, the bright sunlight may seem unbearable at first, but our eyes soon adjust to it. So a sequence of products that goes strong-strong-weak may not lead to as good a discrimination as one that goes weak-weak-strong. This general theory goes by the name "sequential sensitivity analysis." Its predictions have not always been substantiated (Santosa and O'Mahony 2008) and most of the research has been conducted on simple salt solutions, rather than real products. However, the sensory specialist who wants to maximize the sensitivity of the test might want to consider these effects and the possibility of using non-random, i.e. fixed, orders.

## 2.7 Summary and Conclusions

Discrimination testing is one of the three major tools in the sensory evaluation toolbox along with descriptive analysis and consumer affective testing. The discrimination methods are designed to address questions of whether any difference is perceivable, and are often done before descriptive testing is used to characterize the difference, or consumer testing is done to determine if the difference matters. A variety of techniques have been used over the years, most notably the triangle test and duo trio tests for overall differences. The attribute specific tests such as the paired comparison and 3-AFC are more powerful and sensitive to differences than the overall difference tests, but the attribute in question cannot always be specified beforehand. When making a product change in ingredients or processing, it is not always possible to predict the sensory effect, and there may be many of them.

The choice tests are straightforward to conduct, and statistical analysis is simple. Most texts provide common lookup tables that show the minimum number of correct judgments for statistical significance, based on a given panel size. More complex analyses are also available for showing product equivalence, and psychometric models such as Thurstone's abound for the common methods. The latter can provide information on the degree of sensory difference, beyond simple statistical significance. Keep in mind that statistical significance depends in part on the panel size, a more or less arbitrary decision that has nothing to do with the product differences or human perception. Another common index of differences is the proportion of discriminators, determined by the correction for guessing. So there are three pieces of information that can result from any given choice test: the statistical outcome, proportion of discriminators, and the size of the modeled sensory distance or difference. The latter two may be more or less useful to management, given the limitations outlined above. This volume will revisit these themes in subsequent chapters in greater detail.

## References

Amoore, J.E., Venstrom, D., and Davis, A.R. (1968). Measurement of specific anosmia. *Percept. Motor Skills* 26: 143–164.

ASTM (2019a). Standard practice for determining odor and taste thresholds by a forced-choice ascending concentration series method of limits, E-679-19. In: *Annual Book of Standards*, vol. 15.08, 36–42. Conshocken, PA: ASTM International.

ASTM (2019b). Standard practice for defining and calculating individual and group sensory thresholds from forced-choice data sets of intermediate size, E-1432-19. In: *ASTM International Book of Standards*, vol. 15.08, 82–89. Conshocken, PA: ASTM International.

Aust, L.B., Gacula, M.C., Beard, S.A., and Washam, R.W. (1985). Degree of difference test method in sensory evaluation of heterogeneous product types. *J. Food Sci.* 50: 511–513.

Bi, J. (2006a). *Sensory Discrimination Tests and Measurements*. Ames, IA: Blackwell Publishing.

Bi, J. (2006b). Statistical analyses for R-index. *J. Sens. Stud.* 21: 584–600.

Bi, J. (2011). Similarity tests using forced-choice methods in terms of Thurstonian discriminal distance, *d'*. *J. Sens. Stud.* 26: 151–157.

Bi, J. and Ennis, D.M. (2001). Statistical models for the A-not A method. *J. Sens. Stud.* 16: 215–237.

Braun, V., Rogeaux, M., Schneid, N. et al. (2004). Corroborating the 2-AFC and 2-AC Thurstonian models using both a model system and sparkling water. *Food Qual. Prefer.* 15: 501–507.

Brockoff, P.B. (2003). The statistical power of replications in difference tests. *Food Qual. Prefer.* 14: 405–417.

Brockoff, P.B. and Schlich, P. (1998). Handling replications in discrimination tests. *Food Qual. Prefer.* 9: 303–312.

Byer, A.J. and Abrams, D. (1953). A comparison of the triangle and two-sample taste test methods. *Food Technol.* 7: 183–187.

Chacon, R. and Sepulveda, D.R. (2011). Development of an improved two-alternative choice (2AC) sensory test protocol based on the application of the asymmetric dominance effect. *Food Qual. Prefer.* 22: 78–82.

Wolf, M.B. (2020). *Sensory Testing Methods*. Conshohocken, PA: American Society for Testing and Materials.

Chapman, K.W. and Lawless, H.T. (2005). Sources of error and the no-preference option in dairy product testing. *J. Sens. Stud.* 20: 454–468.

Christensen, R.H.B., Cleaver, G., and Brockhoff, P.B. (2011). Statistical and Thurstonian models for the A-Not A protocol with and without sureness. *Food Qual. Prefer.* 22: 542–549.

Dacremont, C. and Sauvegeot, F. (1997). Are replicate evaluations of triangle tests during a session good practice? *Food Qual. Prefer.* 8: 367–372.

Delwiche, J. and O'Mahony, M. (1997). Changes in secreted salivary sodium are sufficient to alter taste sensitivity: use of signal detection measures with continuous monitoring of the oral environment. *Physiol. Behav.* 59: 605–611.

Eggert, J.K. and Zook, K. (1986). *Physical Requirement Guidelines for Sensory Evaluation Laboratories*. ASTM STP 913. Philadelphia, PA: ASTM International.

Ennis, D.M. (1993). The power of sensory discrimination methods. *J. Sens. Stud.* 8: 353–370.

Ennis, D.M. and Bi, J. (1998). The beta-binomial model: accounting for inter-trial variation in replicated difference and preference tests. *J. Sens. Stud.* 13: 389–412.

Ennis, D.M. and Ennis, J.M. (2009). Hypothesis testing for equivalence defined on symmetric open intervals. *Commun. Stat.* 31: 1792–1803.

Ennis, J.M. and Jesionka, V. (2011). The power of sensory discrimination methods revisited. *J. Sens. Stud.* 26: 371–382.

Ferdinandus, A., Oosterom-Kleijngeld, I., and Runneboom, A.J.M. (1970). Taste testing. *MBAA Tech. Quart.* 7: 210–227.

Filipello, F. (1956). A critical comparison of the two-sample and triangular binomial designs. *J. Food Sci.* 21: 235–241.

Finney, D.J. (1949). The adjustment for a natural response rate in probit analysis. *Ann. Appl. Biol.* 36: 187–195.

Frijters, J.E.R. (1979). The paradox of the discriminatory nondiscriminators resolved. *Chem. Senses* 4: 355–358.

Fritz, C. (2009). Methods for determining whether JAR distributions are similar among products (Chi-square, Cochran-Mantel-Haenszel (CMH), Stuart-Maxwell, McNemar). In: *Just-About-Right Scales: Design, Usage, Benefits, and Risks. ASTM Manual MNL63* (ed. L. Rothman and M.J. Parker), 29–37. Philadelphia, PA: ASTM International. Appendix G.

Gacula, M.C., Singh, J., and Altan, S. (2009). *Statistical Methods in Food and Consumer Research*, 2e. Amsterdam: Elsevier/Academic Press.

Harris, H. and Kalmus, H. (1949). The measurement of taste sensitivity to phenylthiourea (P.T.C.). *Ann. Eugenics* 9: 123–135.

Hautus, M.J., Shepherd, D., and Peng, M. (2011). Decision strategies for the a not-A, 2AFC and 2AFC-reminder tasks: empirical tests. *Food Qual. Prefer.* 22: 433–434.

Helm, E. and Trolle, B. (1946). Selection of a taste panel. *Wallerstein Lab. Commun.* 9: 181–194.

Hough, G. (2010). *Sensory Shelf Life Estimation of Food Products*. Boca Raton, FL: CRC Press/ Taylor and Francis.

van Hout, D., Hautus, M.J., and Lee, H.-S. (2011). Investigation of test performance over repeated sessions using signal detection theory: comparison of three nonattribute-specified difference tests 2-AFCR, A-not A and 2-AFC. *J. Sens. Stud.* 26: 311–321.

Huang, Y.-T. and Lawless, H.T. (1998). Sensitivity of the ABX discrimination test. *J. Sens. Stud.* 13: 229–239.

Kamerud, J. and Larson, G. (2010). Use of same–different test and R-index to efficiently compare multiple product differences. Poster presented and the Society of Sensory Professionals Meeting, Napa, CA (October 27–29).

Kemp, S.E., Hollowood, T., and Hort, J. (2009). *Sensory Evaluation. A Practical Handbook.* Oxford: Wiley-Blackwell.

Kim, M.-A., Lee, Y.-M., and Lee, H.-S. (2010). Comparison of *d'* estimates produced by three versions of a duo-trio test for discriminating tomato juices with varying salt concentrations: the effects of the number and position of the reference stimulus. *Food Qual. Prefer.* 21: 504–511.

Lawless, H.T. (1980). A comparison of different methods for assessing sensitivity to the taste of phenylthiocarbamide PTC. *Chem. Senses* 5: 247–256.

Lawless, H.T. (2010). A simple alternative analysis for threshold data determined by ascending forced-choice method of limits. *J. Sens. Stud.* 25: 332–346.

Lawless, H.T. and Heymann, H. (2010). *Sensory Evaluation of Foods, Principles and Practices*, 2e. New York: Springer Publishing.

Lee, H.-S. and Kim, K.-O. (2008). Difference test sensitivity: comparison of three versions of the duo-trio method requiring different memory schemes and taste sequences. *Food Qual. Prefer.* 19: 97–102.

Liggett, R.A. and Delwiche, J.F. (2005). The beta-binomial model: variability in overdispersion across methods and over time. *J. Sens. Stud.* 20: 48–61.

Macmillan, N.A. and Creelman, C.D. (1991). *Detection Theory: A User's Guide.* Cambridge: Cambridge University Press.

MacRae, A.W. (1995). Confidence intervals for the triangle test can give reassurance that products are similar. *Food Qual. Prefer.* 6: 61–67.

Masuoka, S., Hatjopoulos, D., and O'Mahony, M. (1995). Beer bitterness detection: testing Thurstonian and sequential sensitivity analysis models for triad and tetrad methods. *J. Sens. Stud.* 10: 295–306.

Meilgaard, M., Civille, G.V., and Carr, B.T. (2015). *Sensory Evaluation Techniques.* Boca Raton, FL: CRC Press.

O'Mahony, M.A.P.D. (1979). Short-cut signal detection measures for sensory analysis. *J. Food Sci.* 44: 302–303.

O'Mahony, M. (1995). Sensory measurement in food science: fitting methods to goals. *FoodTechnol.* 29: 72–82.

O'Mahony, M. and Goldstein, L.R. (1986). Effectiveness of sensory difference tests: sequential sensitivity analysis for liquid food stimuli. *J. Food Sci.* 51: 1550–1553.

O'Mahony, M.A.P.D.E. and Odbert, N. (1985). A comparison of sensory difference testing procedures: sequential sensitivity analysis and aspects of taste adaptation. *J. Food Sci.* 50: 1055–1058.

O'Mahony, M. and Rousseau, B. (2002). Discrimination testing: a few ideas, old and new. *Food Qual. Prefer.* 14: 157–164.

Pecore, S., Stoer, N., Hooge, S. et al. (2006). Degree of difference testing: a new approach incorporating control lot variability. *Food Qual. Prefer.* 17: 552–555.

Peryam, D.R. and Swartz, V.W. (1950). Measurement of sensory differences. *Food Technol.* 4: 390–395.

Rousseau, B. and Ennis, D.M. (2001). A Thurstonian model for the dual pair (4IAX) discrimination method. *Percept. Psychophys.* 63: 1083–1090.

Rousseau, B. and Ennis, D.M. (2002). The multiple dual-pair method. *Percept. Psychophys.* 64: 1008–1014.

Rousseau, B. and O'Mahony, M. (2000). Investigation of the effect of within-trial retasting and comparison of the dual-pair, same–different and triangle paradigms. *Food Qual. Prefer.* 11: 457–464.

Rousseau, B. and O'Mahony, M. (2001). Investigation of the dual-pair method as a possible alternative to the triangle and same–different tests. *J. Sens. Stud.* 16: 161–178.

Santosa, M. and O'Mahony, M. (2008). Sequential sensitivity analysis for same–different tests: some further insights. *J. Sens. Stud.* 23: 2008.

Schlich, P. (1993). Risk tables for discrimination tests. *Food Qual. Prefer.* 4: 141–151.

Stone, H. and Sidel, J. (1993). *Sensory Evaluation Practices*, 2e. New York: Academic Press.

Watson, C.S., Kellogg, S.C., Kawanishi, D.T., and Lucas, P.A. (1973). The uncertain response in detection-oriented psychophysics. *J. Exp. Psychol.* 92: 180–185.

# 3

# An Overview of Sensory Discrimination Tests

*Stella Lignou and Victoria Norton*

*University of Reading, Reading, UK*

Discrimination testing enables sensory scientists to understand whether two (or more) samples differ in corresponding sensory attributes. This type of testing is commonly used to demonstrate subtle or small differences between stimuli typically centered around two-sample or multi-sample related tests (Kemp et al. 2009; Lawless and Heymann 2010; Meilgaard et al. 2016; Rogers 2017; Stone et al. 2021). There are numerous discrimination tests available; however, depending on the specific context certain tests may be more suitable than others. Accordingly, this chapter will provide an overview of the different discrimination tests commonly utilized by sensory scientists coupled with relevant examples.

Sensory scientists have various factors they need to consider in terms of selecting the right discrimination test. For example, there are two key types of discrimination tests: (1) specified tests focusing on a specific attribute (this attribute will vary depending on the research question) such as *which sample is more mouthdrying?* and (2) unspecified tests which are more general (e.g., *identify the different coded sample*) and where there is no emphasis on a specific attribute. In addition, tests can be both specified and unspecified, namely tetrad and two-out-of-five; accordingly, highlighting the flexibility discrimination testing can provide us with. Discrimination testing requires assessors to make a choice about the difference between samples and in most cases this will be performed as a forced-choice test (e.g., the assessor must select one sample), whereas a non-forced choice test provides an additional option and enables the assessor to select both samples (e.g., no difference) if appropriate. The tests also have different associated tasks such as selecting yes or no, choosing a sample, identifying the odd sample, matching, or sorting samples (Lawless and Heymann 2010). Tests can vary in the number of samples presented from two upwards as well as whether a control or reference sample is utilized. The tests can differ in statistical power; accordingly, the probability of selecting the correct answer may differ, e.g., from 1/2 to 1/10. Sensory discrimination tests are covered by various standards from the

*Discrimination Testing in Sensory Evaluation*, First Edition. Edited by Lauren Rogers, Joanne Hort, Sarah E. Kemp and Tracey Hollowood.

International Organization for Standardization and the ASTM in some cases (e.g., A not A, paired comparison, same-different, triangle, 3-AFC, duo-trio, ranking and Tetrad) but not all (Rogers 2017).

Discrimination testing dates back to the early 19th Century in varying forms; however, it has been on quite the journey and has constantly adapted to meet the current needs. The history of discrimination testing in the sensory science context has been previously reviewed by Rogers (2017) and Meiselman et al. (2022). Currently, there is growing emphasis on discrimination testing efficiency and how best to utilize this to answer the corresponding research question. In addition, discrimination testing can be performed by either naive consumers or trained panelists which demonstrates its versatility and making it a key test within the sensory toolbox. However, deciding whether consumers or trained panelists are more appropriate for any given discrimination test will depend on the specific research question (Ares and Varela 2017). For example, consumers can enable greater ecological validity since they are the main end product users, whereas trained panelists are typically recruited for sensory acuity; however, this subject (consumers vs. panelists) is associated with ongoing discussions (Ares and Varela 2017). Moreover, it should be noted that a combination of consumers and trained panelists should never be utilized in the same test. The main advantage of using trained panelists in discrimination testing can relate to increased power due to panelists' sensory acuity, which is subject to performance monitoring and training as well as their experience with sensory methodology; accordingly, less panelists are often needed. In addition, trained panelists would in most cases be associated with better discriminating abilities and familiarity with sensory attributes (such as astringency) than consumers. Consumers are typically less sensitive to small differences and are linked with larger numbers required to perform any given test; though, if a difference is noticeable, it would most likely be consumer relevant. Regardless of the assessor type, providing clear test instructions and recruiting the appropriate number of assessors is paramount. Statistical tables (see Rogers 2017) can support researchers in understanding the appropriate number of individuals needed to perform such discrimination tests. Typically, the focus is on independent responses and the topic of using individuals as replicates is discussed in another chapter.

The discrimination test determines whether there is a difference or not between samples (performed in suitable bias free environment); this test is considered easy to set up and associated with minimal training. In addition, discrimination testing is useful where there are small differences, strong carry-over and/or non-homogeneous based samples. It should be noted that the statistical power will vary depending on the specific discrimination test. In addition, there are various practical considerations such as research questions, hypotheses, test outcomes, product types, cost implications and assessor types (consumers vs. trained panelists) that all need to be considered in test selection. Moreover, the focus in discrimination testing is primarily on differences, as no detail on the sensory profile can be identified; therefore, descriptive methods are more appropriate in this case (see Stone et al. 2021). For example, if you are developing a new product and are interested in understanding the preferred product compared with the current on-the-market version: a paired comparison test with consumers would be relevant (e.g., *which sample do you prefer?*). In contrast, if the focus was on the extent of differences between the new vs. current product, as well as key contributing attributes, then determining a sensory profile via descriptive methods using a trained panel would be more appropriate.

Discrimination testing can be used for both difference (to identify perceivable differences between samples) and similarity (to determine if samples are similar) with the following considerations: (1) difference: it is important to avoid identifying a difference that does not exist (e.g., type I error) and defining significance at 5% can help to overcome this (in this case alpha risk is more important); and (2) similarity: preventing the type II error (e.g., highlighting the samples are not different yet they are different); accordingly, using more individuals (such as consumers or panelists) can solve this problem by increasing the test power (beta risk is more important in this case) (Kemp et al. 2009). Moreover, it is worth mentioning that no significant difference between products does not always assume similarity. Similarity testing application is typically associated with ingredient or reformulation changes; accordingly, it is important the samples are perceived as similar. As alluded to by Bi (2005) a drawback to similarity testing is the need for a larger sample size (which is higher than difference testing even for the same test). For example, a triangle test can be performed in both difference and similarity formats yet needs 20–40 and 50–100 individuals for difference and similarity outcome respectively (Meilgaard et al. 2016).

In the following pages we will explore some of the commonly used discrimination tests (such as A-not-A, paired comparison, same-different, triangle, 3-AFC, duo-trio, tetrad, ranking, 4-AFC, difference from control, and two-out-of-five) used in sensory science (as summarized in Table 3.1).

**Table 3.1** Summary of discrimination tests commonly utilized in a sensory science context.

| Test | Presented samples (*n*) | Type | Task | Probability | Question | Standards: ISO[a] | Standards: ASTM[b] |
|---|---|---|---|---|---|---|---|
| A not A with reminder |  | Unspecified | Yes\|no | 1/2 | Is the coded sample the same as A or not? | BS ISO 8588:2017 | — |
| Paired comparison (2-AFC) |  | Specified | Choosing | 1/2 | Which sample is the sweetest? | BS EN IS0 5495: 2005+A1:2016 | E2263-12(2018) |
| Same-different |  | Unspecified | Yes\|no | 1/2 | Are the samples the same or different? | — | E2139-05(2018) |
| Triangle |  | Unspecified | Oddity | 1/3 | Identify the different coded sample | BS EN ISO 4120:2021 | E1885-18 |
| 3-AFC |  | Specified | Choosing | 1/3 | Which sample is the more mouthdrying? | BS IS0 13301:2018 | E1432-19 |
| Duo-trio |  | Unspecified | Matching | 1/2 | Which coded sample is the same as the reference? | BS EN ISO 10399:2018 | E2610-18 |
| ABX |  | Unspecified | Matching | 1/2 | Does the third coded sample match the first or second coded sample? | — | — |
| Ranking |  | Specified | Choosing | | Rank the samples in order of bitterness | BS ISO 8587:2006+ A1:2013 | — |
| Difference from control |  | Unspecified | Choosing | | Determine if each of the coded samples is different from the control & record the extent of difference for each sample on the scale. | — | — |
| 4-AFC |  | Specified | Choosing | 1/4 | Which sample is the saltiest? | — | — |

| | | | | | | | |
|---|---|---|---|---|---|---|---|
| Dual-standard |  | Unspecified | Matching | 1/2 | Two questions: which of the coded samples is the same as REF A and which is the same as REF B? | — | — |
| Dual-pair |  | Unspecified | Choosing | 1/2 | Choose the pair which contains the different pair of samples | — | — |
| One-out-of-four |  | Unspecified | Oddity | 1/4 | Which is the odd sample? | — | — |
| Tetrad |  | Both | Sorting | 1/3 | Group the samples into two groups based on similarity (or on a specific attribute) | — | E3009-15e1 |
| Multiple standards |  | Unspecified | Choosing | | Which is the most different sample? | — | — |
| Two-out-of-five |  | Both | Sorting | 1/10 | Identify two groups of samples: two samples which are the same and different from the other three samples. | — | — |

There are several relevant standards for sensory scientists:
[a] ISO relates to International Organisation for Standardisation.
[b] ASTM refers to the American Society for Testing and Materials.
(—) Represents no standard currently available.

## A-NOT A TEST

**BACKGROUND**
The A-not A test requires a familiarization task and is commonly utilized in quality control situations; therefore, is beneficial if the testing cohort are regular product users (Van Hout et al. 2011). This test enables a one sample presentation; therefore, useful with high carry over, temperature controlled or fatiguing samples.

**TEST SUMMARY**
This test involves presenting two samples (A and not-A) and identifying whether the coded sample is the same as A or not.

**PRESENTATION ORDER**
If using a monadic design, there are two possible presentation orders: (1) "A" and (2) "not-A" for this test.

**EXAMPLE QUESTIONNAIRE**
You are provided with two coded samples: "A" or "not-A" (as previously tasted in the preliminary session).
Please taste your samples in order from left to right and decide whether it is "A" or "not-A."
Please cleanse your palate with water and crackers between samples; please DO NOT retaste.
SAMPLE 246:______
SAMPLE 391:______
Any comments:_______________
Thank you ☺

**DATA ANALYSIS**
In brief, the responses for "A" and "not-A" are totaled and a chi-squared test can be utilized to compare responses. It should be noted that if a confidence scale (e.g., not sure to sure) is presented this enables analysis to be conducted via R-index.

**LITERATURE-BASED EXAMPLE**
Recently, the A–not-A test has been successfully utilized in children tasting dairy-based products when focusing on both differences and liking; demonstrating the versatility and the ability to modify such tests depending on the specific population and corresponding research question (Velázquez et al. 2022).

## PAIRED COMPARISION TEST

**BACKGROUND**
The paired comparison is often utilized due to its simplicity and it can easily be adapted to focus on a specific attribute or overall preference; accordingly, it is commonly used in consumer testing and product preference as well as to validate product changes and screen/monitor panelists.

**TEST SUMMARY**
Two samples are presented and assessors are asked to select the more intense sample for a specific attribute (e.g., which sample is sweeter?).

**PRESENTATION ORDER**
There are only two possible presentation orders: (1) AB and (2) BA for this test.

**EXAMPLE QUESTIONNAIRE**
You are presented with two coded samples.
Please taste your samples in order from left to right.
Please cleanse your palate with water and crackers between samples; please DO NOT retaste.
Which sample is sweeter?
CODE:______
Any comments:______________
Thank you ☺

**DATA ANALYSIS**
The number of responses for each sample are calculated and statistical tables (please refer to Rogers 2017) can be used to identify the minimum number of responses needed to reach defined levels of significance (e.g., $p < 0.05$–5% significance level).

**LITERATURE-BASED EXAMPLE**
The 2-AFC test can be useful to demonstrate small differences between samples. For example, Zhou et al. (2016) used the 2-AFC to distinguish between increasing fat levels focusing on identifying the "fattier" sample in a milk and cream-based product.

## SAME-DIFFERENT TEST

**BACKGROUND**

This test is considered simpler and more beneficial than its counterparts (e.g., triangle and duo-trio) for fatiguing and carryover samples (Wolf 2020).

**TEST SUMMARY**

A pair of samples are presented and it determines whether the two samples are the same or different.

**PRESENTATION ORDER**

This test has four typical presentation orders: (1) AA; (2) AB; (3) BB; and (4) BA.

**EXAMPLE QUESTIONNAIRE**

You are presented with two coded samples.
Please taste your samples in order from left to right and decide whether the samples are the SAME or DIFFERENT.
Please cleanse your palate with water and crackers between samples; please DO NOT retaste.
SAMPLES are SAME:______
SAMPLES are DIFFERENT:______
Any comments:________________
Thank you ☺

**DATA ANALYSIS**

Similar to the A-not-A test, responses are totaled, and a chi-squared test can be used to compare responses. Additional analysis using R-index can be utilized if a confidence scale (e.g., not sure to sure) is presented.

**LITERATURE-BASED EXAMPLE**

Chae et al. (2010) used the same-different test to identify subtle differences between pasteurized milks via two different decision approaches (e.g., affective vs. analytical familiarization).

## TRIANGLE TEST

**BACKGROUND**

The triangle test focuses on overall differences and is commonly used in quality control, product development, panelist screening, training and monitoring, and batch-to-batch variation.

**TEST SUMMARY**

This test utilizes three samples, where two samples are the same and one sample is different, and focuses on identifying the different sample (e.g., the odd one out).

**PRESENTATION ORDER**

There are six presentation orders typically followed: (1) AAB; (2) ABA; (3) BAA; (4) BBA; (5) BAB; and (6) ABB.

**EXAMPLE QUESTIONNAIRE**

You are presented with three samples; two are the same and one is different.
Please taste your samples in order from left to right and identify the sample which you think is different.
Please cleanse your palate with water and crackers between samples; please DO NOT retaste.
CODE:______
Any comments:________________
Thank you ☺

**DATA ANALYSIS**

The total correct responses are calculated, and statistical tables (please refer to Rogers 2017) have been commonly used to determine the minimum number of correct responses needed to reach defined levels of significance (e.g., $p < 0.05$ – 5% significance level).

**LITERATURE-BASED EXAMPLE**

The triangle test was used to determine consumers' umami discrimination status (via 24 tests over three sessions) without any apparent learning or fatigue implications as well as demonstrating consistent and reliable results (Hartley et al. 2022).

## 3-AFC TEST

**BACKGROUND**

The 3-AFC test enables presentation of samples varying in strength to subsequently determine an individual or group threshold; therefore, can be used to identify the lowest detectable signal (Wolf 2020).

**TEST SUMMARY**

Three samples are presented where two samples are the same and one sample is different and focuses on the intensity for a specific attribute (e.g., which sample is more mouthdrying?).

**PRESENTATION ORDER**

This test typically utilizes three presentation orders: (1) AAB; (2) ABA; and (3) BAA.

**EXAMPLE QUESTIONNAIRE**

You are presented with three coded samples.

Please taste your samples in order from left to right.

Please cleanse your palate with water and crackers between samples; please DO NOT retaste.

Which sample is more mouthdrying?

CODE:______

Any comments:________________

Thank you ☺

**DATA ANALYSIS**

Calculates the total responses for the different samples and via statistical tables (please refer to Rogers 2017) identifies the minimum number of responses needed to reach defined levels of significance (e.g., $p < 0.05$–5% significance level).

**LITERATURE-BASED EXAMPLE**

Recently, the 3-AFC test has been utilized to determine a mouthdrying detection threshold for whey protein beverages using a trained sensory panel (Norton et al. 2022).

## DUO-TRIO TEST

**BACKGROUND**

The duo-trio test is considered an easy to perform test and is often utilized in fatiguing or carryover-based samples as well as in monitoring panel performance.

**TEST SUMMARY**

Three samples are presented as two coded samples and one reference, where the coded sample most similar to the reference is identified (e.g., which sample matches the reference).

**PRESENTATION ORDER**

The duo-trio has four presentation orders: (1) REF A|AB; (2) REF A|BA; (3) REF B|AB; (4) REF B|BA.

**EXAMPLE QUESTIONNAIRE**

You are presented with three coded samples and one sample is labeled as REFERENCE A.

Please taste the REFERENCE first followed by the other samples in order from left to right. Identify the sample which is the SAME as the reference.

Please cleanse your palate with water and crackers between samples; please DO NOT retaste.

CODE:______

Any comments:__________________

Thank you ☺

**DATA ANALYSIS**

Like the triangle test the total correct responses is calculated. Statistical tables (please refer to Rogers 2017) can determine the minimum number of correct responses needed to reach defined levels of significance (e.g., $p < 0.05$ – 5% significance level).

**LITERATURE-BASED EXAMPLE**

Kim et al. (2010) used the duo-trio test to discriminate tomato juices varying in salt concentrations and focused on design implications such as reference presentation positions.

## TETRAD TEST

### BACKGROUND

There is growing interest in this test due to its cost implications as it has lower sample size and higher power/sensitivity compared with the triangle test (Ennis 2012; Ennis and Rousseau 2012).

### TEST SUMMARY

This test involves presenting four samples (two reference samples and two trial samples) and the samples are grouped into two, based on similarity (or on a specific attribute).

### PRESENTATION ORDER

This test typically uses six presentation orders: (1) AABB; (2) ABAB; (3) ABBA; (4) BBAA; (5) BABA; and (6) BAAB.

### EXAMPLE QUESTIONNAIRE

You are presented with four coded samples.
Please taste your samples in order from left to right. Please GROUP the samples into two groups of two based on similarity.
Please cleanse your palate with water and crackers between samples; please DO NOT retaste.
Please write down the codes in each of your two groups:
GROUP 1 CODES:______ and ______
GROUP 2 CODES:______ and ______
Any comments:________________
Thank you ☺

### DATA ANALYSIS

The total correct responses is calculated, and statistical tables (please refer to Rogers 2017) have been commonly used to determine the minimum number of correct responses needed to reach defined levels of significance (e.g., $p < 0.05$–5% significance level).

### LITERATURE-BASED EXAMPLE

Keefer et al. (2021) used tetrads tests with children to test differences in unflavored and chocolate milks varying in milk fat levels with and without visual cues.

## RANKING TEST

**BACKGROUND**

This test is considered simple, time efficient and requires minimal training; therefore, it is useful for vulnerable populations and applications include: liking/preference, intensity relating to specific attributes and ingredient changes.

**TEST SUMMARY**

Three or more samples (the exact number can vary) are often presented, where the samples are ranked in order of intensity for a specific attribute.

**PRESENTATION ORDER**

The samples should be presented at the time same and follow a balanced presentation order.

**EXAMPLE QUESTIONNAIRE**

You are presented with three coded samples. Please taste your samples in order from left to right.

Rank the samples in order of increasing sweetness (ONE = least sweet and FIVE = most sweet).

Please cleanse your palate with water and crackers between samples; please DO NOT retaste.

| 1st – least sweet | 2nd | 3rd | 4th | 5th – most sweet |
|---|---|---|---|---|
| | | | | |

Any comments:_______________

Thank you ☺

**DATA ANALYSIS**

For each sample the sum rank is calculated and the data is then analyzed using a Friedman test to determine whether there is a significant difference between the rank of the samples tested. If a significant difference is found, the Nemenyi's multiple comparison test is applied to determine which samples are significantly different from each other.

**LITERATURE-BASED EXAMPLE**

Lignou and Oloyede (2021) used a ranking test to determine preference of four meat-based food packages; demonstrating how this test can focus on either a specific attribute or preference depending on the corresponding research question.

## 4-AFC TEST

**BACKGROUND**

The 4-AFC is a variation on the 2-AFC and 3-AFC tests; however, the main difference relates to the number of samples.

**TEST SUMMARY**

Four samples (three samples are the same and one sample is different) are presented and focuses on intensity for a specific attribute (e.g., which sample is most bitter?).

**PRESENTATION ORDER**

This test typically utilizes four presentation orders: (1) AAAB, (2) AABA, (3) ABAA, and (4) BAAA.

**EXAMPLE QUESTIONNAIRE**

You are presented with four coded samples.
Please taste your samples in order from left to right.
Please cleanse your palate with water and crackers between samples; please DO NOT retaste.
Which sample is most bitter?
CODE:______
Any comments:________________
Thank you ☺

**DATA ANALYSIS**

Calculates the total responses for the different samples and via statistical tables identifies the minimum number of responses needed to reach defined levels of significance (e.g., $p < 0.05$–5% significance level).

**LITERATURE-BASED EXAMPLE**

The 4-AFC test has been successfully used for food odor identification and demonstration of age-related differences in an ageing population (de Lichtenberg Broge et al. 2023).

## DIFFFERENCE FROM CONTROL TEST

### BACKGROUND

This test allows multiple sample presentation combined with extent of difference; accordingly, it has various applications such as consumer testing, quality control/ assure, shelf life, product claims, and screening.

### TEST SUMMARY

Three samples are presented as two samples and one control sample to determine the differences (e.g., between samples and the control) as well as record the extent of difference on the provided scale.

### PRESENTATION ORDER

The control should be presented first followed by the test samples simultaneously; all samples should be presented in a balanced order.

### EXAMPLE QUESTIONNAIRE

You are presented with two samples – a control and test samples.

Please evaluate the control sample followed by the test samples in order from left to right.

Determine if the test sample is different from the control and record the extent of difference on the scale below; if no difference is identified select no difference.

Please cleanse your palate with water and crackers between samples; please DO NOT retaste.

CODE:______

| No difference | Very slight difference | Slight difference | Moderate difference | Large difference | Very large difference |
|---|---|---|---|---|---|
| | | | | | |

Any comments:________________

Thank you ☺

### DATA ANALYSIS

The mean difference scores are calculated for every pair by analysis of variance (ANOVA) and Dunnett's multiple comparison test is used to determine which samples are significantly different from the control.

### LITERATURE-BASED EXAMPLE

The difference from control test was utilized to investigate the effects of sodium reduction and flavor enhancement on chicken rice and mee soto broth using trained panelists (Leong et al. 2015).

## TWO-OUT-OF FIVE TEST

**BACKGROUND**

This is considered a more advanced version of the triangle test and is useful for small assessor numbers due to its high power. However, this test is associated with a strong fatigue element; therefore, more suitable for visual, auditory and tactile uses compared with flavor (Meilgaard et al. 2016).

**TEST SUMMARY**

Five samples are presented and select the two samples which are different (or more intense based on a specific attribute) to the other three samples.

**PRESENTATION ORDER**

There are 20 different presentation orders: (1) AAABB; (2) BABBA; (3) ABABA; (4) BAABB; (5) BBBAA; (6) BAAAB; (7) BABAB; (8) BABAA; (9) AABAB; (10) ABBBA; (11) BAABA; (12) ABABB; (13) BBABA; (14) AABBA; (15) ABBAB; (16) BBAAA; (17) ABAAB; (18) BBAAB; (19) ABBAA; and (20) AABBB.

**EXAMPLE QUESTIONNAIRE**

You are presented with five samples, two of which are one product and three are another product.

Please taste your samples in order from left to right.

Please cleanse your palate with water and crackers between samples; please DO NOT retaste.

Identify two groups of samples: two samples which are the same and different from the other three samples.

GROUP 1 CODES:____ and ____

GROUP 2 CODES:____ and ____ and ____

Any comments:________________

Thank you ☺

**DATA ANALYSIS**

Calculates the total responses for the different samples and via statistical tables (please refer to Rogers 2017) identifies the minimum number of responses needed to reach defined levels of significance (e.g., $p < 0.05$–5% significance level).

**LITERATURE-BASED EXAMPLE**

Kang et al. (2021) used the two-out-of-five test to identify appearance differences from processing conditions (minimal vs. intense) with milk powder and reconstituted milk using trained panelists.

In addition, there are a couple of other three and four sample tests which are less commonly used in sensory discrimination testing but still worth mentioning briefly. For example, the ABX test utilizes a two-stage approach where two reference samples (e.g., REFERENCE A and REFERENCE B) are tested followed by third sample (e.g., sample X) (e.g., does sample X match REFERENCE A or B). The one-out-of-four test involves four samples where three samples are the same and one sample is different; the odd sample is selected. The dual-pair test has four samples presented as two matched samples and two unmatched samples, where the pair of different samples needs to be selected. The dual-standard test uses four samples where two references and two samples are presented, where the samples need to be matched to the relevant reference. In addition, the multiple standards test has four samples presented (three reference samples and one trial sample) and the most different sample is selected.

In summary, discrimination testing provides sensory scientists with a range of tests varying in statistical power that can be utilized with different assessor types and research questions; accordingly, highlighting the versatility and easy to perform nature of this test type. Recently, the focus on test efficiency would suggest discrimination testing popularity will continue to increase coupled with its widespread application as demonstrated in this book. Therefore, it could be suggested that the future looks bright for discrimination testing, and it will continue to adapt to meet sensory scientists growing needs.

## References

Ares, G. and Varela, P. (2017). Trained vs consumer panels for analytical testing: fueling a long lasting debate in the field. *Food Qual. Prefer.* 61: 79–86.

Bi, J. (2005). Similarity testing in sensory and consumer research. *Food Qual. Prefer.* 16: 139–149.

Chae, J.E., Lee, Y.-M., and Lee, H.-S. (2010). Affective same-different discrimination tests for assessing consumer discriminability between milks with subtle differences. *Food Qual. Prefer.* 21: 427–438.

Ennis, J.M. (2012). Guiding the switch from triangle testing to tetrad testing. *J. Sens. Stud.* 27: 223–231.

Ennis, J.M. and Rousseau, B. (2012). Reducing costs with tetrad testing. *IFPress* 15: 3–4.

Hartley, I., Orellana, L., Liem, D.G., and Keast, R. (2022). Assessment of the triangle test methodology for determining umami discrimination status. *Chem. Senses* 47: 1–10.

Kang, M., Chen, D., Ruan, R., and Vickers, Z.M. (2021). The effect of intense pulsed light on the sensory properties of nonfat dry milk. *J. Food Sci.* 86: 1–15.

Keefer, H.M., Sipple, L.R., Carter, B.G. et al. (2021). Children's perceptions of fluid milk with varying levels of milk fat. *J. Dairy Sci.* 105: 3004–3018.

Kemp, S.E., Hollowood, T., and Hort, J. (2009). *Sensory Evaluation, A Practical Handbook.* Wiley Blackwell.

Kim, M.A., Lee, Y.-M., and Lee, H.-S. (2010). Comparison of d estimates produced by three versions of a duo-trio test for discriminating tomato juices with varying salt concentrations: the effects of the number and position of the reference stimulus. *Food Qual. Pref.* 21: 504–511.

Lawless, H.T. and Heymann, H. (2010). *Sensory Evaluation of Food: Principles and Practices*, 2e. New York: Springer.

Leong, J., Kasamatsu, C., Ong, E. et al. (2015). A study on sensory properties of sodium reduction and replacement in Asian food using difference-from-control test. *Food Sci. Nutr.* 4 (3): 469–478.

de Lichtenberg Broge, E.H., Wendin, K., Rasmussen, M.A., and Bredie, W.L.P. (2023). Familiarity and identification of everyday food odors in older adults and their influence on hedonic liking. *Food Qual. Prefer.* 103: 104715.

BS ISO 8588:2017. Sensory analysis. Methodology. "A" "not A" test.

BS EN ISO 5495:2005+A1:2016. Sensory analysis. Methodology. Paired comparison test.

BS EN ISO 4120:2021. Sensory analysis. Methodology. Triangle test.

BS ISO 13301:2018. Sensory analysis. Methodology. General guidance for measuring odour, flavour and taste detection thresholds by a three-alternative forced-choice (3-AFC) procedure.

BS EN ISO 10399:2018. Sensory analysis. Methodology. Duo–trio test.

BS ISO 8587:2006+A1:2013. Sensory analysis. Methodology. Ranking.

E2263-12(2018). Standard Test Method for Paired Preference Test.

E2139-05(2018). Standard Test Method for Same–Different Test.

E1885-18. Standard Test Method for Sensory Analysis – Triangle Test.

E1432-19. Standard Practice for Defining and Calculating Individual and Group Sensory Thresholds from Forced-Choice Data Sets of Intermediate Size.

E2610-18. Standard Test Method for Sensory Analysis – Duo–Trio Test.

E3009-15e1. Standard Test Method for Sensory Analysis – Tetrad Test.

Lignou, S. and Oloyede, O.O. (2021). Consumer acceptability and sensory profile of sustainable paper-based packaging. *Foods* 10: 990.

Meilgaard, M., Civille, G., and Carr, B. (2016). *Sensory Evaluation Techniques*. Boca Raton, FL: Taylor & Francis.

Meiselman, H.L., Jaeger, S.R., Carr, B.T., and Churchill, A. (2022). Approaching 100 years of sensory and consumer science: developments and ongoing issues. *Food Qual. Prefer.* 100: 104614.

Norton, V., Lignou, S., Faka, M., and Methven, L. (2022). Individual variation in mouthfeel sensitivity: investigating influences of whey protein content, consumer age, food format and fat addition. *Food Qual. Prefer.* 101: 104638.

Rogers, L. (2017). *Discrimination Testing in Sensory Science: A Practical Handbook*. Cambridge: Woodhead Publishing.

Stone, H., Bleibaum, R.N., and Thomas, H.A. (2021). *Sensory Evaluation Practices*, 5e. San Diego: Academic Press.

Van Hout, D., Hautus, M.J., and Lee, H.S. (2011). Investigation of test performance over repeated sessions using signal detection theory: comparison of three nonattribute-specified difference tests 2-AFCR, A–NOT-A and 2-AFC. *J. Sens. Stud.* 26: 311–321.

Velázquez, A.L., Vidal, L., Varela, P., and Ares, G. (2022). Can children use the A–not A test? *J. Sens. Stud.* 37 (4): e12749.

Wolf, M.B. (2020). *Sensory Testing Methods*, 3e. ASTM International Manual.

Zhou, X., Shen, Y., Parker, J.K. et al. (2016). Relative effects of sensory modalities and importance of fatty acid sensitivity on fat perception in a real food model. *Chemosens. Percept.* 9: 105–119.

# 4

# Determining Individual Variation in Ability to Discriminate: Factors Affecting Responsiveness and Performance

*Sara Spinelli, Caterina Dinnella and Erminio Monteleone*

*SensoryLab, Department of Agriculture, Food, Environment and Forestry (DAGRI), University of Florence, Florence, Italy*

## 4.1 Introduction

Individuals vary substantially in their chemo- and somatosensory perception. Fungiform papillae density (FPD: FP/cm$^2$) and responsiveness to the bitter taste of 6-*n*-propylthiouracil (PROP status) are the most studied phenotypic markers of responsiveness to oral stimulations. However, while the results related to FDP are somewhat controversial, the bitterness responsiveness to PROP has been consistently reported as a marker of oral responsiveness. Thermal taster status (TTS) is another marker of individual variation in oral sensitivity. Thermal tasters (TT) perceive a taste when a small area of the tongue is warmed and/or cooled and are reported to be more responsive to orosentations than thermal non-tasters (TnT). Furthermore, individuals have been found to differ in their tactile acuity, affecting somatosensory perception of foods. Other factors affect individual oral responsiveness, including not only age and gender but also psychological traits. In this chapter we present an overview of different factors affecting oral and smell responsiveness and the conditions and variables that may influence an individual ability to discriminate, including those related to involvement, knowledge, expertise and environment.

## 4.2 Overview of the Factors Affecting Individual Differences in Oral and Olfactory Responsiveness

### 4.2.1 Age and Gender

It is well established that sensory acuity diminishes with age and even if the process is continuous (for example, due to the decrease in the fungiform papillae number, see Section 4.2.2), actual losses in the sense of taste are more commonly perceived beyond the age of 60 years (Hoffman et al. 2009; Doets and Kremer 2016). The extent of the decline is

*Discrimination Testing in Sensory Evaluation*, First Edition. Edited by Lauren Rogers, Joanne Hort, Sarah E. Kemp and Tracey Hollowood.

not clear, but thresholds of taste identification and taste detection appear to be increased in older adults in 70–100% of studies, depending on the taste quality (Methven et al. 2012). Furthermore, decrements in olfactory function with increasing age have been observed, estimating an effect in over half of those between the ages of 65 and 80 years and in over three quarters of those over the age of 80 years (Doty and Kamath 2014). Multiple factors contribute to such age-related loss, including altered nasal engorgement, increased propensity for nasal disease, but also cumulative damage to the olfactory epithelium from viral and other environmental insults. It should also be considered that given the possible long-term effect on taste/olfactory loss or impairment of Covid-19, we may expect alterations in smell/taste perception also in younger individuals even beyond the pandemic (Xydakis et al. 2021).

Sex/gender has also been found to influence taste/olfactory acuity. A higher proportion of women have been reported to have a higher oral (Section 4.2.3) and odor sensitivity (Section 4.2.5). Based on this general heightened taste response in women signaled by their highest PROP responsiveness, we may have expected a heightened taste response also to tastants in aqueous solutions or in foods, but the evidence is mixed. In fact, even though it has been reported that the intensity of high concentrations of sweet, sour, and bitter tastants are rated higher by women than by men (Bajec and Pickering 2008; Michon et al. 2009), overall the evidence is not strong, partly due to a lack of studies, as well as a lack of sufficient statistical power. Recent studies with large sample sizes still report controversial results, with some failing to find systematic differences by gender in taste responses to basic tastants in solutions (Tuorila et al. 2017; Dinnella et al. 2018), but others finding that women are more responsive (Fischer et al. 2013a). For example, the case of saltiness results are mixed: some studies reported that men are less sensitive at threshold than women, consistent with the literature on potential roles of sex hormones on salt perception (Duffy et al. 1998; Okoro et al. 1998; Hayes et al. 2010); but other studies on suprathreshold responses did not (Dinnella et al. 2018). We may speculate that these differences could be related to different tastant concentrations selected in the different studies and thus that sex- and gender-differences are concentration-specific. Something similar has been hypothesized for odor responsiveness: in one small study, women were found to perceive nasal irritation induced by amylacetate as more intense, but this was only true at lower concentrations (Claeson and Nordin 2011). On the other hand, gender differences were consistently reported in response to somatosensory sensations stimulating the trigeminal nerve, such as pungency, with women reporting heightened perception, and also lower liking and consumption (Byrnes and Hayes 2015; Spinelli et al. 2018).

### 4.2.2 Fungiform Papillae

Fungiform papillae (FP) are considered the key anatomic structures for oral stimuli detection and transduction (Mistretta and Bradley 2021). They carry the taste buds which respond to chemical stimulation, are innervated by chorda tympani and trigeminal nerve fibers which respond to chemesthetic agents (pungency and spiciness) and the trigeminal neurons that surround the taste buds (Whitehead et al. 1985; Whitehead and Kachele 1994) are responsible for perception of food textural attributes (Engelen and Van Der Bilt 2008). Considerable individual variability has been reported in FPD, from 0 to over 200 papillae/$cm^2$

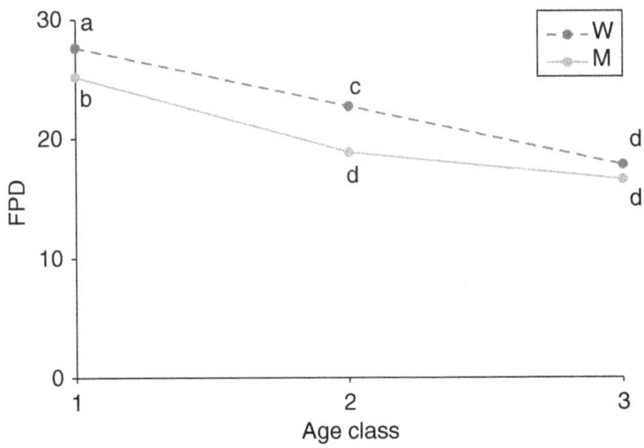


**Figure 4.1** Gender (W: women, M: men) and age class (1: 18–30 y.o.; 2: 31–45 y.o.; and 3: 46–60 y.o.) interactions on fungiform papillae density (FPD) ($n$ = 1166). Different letters indicate significantly different values ($P \leq 0.005$) (adapted from Robino et al. 2022).

(Piochi et al. 2018). Gender and age significantly affect FPD (Shahbake et al. 2005; Feeney and Hayes 2014; Webb et al. 2015; Dinnella et al. 2018). Women generally show a higher FPD mean value than men. FPD mean values significantly decreased with age, and the decline is more evident in men than in women (Dinnella et al. 2018) (Figure 4.1). FP also vary in size and shape (Essick et al. 2003). Recently a classification of subjects based on differences in both FP density and diameter was proposed: high density and large diameter, high density and small diameter, low density and large diameter, and low density and small diameter. Individual variation in oral responsiveness significantly varied according to FP groups, with the highest responsiveness associated to the pattern characterized by low density and small size FP (Piochi et al. 2019).

FP anatomic features suggest that their density underlies the intensity of food sensory properties and, according to the spatial summation theory, the greater the number of FP, the greater the perception of oral stimuli intensity. Positive relationships between FPD and responsiveness to taste stimuli at supra-threshold concentration have been reported, finding that subjects with higher FPDs perceived greater intensity of taste stimuli (Smith 1971; Miller and Reedy 1990; Stein et al. 1994; Bartoshuk 2000; Delwiche et al. 2001; Yackinous and Guinard 2002). However, more recent studies on large population samples have suggested a lack of association between the perception of prototypical taste solution intensity and FPD (Fischer et al. 2013a, b; Feeney and Hayes 2014; Webb et al. 2015; Dinnella et al. 2018). Conflicting results also are reported in the relationships between FPD and perception of tactile sensations. Positive associations have been reported between FPD and the perception of texture related sensations such as creaminess in milk (Hayes and Duffy 2007), hardness and crunchiness in biscuits (Zhou et al. 2021). FPD has also been positively associated to lingual tactile acuity such as oral spatial stereognosis (Essick et al. 2003; Bangcuyo and Simons 2017) and point-pressure measures (Zhou et al. 2021). On the other hand, other studies failed to find significant associations between FPD and food texture attributes, such as astringency in bread (Bakke and Vickers 2008) and fat content (Nachtsheim and Schlich 2013). These results taken together indicate that caution should be applied in considering FP density as an indicator of oral responsiveness, and other factors such as FP size and size uniformity might also play a role (Piochi et al. 2019).

### 4.2.3 Responsiveness to PROP

Individual capacity to perceive bitter thiourea compounds such as PROP (6-*n*-propylthiouracil) varies and three groups of individuals can be identified: (1) "supertasters" (ST) those reporting very high bitter intensity perception induced by PROP, (2) "medium tasters" (MT) those who perceive moderate intensity induced by PROP, and (3) "non-tasters" (NT) with low or null bitter intensity perception induced by PROP (Bartoshuk et al. 1994).

Individual differences in PROP perception have frequently been correlated with the intensity of other oral stimuli. Several studies have demonstrated that responsiveness to PROP is positively associated with responsiveness to chemosensory stimulation in taste model solutions (Prescott et al. 2001; Hayes et al. 2008; Fischer et al. 2014; Webb et al. 2015) and real food (Dinehart et al. 2006; Zhao and Tepper 2007; Masi et al. 2015; Dinnella et al. 2018; Spinelli et al. 2018; Robino et al. 2022). The perception of PROP bitterness was found to be positively associated to the perception of tactile/texture-related sensations such as astringency (Pickering and Robert 2006; Melis et al. 2017), creaminess (Kirkmeyer and Tepper 2003), and roughness (Bakke and Vickers 2008). Furthermore, PROP responsiveness was reported to increase discrimination among foods and beverages with systematic variations in tastes, oral irritants (Prescott et al. 2004), and textures (de Wijk et al. 2007; Donovan et al. 2016).

The specific mechanism explaining the heightened responsiveness to different oral stimuli of PROP tasters remains unclear. Genetic variation in the *TAS2R38* gene is mainly responsible for individual differences in PROP perception (Kim et al. 2003). Three single nucleotide polymorphisms of *TAS2R38* lead to three amino acid substitutions (A49P, A262V and V291I) that define two common haplotypes, physical groupings of genomic variants (or polymorphisms) that tend to be inherited together): PAV (considered the "taster haplotype") and AVI (considered the "non-taster" haplotype). PROP NT are typically homozygous (i.e., they carry the same haplotype) for the AVI haplotype (AVI/AVI), while a large genotypic overlap has been found between medium and supertasters. However, the incomplete correspondence between *TAS2R38* genotype and PROP phenotype was reported in recent studies with a few AVI homozygous individuals responding to PROP and some PAV homozygous individuals blind to this stimulus (Nolden et al. 2020; Robino et al. 2022). Additional factors other than TAS2R38 polymorphism have been related to the variability in responsiveness to PROP which include the polymorphisms in the gustin (CA6) gene involved in taste bud growth and maintenance (Padiglia et al. 2010; Calò et al. 2011), FPD (Bartoshuk et al. 1994; Tepper and Nurse 1998; Essick et al. 2003; Piochi et al. 2019) and salivary protein composition (Melis et al. 2017). Demographic factors also affect responsiveness to PROP, with PROP bitterness ratings significantly higher in women than in men and PROP responsiveness that declines with age (Mennella et al. 2010; Tepper et al. 2017; Barragán et al. 2018; Dinnella et al. 2018) (Figure 4.2).

The association of PROP responsiveness to FPD has been extensively investigated. Studies on large population samples tend to agree on the lack of direct association between

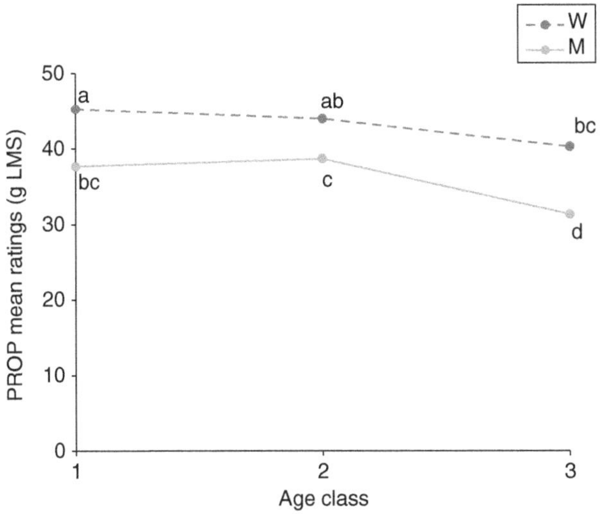


**Figure 4.2** Gender (W: women, M: men) and age class (1: 18–30 y.o.; 2: 31–45 y.o.; and 3: 46–60 y.o.) interactions on PROP mean ratings ($n$ = 1166). Different letters indicate significantly different values ($P \leq 0.005$) (adapted from Robino et al. 2022).

responsiveness to PROP and variation in FPD (Fischer et al. 2013a, b; Feeney and Hayes 2014; Garneau et al. 2014; Robino et al. 2022). A complex interplay between PROP responsiveness and FPD in modulating the intensity to suprathreshold tastes and somatosensory stimuli has been observed (Dinnella et al. 2018). Recent findings corroborate the "central gain" theory (Green and George 2004) which hypothesize that individual differences in signal central processing might contribute to individual variations in responsiveness to orosensory stimuli.

### 4.2.4 Thermal Taster Status

Thermal taster status (TTS) is defined based on the variation in orosensory perception reported by some individuals in response to warming or cooling of the tongue surface (Cruz and Green 2000). Thermal tasters (TT) experience tastes and flavors in addition to changes in tongue temperature. The modalities of tongue thermal stimulation and the location of the stimulated region influence the quality of thermally evoked tastes (Skinner et al. 2018; Thibodeau et al. 2020). Thermal non-tasters (TnT) only perceive temperature modification during warming or cooling of the tongue. A further group of non-classifiable individuals was identified (NN) who report non-reproducible variation in orosensation evoked by tongue thermal stimulation that possibly consists in misclassified TT or TnT individuals (Thibodeau et al. 2019).

TTS is considered a marker of individual variation in orosensory perception, as intensities from suprathreshold taste, astringent, metallic and pungent solutions were rated higher by TTs than TnT (Green and George 2004; Bajec and Pickering 2008; Yang et al. 2014). Moreover, significant associations have been reported between TTS and perception of taste and flavors in alcoholic beverages, with TT reporting intensity of wine and beer sensory attributes higher than TnT (Pickering et al. 2009; Pickering et al. 2010).

Multiple mechanisms have been hypothesized to explain the diversity of sensory modalities affected by the thermal taste status, and either central and peripheral mechanisms appear to contribute to individual variability in TTS (Skinner et al. 2018; Thibodeau et al. 2020). It has been hypothesized that thermally elicited sensations may be evoked by the direct activation of temperature-sensitive taste transduction pathways (Thibodeau et al. 2020). Furthermore, the cross-wiring of receptors in TT (i.e., the activation of the temperature pathway can activate the pathway of taste) was hypothesized to account for the differences observed between TT and TnT in the activation of the taste brain regions (Hort et al. 2016).

### 4.2.5 Oral Tactile Acuity

Oral tactile acuity measures have been recently employed to study individual differences in food texture perception. The most common measures of oral tactile acuity include: the point-pressure method with the Von Fryer monofilaments (Yackinous and Guinard 2001; Breen et al. 2019; Etter et al. 2020; Zhou et al. 2021), the identification of letters and shapes (oral stereognosis) (Essick et al. 1999, 2003; Lukasewycz and Mennella 2012; Steele et al. 2014; Bangcuyo and Simons 2017), the identification of two point distance (Olarte Mantilla et al. 2022), and grating orientation (spatial resolution) (Phillips and Johnson 1981; Wohlert 1996). Limitations have been observed for all these measures. The majority of subjects are able to detect the lowest point-pressure test force thus indicating the strong "floor effect" of this method (Santagiuliana et al. 2019) while the familiarity with the tactile cues (i.e., the alphabet letters) impairs the reliability of oral stereognosis measures (Shupe et al. 2018). Scientific literature regarding the association of oral tactile measures and food texture perception is still relatively scarce. However, consistently with the hypothesis that high oral somatosensory responsiveness associates to heightened perception of food texture-related sensations, positive relationships have been found between responses from oral tactile acuity measures and particle perception in yogurt (Olarte Mantilla et al. 2022) and in chocolate (Breen et al. 2019), and to increased hardness perception in biscuits (Zhou et al. 2021).

### 4.2.6 Psychological Traits

Recent studies have showed that variations in psychological traits affect not only the acceptability of foods, but also the perception of sensory properties. A heightened perception of pungency, both in aqueous solution and in a model food, has been reported in individuals that were higher in food neophobia, sensitivity to disgust and sensitivity to punishment (Spinelli et al. 2018). Food Neophobia, defined as the reluctance to try and eat unfamiliar foods, is generally quantified using the 10-item instrument developed by Pliner and Hobden (1992). The sensitivity to punishment subscale is formed by a set of items reflecting situations which describe individual differences in reactivity and responsivity to the Behavioral Inhibition System (BIS) and has been measured through a subscale of the Sensitivity to Punishment and Sensitivity to Reward Questionnaire (Torrubia et al. 2001). The responsivity to core-visceral disgust (rotten food, vermin, body fluids) can be measured using the 8-item short form of the Disgust Sensitivity Scale (Disgust Scale-Short form, Haidt et al. 1994; Haidt 2004; Olatunji et al. 2007; Inbar et al. 2009). A lower perceived intensity of sweetness and a heightened intensity of bitterness and astringency was associated to a higher sensitivity to disgust in the assessment of chocolate pudding samples differing in their sucrose content (Spinelli et al. 2021). These findings suggest that some psychological traits impact acceptability of foods, but also the perception of specific sensory properties, such as bitterness, pungency and astringency. Further research is needed to better understand this issue to verify if certain psychological traits may lead to variations in perception that impact, in turn, food acceptability. Another alternative possibility would be that the variation in perception contributes to the psychological trait.

### 4.2.7 Individual Differences in Odor Sensitivity

Large inter-individual variability in odor sensitivity has been observed. Along a continuum, olfactory deficits can be categorized as partial impairment (mild, moderate, or severe hyposmia; lower sensitivity to odorants and/or decreased identification) or as a total deficit (anosmia). In addition to these quantitative smell problems, there are qualitative deficiencies that can be observed, such as parosmia (distorted and incorrect olfactory perception induced by smelling something) and phantosmia (phantom smell) (Manesse et al. 2021). Olfactory impairment is common among older adults and a recent study reports that the overall prevalence of anosmia was 22.3% among "African Americans" and 10.4% among "European Americans" associated with older age and male sex in USA (Dong et al. 2017). Factors other than age may contribute to olfactory impairments such as acute or chronic infections (including Covid-19), tumors, head traumas, allergies, pollutants, medications, etc. However, the prevalence of olfactory dysfunction has also been found to vary considerably at a younger age. In a recent study conducted in France on a sample of 3685 male and female individuals aged 4–89 years old, the overall prevalence of olfactory dysfunction, as measured by psychophysical testing, was 17% and varied with age and gender: from 10% in 4- to 12-year-old children to >30% after 60 years of age (Manesse et al. 2021); overall, women were affected less often than men. On the other hand, the prevalence of qualitative smell disorders (parosmia and phantosmia) was much lower (1%). Furthermore, specific anosmia exists, namely anosmia to a specific odor (e.g., some people cannot detect the odor of isobutyric acid even though their sense of smell may be otherwise not decreased). For a description of the most common clinical methodologies to measure smell sensitivity used also in sensory tests we refer the reader to Doty (2015) and Thomas-Danguin et al. (2003).

Some studies reported some association between oral and olfactory responsiveness, thus suggesting that there may be a common underlying factor determining some individual differences in odor, taste and trigeminal sensitivity. There is some evidence that odor responsiveness to odorants that stimulate the trigeminal nerve is related to oral responsiveness (Piochi et al. 2021), even if a general chemical sensitivity that spans chemosensory modalities seems unlikely due to mixed results (Lundström et al. 2012).

Several odors have been proven to have a bimodal sensitivity distribution, but only a few of these are bimodally distributed such that two groups exhibit different sensitivity. This is the case, for example, for androstadienone (proposed as a human pheromone and that can, among other places, be found in male axillary secretion) that is perceived at a higher intensity by a subgroup of individuals; importantly this higher responsiveness is not associated with a higher general olfactory acuity or higher trigeminal sensitivity (Lundström et al. 2003). In the case of other bimodal odors, bimodality seems to depend instead on a group of individuals with specific anosmia to the particular odorant, although the choice of stimulus concentration range is of course crucial for this conclusion (Lundström et al. 2003).

## 4.3 Factors Affecting Consumer and Trained Panelists Ability to Discriminate

### 4.3.1 Training and Familiarity

In addition to inter-individual differences in taste and odor responsiveness, other factors may influence individual responsiveness and ability to discriminate. Intensive training in recognition and evaluation of sensory properties is certainly one of these factors. It is well known that the ability of trained panelists to discriminate is higher when compared to that of consumers with no training, and this is why trained panelists are recommended when sensory differences among products are small (see Chapter 2). Another solution is using semi-trained panels of consumers with a variety of descriptive and similarity methodologies that have been indicated as "rapid methods" as they are faster for data collection (and less resource-consuming) than classic descriptive analysis with reliable results (e.g., Flash Profile, Check-All-That-Apply, Sorting, Projective Mapping, Valentin et al. 2012; Varela and Ares 2012). When a sample of consumers is used instead of trained panelists the number of subjects must be increased (the number changes depending on the methodology). In some cases, a semi-trained panel has also been suggested. In this case a small group of consumers (the number depends on the methodology) participate in a short training session (Valentin et al. 2012; Varela and Ares 2012). For example in the case of CATA it has been suggested to work with 30 consumers that participate in a two-hour training session including the definition of the sensory descriptors and the evaluation of the different physical references (Veríssimo et al. 2021).

However, consumers without any specific training may be entirely appropriate for discrimination tasks when making judgments about what other consumers might perceive in products (Prescott and Monteleone 2015). In some cases, trained panelists may overestimate differences that go unnoticed for consumers. This may impact the decision to reject samples due to off-flavors that might not be detected by consumers, with economic consequences. The consumer rejection threshold method (also known as the rejection threshold method) was developed when it was realized that this was happening in the wine sector, with extremely sensitive trained panels rejecting wines that had become infected with cork taint (trichloranisol, TCA) while this taint was not perceived by the consumers (Prescott et al. 2005). This method was established by presenting pairs of wines that were identical save for the addition of progressively higher quantities of TCA. Consumers were just asked to try both wines and choose which one they preferred. The consumer rejection threshold represented the TCA concentration at which the wine without added TCA became consistently preferred. This study generated a threshold value that was significantly higher than the detection threshold values commonly provided by trained panels, implying that batches of wine were being rejected unnecessarily due to low TCA levels, despite the fact that the untrained wine drinker would find them perfectly acceptable. The importance of contextualizing the results of discrimination tests to make business decisions have been highlighted and some guidelines and scales to support in this task are available (Castura and Franczak 2017).

### 4.3.2 Affective Priming

While familiarization and training can enhance an individual's ability to discriminate, other aspects play a role in sensory perception that are related to attention and other cognitive processes such as memory and emotions. It has been demonstrated that consumers are highly sensitive to product variations when they are first emotionally primed, thus challenging the common idea that an affective response to foods (e.g., liking) is necessarily at odds with perceptual sensitivity (Prescott and Monteleone 2015). A technique known as the authenticity test (AT), first ideated by Mojet and Koster (1986) and then replicated in several different contexts (Wolf Frandsen et al. 2007; Boutrolle et al. 2009; Chae et al. 2010), has been used to this aim. In the AT information about products provided to very familiar consumers are manipulated in order to induce an emotional state of alarm/warning/concern about the products. Usually, a contrast is established between the "genuine" (basically standard) version of their product and another version that the consumer is told is created from inferior materials, is a foreign import, or is made in unconventional ways. In reality the product that is evaluated together with the familiar product is, for example, a competitor product of no inferior quality. The first study using the AT reported that beer drinkers primed with a story about proposed changes in the brewing processes of their favorite lager beer resulted in an increase in the discriminability of this beer from competitor brands, compared to a simple (analytical) discrimination test on the same lagers (Mojet and Koster 1986).

This method shows that not only training, but also familiarity with the product can affect an individual's ability to discriminate. In fact, for the AT to work, the product must be highly familiar to the consumers. Furthermore, these findings indicate that an emotional state may improve the ability to discriminate. When the consumers are put in a situation of potential "alarm" or "threat" for a product, discrimination ability is higher probably due to increased attention and activation of an implicit memory, invoking memory for what is not right (Prescott 2018). It has also been demonstrated that the AT is not a mere proxy for preference as, for some subjects, the sense of authenticity inferred from the sensory perception of the products can be different to liking (Boutrolle et al. 2009). The cognitive mechanisms behind the AT should be further explored to be fully understood, but its implications are very interesting, suggesting a wider use of this methodology with consumers. For details about the practical applications of this methodology we refer the reader to Prescott (2018).

### 4.3.3 Panel Composition

Panel composition represents another point to consider in the interpretation of results from discrimination tests. Random noise due to individual variability might cover subtle differences among samples. Thus, the within-subject design, when applicable, seems the more advisable experimental design as it is the design that ensures wider control on individual variability. It is also extremely important to not mix trained panelists with experts (e.g., internal employees) or consumers. As mentioned before (Sections 4.3.1 and 4.3.2), these types of assessors differ greatly in their familiarity with the product and with the methodology and have different levels of product and sensory knowledge. The choice of different types of subjects should be based on the objectives of the study. Where consumers are preferred for the tests, it is highly recommended to recruit them based on a homogeneous familiarity level with the product/product category because their ability to discriminate can change accordingly.

Regarding the number of panelists, any discrimination test's statistical power will grow as the number of assessors increases, giving greater confidence in the accuracy of the findings. This is crucial and there are published tables (known as critical value tables; see Rogers 2017) on the bare minimum of assessors necessary to establish a statistical significance for each test.

### 4.3.4 Environmental Conditions

Environmental conditions can hinder assessor discrimination ability when samples with subtle differences are evaluated. A large source of noise could be related to the presence of odors in the work environments, very frequent in a flavor manufacturer company (e.g., when tests are performed with employees). Strict control of ambient odor while running sensory evaluation is highly recommended to control for olfactory adaptation and consequent changes in odor and flavor perception (Lawless and Heymann 2010). Moreover, the presence of people engaged in different activities may hold assessor attention during evaluation thus preventing the perception of differences among samples, especially when they are subtle. Thus, performing the test in a quiet room, preferably in individual booths, is highly recommended but other options are also available (see Section 4.4). Whatever the environment chosen for the test, it is crucial to follow all the guidelines and standards for sensory testing in order to control the most typical psychological errors (see "Principles of good practices," in Lawless and Heymann 2010, International Sensory Science Standards, ASTM, Rogers 2017).

## 4.4 Future Developments

A larger involvement of consumers in discrimination testing can be foreseen in the future. With the increase of the spread of online testing, also due to the restrictions of the Covid-19 pandemic, we are developing procedures for analytical testing out of the lab but in semi-controlled conditions. This is the case for example of "remote testing," namely sensory testing conducted at home or in the office (when internal personnel are involved) but in assisted and controlled conditions (Dinnella et al. 2022). In remote testing the panel leader guides the evaluation step by step following the same principles as for laboratory testing, using videoconference systems. The procedure was found to be effective both in training and in evaluation tasks with results comparable to the ones of traditional lab tests and was applied to several sensory methodologies including discrimination tests. This suggests that remote sensory testing, if appropriately conducted following meticulously the indicated sensory protocols, is a promising technique that can be applied even beyond the global pandemic. Furthermore, when the objective of the study is to understand if consumers detect a difference, the use of real contexts and evoked scenarios (even manipulated as in the ATs) to promote a more immersive experience should be considered.

## 4.5 Summary

This chapter presents an overview of the factors affecting individual differences in oral and smell responsiveness including age and gender, FP, TTS, oral tactile and taste acuity, and psychological traits. Furthermore, other factors affecting discriminating ability in consumers and trained panelists has been highlighted, focusing on training and familiarity, but also affective priming, panel composition, and environmental conditions.

## Acknowledgments

The authors wish to acknowledge Lauren Rogers who provided thoughtful feedback on an earlier draft of this chapter.

## References

Bajec, M.R. and Pickering, G.J. (2008). Thermal taste, PROP responsiveness, and perception of oral sensations. *Physiol. Behav.* 95: 581–590.

Bakke, A. and Vickers, Z. (2008). Relationships between fungiform papillae density, prop sensitivity and bread roughness perception. *J. Texture Stud.* 39: 569–581.

Bangcuyo, R.G. and Simons, C.T. (2017). Lingual tactile sensitivity: effect of age group, sex, and fungiform papillae density. *Exp. Brain Res.* 235: 2679–2688.

Barragán, R., Coltell, O., Portolés, O. et al. (2018). Bitter, sweet, salty, sour and umami taste perception decreases with age: sex-specific analysis, modulation by genetic variants and taste-preference associations in 18 to 80 year-old subjects. *Nutrients* 10: 1539.

Bartoshuk, L.M. (2000). Comparing sensory experiences across individuals: recent psychophysical advances illuminate genetic variation in taste perception. *Chem. Senses* 25: 447–460.

Bartoshuk, L.M., Duffy, V.B., and Miller, I.J. (1994). PTC/PROP tasting: anatomy, psychophysics, and sex effects. *Physiol. Behav.* 56: 1165–1171.

Boutrolle, I., Delarue, J., Köster, E. et al. (2009). Use of a test of perceived authenticity to trigger affective responses when testing food. *Food Qual. Prefer.* 20: 418–426.

Breen, S.P., Etter, N.M., Ziegler, G.R., and Hayes, J.E. (2019). Oral somatosensory acuity is related to particle size perception in chocolate. *Sci. Rep.* 9: 7437.

Byrnes, N.K. and Hayes, J.E. (2015). Gender differences in the influence of personality traits on spicy food liking and intake. *Food Qual. Prefer.* 42: 12–19.

Calò, C., Padiglia, A., Zonza, A. et al. (2011). Polymorphisms in TAS2R38 and the taste bud trophic factor, gustin gene co-operate in modulating PROP taste phenotype. *Physiol. Behav.* 104: 1065–1071.

Castura, J.C. and Franczak, B.C. (2017). Statistics for use in discrimination testing. In: *Discrimination Testing in Sensory Science: A Practical Handbook* (ed. L. Rogers), 31–65. Elsevier.

Chae, J.E., Lee, Y.M., and Lee, H.S. (2010). Affective same–different discrimination tests for assessing consumer discriminability between milks with subtle differences. *Food Qual. Prefer.* 21: 427–438.

Claeson, A.S. and Nordin, S. (2011). Gender differences in nasal chemesthesis: a study of detection and perceived intensity. *Chemosens. Percept.* 4: 25–31.

Cruz, A. and Green, B.G. (2000). Thermal stimulation of taste. *Nature* 403: 889–892.

Delwiche, J.F., Buletic, Z., and Breslin, P.A.S. (2001). Relationship of papillae number to bitter intensity of quinine and PROP within and between individuals. *Physiol. Behav.* 74: 329–337.

Dinehart, M.E., Hayes, J.E., Bartoshuk, L.M. et al. (2006). Bitter taste markers explain variability in vegetable sweetness, bitterness, and intake. *Physiol. Behav.* 87: 304–313.

Dinnella, C., Monteleone, E., Piochi, M. et al. (2018). Individual variation in PROP status, fungiform papillae density, and responsiveness to taste stimuli in a large population sample. *Chem. Senses* 43: 697–710.

Dinnella, C., Pierguidi, L., Spinelli, S. et al. (2022). Remote testing: sensory test during Covid-19 pandemic and beyond. *Food Qual. Prefer.* 96: 104437.

Doets, E.L. and Kremer, S. (2016). The silver sensory experience – a review of senior consumers' food perception, liking and intake. *Food Qual. Prefer.* 48: 316–332.

Dong, J., Pinto, J.M., Guo, X. et al. (2017). The prevalence of anosmia and associated factors among U.S. black and white older adults. *J. Gerontol. Ser. A Biol. Sci. Med. Sci.* 72: 1080–1086.

Donovan, J.D., Keller, K.L., and Tepper, B.J. (2016). A brief task to assess individual differences in fat discrimination. *J. Sens. Stud.* 31: 296–305.

Doty, R.L. (2015). Olfactory dysfunction and its measurement in the clinic. *World J. Otorhinolaryngol. Head Neck Surg.* 1: 28–33.

Doty, R.L. and Kamath, V. (2014). The influences of age on olfaction: a review. *Front. Psychol.* 5: 20.

Duffy, V.B., Bartoshuk, L.M., Striegel-Moore, R., and Rodin, J. (1998). Taste changes across pregnancy. In: *Annals of the New York Academy of Sciences*, vol. 855, 805–809. New York Academy of Sciences.

Engelen, L. and Van Der Bilt, A. (2008). Oral physiology and texture perception of semisolids. *J. Texture Stud.* 39: 83–113.

Essick, G.K., Chen, C.C., and Kelly, D.G. (1999). A letter-recognition task to assess lingual tactile acuity. *J. Oral Maxillofac. Surg. Off. J. Am. Assoc. Oral Maxillofac. Surg.* 57: 1324–1330.

Essick, G.K., Chopra, A., Guest, S., and McGlone, F. (2003). Lingual tactile acuity, taste perception, and the density and diameter of fungiform papillae in female subjects. *Physiol. Behav.* 80: 289–302.

Etter, N.M., Breen, S.P., Alcala, M.I.M. et al. (2020). Assessment of midline lingual point-pressure somatosensation using Von Frey hair monofilaments. *J. Vis. Exp. JoVE* 21 (156): 2020.

Feeney, E.L. and Hayes, J.E. (2014). Regional differences in suprathreshold intensity for bitter and umami stimuli. *Chemosens. Percept.* 7: 147–157.

Fischer, M.E., Cruickshanks, K.J., Schubert, C.R. et al. (2013a). Taste intensity in the Beaver Dam Offspring Study. *Laryngoscope* 123: 1399–1404.

Fischer, M.E., Cruickshanks, K.J., Schubert, C.R. et al. (2013b). Factors related to fungiform papillae density: the beaver dam offspring study. *Chem. Senses* 38: 669–677.

Fischer, M.E., Cruickshanks, K.J., Pankow, J.S. et al. (2014). The associations between 6-n-propylthiouracil (PROP) intensity and taste intensities differ by TAS2R38 haplotype. *J. Nutrigenet. Nutrigenom.* 7: 143–152.

Garneau, N.L., Nuessle, T.M., Sloan, M.M. et al. (2014). Crowdsourcing taste research: genetic and phenotypic predictors of bitter taste perception as a model. *Front. Integr. Neurosci.* 8: 33.

Green, B.G. and George, P. (2004). "Thermal taste" predicts higher responsiveness to chemical taste and flavor. *Chem. Senses* 29: 617–628.

Haidt, J. (2004). The Disgust Scale Home Page. https://pages.stern.nyu.edu/~jhaidt/disgustscale.html

Haidt, J., McCauley, C., and Rozin, P. (1994). Individual differences in sensitivity to disgust: a scale sampling seven domains of disgust elicitors. *Personal. Individ. Differ.* 16: 701–713.

Hayes, J.E. and Duffy, V.B. (2007). Revisiting sugar-fat mixtures: sweetness and creaminess vary with phenotypic markers of oral sensation. *Chem. Senses* 32: 225–236.

Hayes, J.E., Bartoshuk, L.M., Kidd, J.R., and Duffy, V.B. (2008). Supertasting and PROP bitterness depends on more than the TAS2R38 gene. *Chem. Senses* 33: 255–265.

Hayes, J.E., Sullivan, B.S., and Duffy, V.B. (2010). Explaining variability in sodium intake through oral sensory phenotype, salt sensation and liking. *Physiol. Behav.* 100: 369–380.

Hoffman, H.J., Cruickshanks, K.J., and Davis, B. (2009). Perspectives on population-based epidemiological studies of olfactory and taste impairment. *Ann. N. Y. Acad. Sci.* 1170: 514–530.

Hort, J., Ford, R.A., Eldeghaidy, S., and Francis, S.T. (2016). Thermal taster status: evidence of cross-modal integration. *Hum. Brain Mapp.* 37: 2263–2275.

Inbar, Y., Pizarro, D.A., and Bloom, P. (2009). Conservatives are more easily disgusted than liberals. *Cognit. Emot.* 23: 714–725.

Kim, U.-K., Jorgenson, E., Coon, H. et al. (2003). Positional cloning of the human quantitative trait locus underlying taste sensitivity to phenylthiocarbamide. *Science* 299: 1221–1225.

Kirkmeyer, S.V. and Tepper, B.J. (2003). Understanding creaminess perception of dairy products using free-choice profiling and genetic responsivity to 6-n-propylthiouracil. *Chem. Senses* 28: 527–536.

Lawless, H.T. and Heymann, H. (2010). *Sensory Evaluation of Food – Principles and Practices.* New York: Springer.

Lukasewycz, L.D. and Mennella, J.A. (2012). Lingual tactile acuity and food texture preferences among children and their mothers. *Food Qual. Prefer.* 26: 58–66.

Lundström, J.N.J.N., Hummel, T., and Olsson, M.J.M.J. (2003). Individual differences in sensitivity to the odor of 4,16-androstadien-3-one. *Chem. Senses* 28: 643–650.

Lundström, J.N., Gordon, A.R., Wise, P., and Frasnelli, J. (2012). Individual differences in the chemical senses: is there a common sensitivity? *Chem. Senses* 37: 371–378.

Manesse, C., Ferdenzi, C., Mantel, M. et al. (2021). The prevalence of olfactory deficits and their effects on eating behavior from childhood to old age: a large-scale study in the French population. *Food Qual. Prefer.* 93: 104273.

Masi, C., Dinnella, C., Monteleone, E., and Prescott, J. (2015). The impact of individual variations in taste sensitivity on coffee perceptions and preferences. *Physiol. Behav.* 138: 219–226.

Melis, M., Yousaf, N.Y., Mattes, M.Z. et al. (2017). Sensory perception of and salivary protein response to astringency as a function of the 6-n-propylthioural (PROP) bitter-taste phenotype. *Physiol. Behav.* 173: 163–173.

Mennella, J.A., Pepino, M.Y., Duke, F.F., and Reed, D.R. (2010). Age modifies the genotype-phenotype relationship for the bitter receptor TAS2R38. *BMC Genet.* 11: 60–69.

Methven, L., Allen, V.J., Withers, C.A., and Gosney, M.A. (2012). Ageing and taste. *Proc. Nutr. Soc.* 71: 556–565.

Michon, C., O'Sullivan, M.G., Delahunty, C.M., and Kerry, J.P. (2009). The investigation of gender-related sensitivity differences in food perception. *J. Sens. Stud.* 24: 922–937.

Miller, I.J. and Reedy, F.E. (1990). Variations in human taste bud density and taste intensity perception. *Physiol. Behav.* 47: 1213–1219.

Mistretta, C.M. and Bradley, R.M. (2021). The fungiform papilla is a complex, multimodal, oral sensory organ. *Curr. Opin. Physiol.* 20: 165–173.

Mojet, J. and Koster, E.P. (1986). Onderzoek naar de waardering van drie laag-alcoholische bieren (Research on the appreciation of three low-alcohol beers). Report Psychological Laboratory, Utrecht University.

Nachtsheim, R. and Schlich, E. (2013). The influence of 6-n-propylthiouracil bitterness, fungiform papilla count and saliva flow on the perception of pressure and fat. *Food Qual. Prefer.* 29: 137–145.

Nolden, A.A., McGeary, J.E., and Hayes, J.E. (2020). Predominant qualities evoked by quinine, sucrose, and capsaicin associate with PROP bitterness, but not TAS2R38 genotype. *Chem. Senses* 45: 383–390.

Okoro, O.E., Uroghide, G.E., Jolayemi, T.E. et al. (1998). Studies on taste thresholds in a group of adolescent children in rural Nigeria. *Food Qual. Prefer.* 9: 205–210.

Olarte Mantilla, S.M., Shewan, H.M., Shingleton, R. et al. (2022). Oral physiology, sensory acuity, product experience and personality traits impact consumers' ability to detect particles in yoghurt. *Food Qual. Prefer.* 96: 104391.

Olatunji, B.O., Williams, N.L., Tolin, D.F. et al. (2007). The disgust scale: item analysis, factor structure, and suggestions for refinement. *Psychol. Assess.* 19: 281–297.

Padiglia, A., Zonza, A., Atzori, E. et al. (2010). Sensitivity to 6-n-propylthiouracil is associated with gustin (carbonic anhydrase VI) gene polymorphism, salivary zinc, and body mass index in humans. *Am. J. Clin. Nutr.* 92: 539–545.

Phillips, J.R. and Johnson, K.O. (1981). Tactile spatial resolution. II. Neural representation of bars, edges, and gratings in monkey primary afferents. *J. Neurophysiol.* 46: 1192–1203.

Pickering, G.J. and Robert, G. (2006). Perception of mouthfeel sensations elicited by red wine are associated with sensitivity to 6-N-propylthiouracil. *J. Sens. Stud.* 21: 249–265.

Pickering, G.J., Moyes, A., Bajec, M.R., and Decourville, N. (2009). Thermal taster status associates with oral sensations elicited by wine. *Aust. J. Grape Wine Res.* 16: 361–367.

Pickering, G.J., Bartolini, J.A., and Bajec, M.R. (2010). Perception of beer flavour associates with thermal taster status. *J. Inst. Brew.* 116: 239–244.

Piochi, M., Dinnella, C., Prescott, J., and Monteleone, E. (2018). Associations between human fungiform papillae and responsiveness to oral stimuli: effects of individual variability, population characteristics, and methods for papillae quantification. *Chem. Senses* 43: 313–327.

Piochi, M., Pierguidi, L., Torri, L. et al. (2019). Individual variation in fungiform papillae density with different sizes and relevant associations with responsiveness to oral stimuli. *Food Qual. Prefer.* 78.

Piochi, M., Dinnella, C., Spinelli, S. et al. (2021). Individual differences in responsiveness to oral sensations and odours with chemesthetic activity: relationships between sensory modalities and impact on the hedonic response. *Food Qual. Prefer.* 88: 104112.

Pliner, P. and Hobden, K. (1992). Development of a scale to measure the trait of food neophobia in humans. *Appetite* 19: 105–120.

Prescott, J. (2018). Affect-based discrimination methods. Methods in consumer research. *N. Approaches Class. Methods* 1: 241–259.

Prescott, J., & Monteleone, E. (2015). Consumer perceptions of food and beverage flavour. In Parker J.K., Elmore J.S. and Methven L. Flavour Development, Analysis and Perception in Food and Beverages (pp. 369–385). Woodhead Publishing.

Prescott, J., Ripandelli, N., and Wakeling, I. (2001). Binary taste mixture interactions in PROP non-tasters, medium-tasters and super-tasters. *Chem. Senses* 26: 993–1003.

Prescott, J., Soo, J., Campbell, H., and Roberts, C. (2004). Responses of PROP taster groups to variations in sensory qualities within foods and beverages. *Physiol. Behav.* 82: 459–469.

Prescott, J., Norris, L., Kunst, M., and Kim, S. (2005). Estimating a "consumer rejection threshold" for cork taint in white wine. *Food Qual. Prefer.* 16: 345–349.

Robino, A., Concas, M.P., Spinelli, S. et al. (2022). Combined influence of TAS2R38 genotype and PROP phenotype on the intensity of basic tastes, astringency and pungency in the Italian taste project. *Food Qual. Prefer.* 95: 104361.

Rogers, L. (2017). *Discrimination Testing in Sensory Science. A Practical Handbook.* Woodhead Publishing Series in Food Science, Technology and Nutrition.

Santagiuliana, M., Marigómez, I.S., Broers, L. et al. (2019). Exploring variability in detection thresholds of microparticles through participant characteristics. *Food Funct.* 10: 5386–5397.

Shahbake, M., Hutchinson, I., Laing, D.G., and Jinks, A.L. (2005). Rapid quantitative assessment of fungiform papillae density in the human tongue. *Brain Res.* 1052: 196–201.

Shupe, G.E., Resmondo, Z.N., and Luckett, C.R. (2018). Characterization of oral tactile sensitivity and masticatory performance across adulthood. *J. Texture Stud.* 49: 560–568.

Skinner, M., Eldeghaidy, S., Ford, R. et al. (2018). Variation in thermally induced taste response across thermal tasters. *Physiol. Behav.* 188: 67–78.

Smith, D.V. (1971). Taste intensity as a function of area and concentration: differentiation between compounds. *J. Exp. Psychol.* 87: 163–171.

Spinelli, S., De Toffoli, A., Dinnella, C. et al. (2018). Personality traits and gender influence liking and choice of food pungency. *Food Qual. Prefer.* 66: 113–126.

Spinelli, S., Prescott, J., Pierguidi, L. et al. (2021). Phenol-rich food acceptability: the influence of variations in sweetness optima and sensory-liking patterns. *Nutrients* 13: 866.

Steele, C.M., Hill, L., Stokely, S., and Peladeau-Pigeon, M. (2014). Age and strength influences on lingual tactile acuity. *J. Texture Stud.* 45: 317–323.

Stein, N., Laing, D.G., and Hutchinson, I. (1994). Topographical differences in sweetness sensitivity in the peripheral gustatory system of adults and children. *Dev. Brain Res.* 82: 286–292.

Tepper, B.J. and Nurse, R.J. (1998). PROP taster status is related to fat perception and preference. *Ann. N. Y. Acad. Sci.* 855: 802–804.

Tepper, B.J., Melis, M., Koelliker, Y. et al. (2017). Factors influencing the phenotypic characterization of the oral marker. *PROP. Nutr.* 9: 1275.

Thibodeau, M., Saliba, A., Bajec, M., and Pickering, G. (2019). Examination and validation of classification schema for determining thermal taste status. *Chemosens. Percept.* 12: 69–89.

Thibodeau, M., Bajec, M., Saliba, A., and Pickering, G. (2020). Homogeneity of thermal tasters and implications for mechanisms and classification. *Physiol. Behav.* 227: 113160.

Thomas-Danguin, T., Rouby, C., Sicard, G. et al. (2003). Development of the ETOC: a European test of olfactory capabilities. *Rhinology* 41: 142–151.

Torrubia, R., Ávila, C., Moltó, J., and Caseras, X. (2001). The Sensitivity to Punishment and Sensitivity to Reward Questionnaire (SPSRQ) as a measure of Gray's anxiety and impulsivity dimensions. *Personal. Individ. Differ.* 31: 837–862.

Tuorila, H., Keskitalo-Vuokko, K., Perola, M. et al. (2017). Affective responses to sweet products and sweet solution in British and Finnish adults. *Food Qual. Prefer.* 62: 128–136.

Valentin, D., Chollet, S., Lelièvre, M., and Abdi, H. (2012). Quick and dirty but still pretty good: a review of new descriptive methods in food science. *Int. J. Food Sci. Technol.*.

Varela, P. and Ares, G. (2012). Sensory profiling, the blurred line between sensory and consumer science. A review of novel methods for product characterization. *Food Res. Int.* 48: 893–908.

Veríssimo, C.M., Macêdo Morais, S., Andrade Lima, L.L. et al. (2021). A short training as an enhancer of sensory ability: the case of red wine consumers. *J. Sens. Stud.* 36: e12629.

Webb, J., Bolhuis, D.P., Cicerale, S. et al. (2015). The relationships between common measurements of taste function. *Chemosens. Percept.* 8: 11–18.

Whitehead, M.C. and Kachele, D.L. (1994). Development of fungiform papillae, taste buds, and their innervation in the hamster. *J. Comp. Neurol.* 340: 515–530.

Whitehead, M.C., Beeman, C.S., and Kinsella, B.A. (1985). Distribution of taste and general sensory nerve endings in fungiform papillae of the hamster. *Am. J. Anat.* 173: 185–201.

de Wijk, R.A., Dijksterhuis, G., Vereijken, P. et al. (2007). PROP sensitivity reflects sensory discrimination between custard desserts. *Food Qual. Prefer.* 18: 597–604.

Wohlert, A.B. (1996). Tactile perception of spatial stimuli on the lip surface by young and older adults. *J. Speech Hear. Res.* 39: 1191–1198.

Wolf Frandsen, L., Dijksterhuis, G.B., Brockhoff, P.B. et al. (2007). Feelings as a basis for discrimination: Comparison of a modified authenticity test with the same–different test for slightly different types of milk. *Food Qual. Prefer.* 18: 97–105.

Xydakis, M.S., Albers, M.W., Holbrook, E.H. et al. (2021). Post-viral effects of COVID-19 in the olfactory system and their implications. *Lancet Neurol.* 20: 753–761.

Yackinous, C. and Guinard, J.X. (2001). Relation between PROP taster status and fat perception, touch, and olfaction. *Physiol. Behav.* 72: 427–437.

Yackinous, C.A. and Guinard, J.-X. (2002). Relation between PROP (6-n-propylthiouracil) taster status, taste anatomy and dietary intake measures for young men and women. *Appetite* 38: 201–209.

Yang, Q., Hollowood, T., and Hort, J. (2014). Phenotypic variation in oronasal perception and the relative effects of PROP and Thermal Taster Status. *Food Qual. Prefer.* 38: 83–91.

Zhao, L. and Tepper, B.J. (2007). Perception and acceptance of selected high-intensity sweeteners and blends in model soft drinks by propylthiouracil (PROP) non-tasters and super-tasters. *Food Qual. Prefer.* 18: 531–540.

Zhou, X., Yeomans, M., Thomas, A. et al. (2021). Individual differences in oral tactile sensitivity and gustatory fatty acid sensitivity and their relationship with fungiform papillae density, mouth behaviour and texture perception of a food model varying in fat. *Food Qual. Prefer.* 90: 104116.

# 5

# Similarity or Equivalence Testing

*Linda Lopez*

*S&T, Givaudan International, Kemptthal, Switzerland*

## 5.1 Introduction

There are many times when businesses are not interested in making a difference to their product's sensory characteristics but rather to ensure these characteristics stay the same. The obvious first case is in quality control, where daily production needs to consistently deliver the same sensory experience. In a manufacturing context, production of a given product line might happen at different frequencies daily, weekly, or monthly or even more randomly as sales volume might vary. The differences in production frequencies introduces some variability in personnel, cleaning procedure, machinery setting, and/or natural raw material variability. Sensory quality checks are essential to ensure product consistency by measuring how similar the recently produced batch is to a reference.

Stability during shelf life is another case where businesses are keen to ensure a consistent product. Most foods come with a "Use by" or a "Best by" date. As opposed to the "use by" date that is an indication of when the product might no longer be safe to eat, the "best by" date is determined by food manufacturers to deliver the best physical and sensory quality of their products. That means they need to determine when the product sensory characteristics begin to change in relation to the fresh product.

Another example can be seen with changes in product formulations, such as qualifying a new raw material supplier, ingredient, or recipe. Large food manufacturers, producing millions of tons of products each year, will often have several raw material suppliers, and so they need to ensure these different sources of raw material will not impact their consumers' sensory experiences.

With our society becoming more and more environmentally conscious, so do companies' efforts to reduce carbon footprint. This includes changes in processes, machinery, raw materials, manufacturing location, or transport conditions. All these changes can also impact sensory quality.

*Discrimination Testing in Sensory Evaluation*, First Edition. Edited by Lauren Rogers, Joanne Hort, Sarah E. Kemp and Tracey Hollowood.

Finally, whilst sensory similarity is often used to reduce the risk of changes, it can also be sought for innovation and marketing purposes, such as when claiming "as good as brand X" or "less sugar, same great taste." Sensory claims are regulated in many countries, guidance on how to set sensory/consumer tests to substantiate the claims are set by international standards (2021).

For a business, these changes will represent different risks and/or opportunities, some easy to quantify in financial terms such as the cost-saving opportunity or level of consumer complaints, others less so. For well-known branded food manufacturers with iconic products that are well recognized by consumers, the risk of any recipe change presents not only a risk of consumer complaints but more broadly a risk to the company's reputation. The omnipresence of social media and the rise in consumer expectations means that any product issue can quickly turn into a public relations nightmare. Major brands such as Cadbury's, McDonald's, Coca-Cola, and many others have had some of their recipe changes highlighted and debated in the media (Bryant 2021). This can impact the company's reputation and ultimately, its financial key performance indicators. When done correctly reformulation can have positive impact such as that demonstrated by salt reduction in many food categories (Jargon 2014).

These different needs for similarity testing also require differing turnaround times. In a manufacturing context, when a batch is on hold, a similarity test should be able to be delivered in a matter of hours. But for a global food manufacturer looking to reformulate an iconic product, for example to replace their artificial flavors, a more robust testing program might be required and include different levels of testing (both with expert panels and naïve consumers) that would take weeks or even months to complete. It is therefore important for a sensory professional to understand the business context in order to recommend the best approach for conducting the sensory studies.

Whilst many of the considerations already discussed in this book will also apply when looking at similarity approaches, a key difference is that similarity testing requires minimizing a different risk. In statistical terms, when setting up a test, the null hypothesis $H_0$ is typically "The products are not different," as scientists are looking to prove an effect. In setting up a statistical test, one needs to consider the results obtained as a result of two scenarios: the results are in line with the actual situation (i.e. results show a difference when there is one or do not show a difference when there is not one) or the results are not in line with the actual situation, these are the risks of making a wrong conclusion. In this case there are two types of risks, the first type is the risk of concluding there is a difference when there is not one. This is the well-known alpha risk that scientists are often trained to look at as many tests are looking to prove an effect. The second risk is the risk of concluding there is no difference when there is one. This is called the beta risk and is the one that sensory scientists interested in similarity testing will have to control (Figure 5.1).

Failing to control this risk can have incredibly large consequences particularly in the case of well-known branded products where consumers are used to a consistent and distinctive sensory experience. A perceptible difference could be rejected by consumers resulting in decreased sales and reputational damage to the brand. Increasing panel numbers to increase the power of the test needs to be considered carefully, as with a large enough panel size, a significant difference will ultimately always be found but might not be meaningful to the consumer.

| | | Test results: Difference | Test results: No difference |
|---|---|---|---|
| Actual situation | Difference exists | Correct conclusion | "Miss" Type II error Beta-risk |
| Actual situation | No difference | "False alarm" Type I error Alpha-risk | Correct conclusion |

**Figure 5.1** Statistical decision matrix showing the two types of error when setting up a statistical test.

In this chapter we will first look at pragmatic approaches to setting up similarity testing giving the sensory professional a checklist to consider using when faced with similarity questions, including method, panel type and calibration, availability of historical results and correlation with consumer test results. Then we will dive into the evolution of statistical approaches to similarity testing from the power approach and interval testing to Thurstonian scaling applied to various discrimination test methods. Finally, we will look at different cases of application of sensory similarity testing for sensory claim substantiation or product reformulation looking at the impact of panel type, product type and test set-up.

## 5.2 Practical and Common-sense Approach for Similarity Testing

For industry professionals, changing their sensory methodology can be a daunting task that will receive little management support unless a quantifiable benefit can be demonstrated. Large companies with well-established sensory programs might be wary of change if no problem has been raised. There might also be high pressure to implement a change project within a certain time frame requiring the sensory professional to work with methods that can fit within the time plan. Unfortunately, what those companies often miss is the incremental degradation that can occur with sequential changes as illustrated in Figure 5.2. A common way to test for similarity is to test the new recipe against the current one (the target). When recipe or production changes occur often, a common occurrence in the food industry, every subsequent change can be validated against the target but lead to a significant shift in the sensory perception compared to the original product.

Smaller companies might not see the need for a dedicated sensory program when they are used to more pragmatic solutions, for example where any change in product is approved by tasting around a table or by the company manager. Yet, for growing companies extending their production capacities, good sensory methods for similarity testing will be crucial to maintain the sensory distinctiveness of their successful products.

Hence before embarking in a program of methodology change to apply a state-of-the-art similarity test method, it is worth checking whether the existing approach (methodology, panel, environmental setting, etc.) is good enough to accept a "no significant difference" as

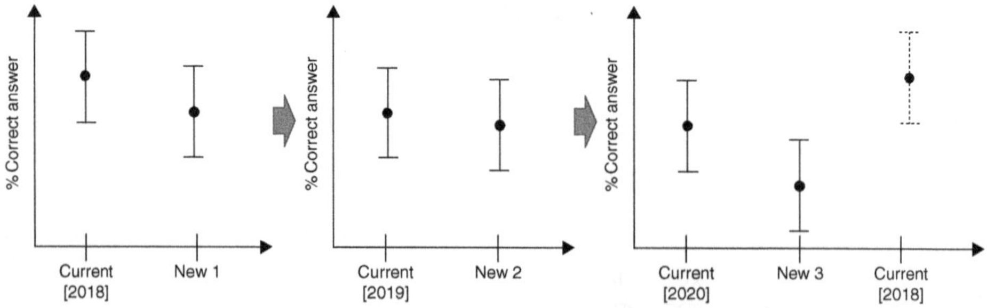


**Figure 5.2** Incremental sensory degradation with sequential changes. In this hypothetical example, a recipe change was validated in 2018, then another in 2019 and another again in 2020. Every time, the recipe change was validated against the "current" recipe by a difference test. Subsequent changes can lead to a significant difference in sensory characteristics of the product compared the original 2018 recipe.

a similarity result. Lawless and Heymann have highlighted a list of prerequisites to take into consideration on which we have built (Lawless and Hildegard 2010, Chapter 5, Similarity, equivalence testing, and discrimination theory).

We will look at protocol consistency (test method and instructions, panel size and type, testing facility), the use of already available data, validation testing, panel screening, and the use of other type of data (such as descriptive or consumer data).

## 5.2.1 Protocol Consistency

It is important to first assess the basics: is the protocol used (i.e. sensory discrimination test/method, known population of panelists, specific test conditions, type of product tested, etc.) robust and consistent enough for the particular change to assess? The following elements address practical considerations that can serve as a simple check list.

### 5.2.1.1 Specific Quality Control Protocol Consideration

When looking at quality control, we are looking at small changes (no difference in recipe or process) and smaller risk (the volume and sale value of a production batch) whilst being limited by the turnaround time for the test (minutes or hours) and cost. Often only a limited number of panelists might be available and there will be variation in panelists between shifts. The type of test would need to be adapted to ensure testing "fatigue" can be controlled. Whilst quality control presents many constraints, it also has advantages with the daily frequency of testing helping to train and make panelists attune to small changes.

The sensory scientist in this case can still greatly influence the quality of the panel by:

- Pre-screening panel members for sensory abilities and motivation.
- Training the panel members on basic sensory concepts and specific sensory methodology.
- Selecting and validating appropriate references with another panel.
- Checking that the panel adhere to general sensory principles (e.g. palate cleansing).
- Setting up appropriate testing frequencies to enable correct batch discrimination.

The following sections apply to general sensory panel management as well as quality control panels.

#### 5.2.1.2 Test Method Choice and Detailed Operating Instructions

The choice of the test method needs to take into consideration several factors:

- How much fatigue might the products to be tested cause? The tetrad test has been often documented as being a more powerful method than others, but in some cases where tasting might cause fatigue, such as with a product with lingering characteristics, a triangle or 2-AFC test might be preferred.
- How heterogeneous are the products from assessment to assessment? A triangle test using only one assessment might produce very un-reliable results if the product to be tested is heterogeneous from assessment to assessment, such as some multicomponent snack food or sauces. Dealing with issues of heterogeneity will be further discussed in Section 5.4.2.2.
- How much training is necessary? Some methods, such as degree of difference, would require specific training sessions and knowledge about the product categories to be tested, whilst a triangle test would only require instructions to be given on the day of testing after initial test familiarization.
- What panelists are available? Test methods and instruction might be very well followed by a dedicated panel whose job description it is to do product testing. An employee panel, where panelists participate in product testing on a voluntary basis might be more sensitive to task length and complexity of instructions. The method choice would also in this case need to take into account any biases that employees might have (e.g. knowledge of the product, project, etc.).

A test method is often well defined but not always precisely enough documented. This can lead to deviation in how the method is practically executed. For example, have the task instructions been properly described and executed? The impact of task instructions has been reported in many articles, for example, Ennis et al. (2012) have shown that task instructions can have an impact on test sensitivity.

#### 5.2.1.3 Panel Size and Type

One of the most critical choices to make is whether the difference or similarity test needs to be carried out by expert or naïve panelists. This is heavily debated in the sensory community (Moskowitz 1998). On one hand, supporters of expert panels argue that an expert panel is more sensitive to change and will be able to detect differences earlier than an untrained panel, thus providing an "insurance policy" for the company. On the other hand, supporters of consumer panels argue that what the consumers think is what matters and expert panels are not a true representation of "real life," therefore leading to potential missed opportunities. Both approaches have their merits. The way the data can be used, and the decision made from the results would be different depending on the approach and the level of information available to link expert panel data to "real life" consumer data. This is further discussed in Section 5.2.4.

Panelists' numbers also need to be consistent so that the strength of the statistical information of each test can be appropriately compared. Is there a defined minimum number below which tests can no longer be carried out, is it adhered to? Attendance records can drop during holiday season, what solution has been put in place to minimize the impact? To address this issue, some panels have planned closures for a set number of weeks during the summer or winter holiday season.

#### 5.2.1.4 Testing Facility

Whilst the sensory lab or consumer testing booths have provided a great consistency for sensory testing, during the pandemic the need to move a large number of tests to in-home testing has led to more questions being asked about the role of the testing facility. In some cases, with the use of cloud-based data recording systems, moving testing to in-home had no impact on test results quality and facilitated better attendance (Dinnella et al. 2022).

Product characteristics are a major factor to consider in choosing the right testing facility. Many products require extensive preparation and controlled temperature for testing such as cooking sauces or hot drinks, hence a dedicated ISO-certified sensory facility might be preferred to be able to control this inherent variability. Whilst for many personal care products such as shampoo, in home testing might still be a preferred choice to represent a more realistic product usage that is also the most convenient for both panelist and panel leader.

The testing facility and equipment also has a crucial role to play in helping to blind products. The ability to test under different light conditions can help mask any color differences that can otherwise impact perception.

### 5.2.2 Availability of Historical Data

What businesses are ultimately concerned with is loss of sales, but since a lot of other factors than sensory can influence sales data, it is notoriously difficult to directly correlate sales to product changes. There are fortunately other types of data that can be used to understand the impact of a change such as hedonic consumer tests, consumer complaints data (more prominent in Western countries) or descriptive analysis. Tying up discrimination test data with these results can help provide a track record of the protocol, and, if it matches well enough, this can provide management with confidence in their validation tool.

Additionally, the availability of historical discrimination and similarity test data can be used to review and to calculate the average statistical power of the methodology and therefore the risk that the company might be taking by making decisions based on such data.

### 5.2.3 Regular Validation Tests

Availability of known defective or different samples as well as known equivalent samples can help to provide calibration validation tests to further demonstrate that the method can pick up differences and does not raise false alarms. Knowledge about the product category is required to identify suitable validation samples.

A common practice in quality control is to retain samples as references (both pass or failed) when product shelf life allows. New production batches can be compared to reference samples and become a new reference for future production.

There also exist some companies providing flavor standards for use in panel proficiency evaluation (WQPMag 2016).

### 5.2.4 Panel Screening and Correlation to Descriptive or Consumer Data

As already discussed, both expert or naïve consumers can be used for similarity and difference testing. Either approach is good as long as the relationship between the expert panel and consumer impact is known. Rousseau has provided a model to look specifically at differences in sensitivities of expert and consumer panels (Rousseau 2004). By doing 10 studies using the same products in consumer and expert panels, Rousseau was able to compare the perceptual variance, $d'$, and showed the higher sensitivity of the expert panel as well as the shape of the relationship between the trained and consumer panels. Once this relationship is known, the impact on consumer liking would need to be studied with a similar sample set in order to determine a true threshold level for consumer acceptability. Figure 5.3 represents the typical relationship expected between an expert trained panel and a naïve consumer population. As an expert panel is more sensitive than a consumer panel, they should discriminate more and therefore generate greater $d'$ values compared to a consumer panel when testing the same products for difference.

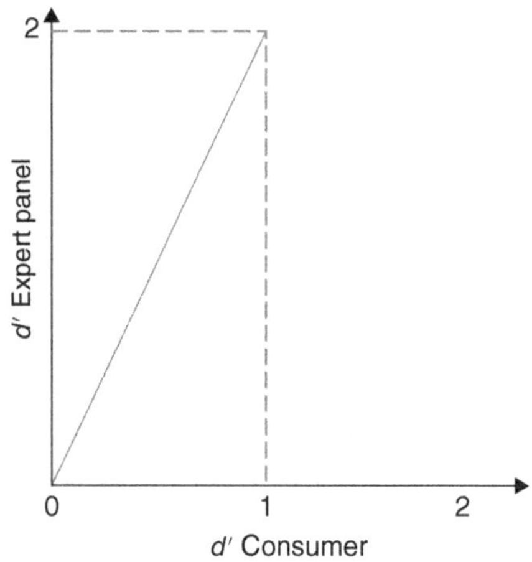


**Figure 5.3** Theoretical relationship between expert and naïve consumer panel as represented by $d'$ variance. The slope of the curve can represent the degree of specialization of a trained panel compared to a general consumer population

Furthermore, in some situations, descriptive data might also be available together with discrimination data. Correlating discrimination data with the number of significantly different attributes can provide valuable information on the discrimination power of the test. Mapping out the descriptive profile of products over time is a good way to identify product drift over time whilst preventing sensory degradation (Pecore and Kellen 2002).

Relying on $p$-values suggesting no significant difference can miss changes that could potentially impact consumers. Rousseau has shown that products that passed a triangle test were subsequently rejected in a consumer preference test (Rousseau and Ennis 2017). Ennis showed that products that would have been rejected by looking solely at $p$-values generated by an expert panel, can be accepted after computing the probability of rejection and consumer relevant confidence at different test power (Ennis and Ennis 2018). In this case study, a tetrad test was carried out to measure if an ingredient change on a candy had any impact on sensory perception. The test showed a significant $p$-value, but a $d'$ value lower than 1. To understand if this $d'$ value is below the consumer relevant difference $\tau_R$, an unknown value; the authors used the theory for multiplicative comparisons. A same-different test was carried out with 200 consumers to evaluate $\delta$ and $\tau_R$ and their variance. By knowing the increased sensitivity of the expert panel, the $\tau_R$ value and variance estimated can be re-scaled to the expert sensitivity. The comparison of the rescaled $\tau_R$ to $d'$ lead to calculating the confidence interval which returned a value of 98% leading to a high confidence level to accept the change when the initial test would have rejected it.

These effects (accepting a change when it is consumer relevant or rejecting a change when it is not consumer relevant) call for a different approach to be considered when carrying out similarity testing.

## 5.3 Evolution of Similarity Testing Methodology

### 5.3.1 Test for Significant Similarity

An early approach to the problem of demonstrating product similarity or equivalence was presented by Meilgaard et al. (2006). It is based on the fact that we do not have to test against the chance performance level in applying the binomial test on proportions. Instead, we can test against a more "relevant" level of expected proportion correct as represented in Figure 5.4. These types of testing were set in order to limit the beta risk usually to 5 or 10% and are sometimes referred to as the "power approach."

The alternative level can be set in order to be more conservative, for example only allowing 10% of discriminators, or broader to allow 30% or 40% depending on the type of change and business context. A significant recipe change on a well-known product might call for a very conservative approach to be taken, whilst a more minor ingredient supplier change on an own label product might call for a higher tolerance level. The proportion of discriminators for a significant difference to be found can be then worked out using Abbotts' formula (Eq. 5.1) (Finney 1971). Tables showing critical values for significant similarity for paired comparison can be found in ASTM standard (2020), triangle and duo-trio tests can also be found in Lawless and Hildegard (2010) and for additional types of tests in Meilgaard et al. (2006).

$$P_{observed} = P_{adjusted} + P_{chance}\left(1 - P_{adjusted}\right) \quad (5.1)$$

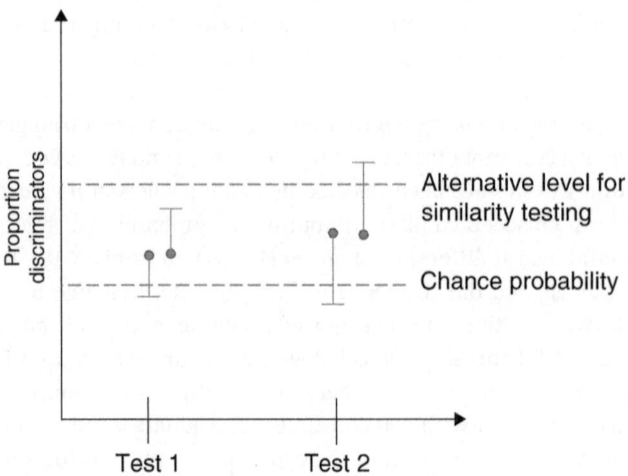


**Figure 5.4** Representation of an alternative level test for significant similarity where the statistical test is no longer carried out against the chance probability but against an alternative level set for similarity. Error bars represent 95% confidence interval one tail. In this case, the products tested in test 1 would have no statistical differences as the error bar overlaps with the chance level and, tested against a higher level for similarity, would have been found significantly similar as the error bar does not overlap with the alternative level for similarity testing. In test 2, the products would again have found to be not significantly different, but would not have found to be significantly similar as the error bar does cross over the higher level set for similarity.

To illustrate this, here is a short example. In this example, we are conducting a triangle test with 60 panelists. The results show that 21 choose the correct (different) sample. First, we look in the tables for difference testing. The critical value for difference is 27 at 5% probability level, as we are below this value, we can conclude that the samples are not significantly different. To determine if we have significant similarity, we first set our similarity criteria at no more than 30% discriminators in the population with a beta risk of 10%. Then, we look at tables in Rogers (2017) and identify that for our criterion the critical value is 25. Because we are below this value, we can conclude that the products are significantly similar.

### 5.3.2 The Two One-Sided Test Approach (TOST) and Interval Testing

Testing for equivalence through a non-significant test, even at a higher level has not been accepted by the scientific community concerned with bioequivalence. One of the main pitfalls of the "power method" is that for a sample size big enough, the chance of finding equivalence becomes negligible. As the sample size increases, the standard deviation decreases, hence the probably of finding a significant difference increases and the probability of finding similarity decreases, although this size of difference might be meaningless for consumers.

Instead, interval testing commonly used in drug delivery studies has been described as the methodology of choice. Bi has proposed a procedure based on this approach applied to different situations in sensory and consumer testing: hedonic or intensity rating (Bi 2005).

In this approach the null hypothesis is that the difference of two treatments is equal to or larger than a specified allowed difference and the alternative hypothesis is that the difference is smaller than the specified value, i.e.:

$$H_0 : \left|\mu_1 - \mu_2\right| \geq \Delta_0 \tag{5.2}$$

$$H_1 : -\Delta_0 < \mu_1 - \mu_2 < \Delta_0$$

$H_0$ represents the null hypothesis and $H_1$ represents the alternative hypothesis. $\mu_1 - \mu_2$ is the confidence interval. $\Delta_0$ is the allowed difference specified.

If the null hypothesis is rejected, the alternative hypothesis (i.e. similarity between the two products for comparison) can be concluded. In this type of testing approach, Type I error, $\alpha$, is the probability of concluding similarity when the two treatments are in fact different. An $\alpha = 0.05$ or 0.1 is usually selected. Type II error, $\beta$, is the probability of failing to conclude similarity when the two treatments are similar. The power of a similarity test, $1 - \beta$, is then the probability of correctly rejecting the null hypothesis of difference and accepting the alternative hypothesis of similarity when the two treatments are similar. A $\beta = 0.1$ or 0.2, i.e., $1 - \beta = 0.9$ or 0.8 is usually selected.

The confidence interval approach can also be used for similarity evaluation. If the confidence interval $\mu_1 - \mu_2$ is within the allowed specified difference $(-\Delta_0, \Delta_0)$, similarity can be concluded.

### 5.3.3 Thurstonian Scaling

A common issue in discrimination testing is that the proportion of discriminators is not constant across methodology as illustrated by Rousseau (2021). Figure 5.5 shows how different proportion correct and proportion discriminators have been reported by different

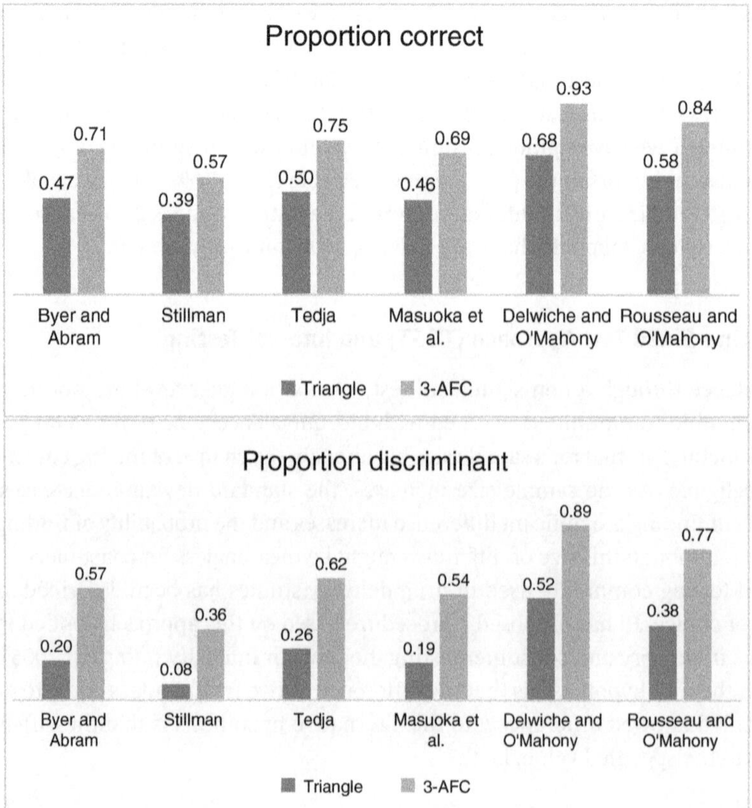


**Figure 5.5** Comparison of proportion correct and proportion of discriminants from various studies using triangle and 3-AFC (adapted from Rousseau 2021).

authors for triangle and 3-AFC tests. This makes sense as each method represent a different proportion of chance responses as well as a different mental load on the participants.

This issue pushed researchers in the sensory field to look at methods used in psychophysics for a method that could be more universal.

Thurstone formulated a law of comparative judgment in 1927 that provided the basis for this measurement and this is explained in detail below (Thurstone 1927).

#### 5.3.3.1 The Theory and Formulae

This approach is derived from the signal detection theory. This theory has been developed in psychophysics and experimental psychology, particularly in the measure of thresholds. In this type of experiment, participants (also known as receivers) are asked to say whether they perceive (see, hear, touch, taste, or smell) a signal or not. The signal intensity is raised until the receiver replies "yes, now I do." Researchers realized that a lot of other factors than sensory characteristics can influence the respondent's reply.

In a classic signal detection experiment, two levels of a stimulus are to be evaluated: a blank and weak signal close to threshold. These stimuli are presented one at a time and the receiver responds "Yes, this is the signal" or "No, this is a blank." The responses can then

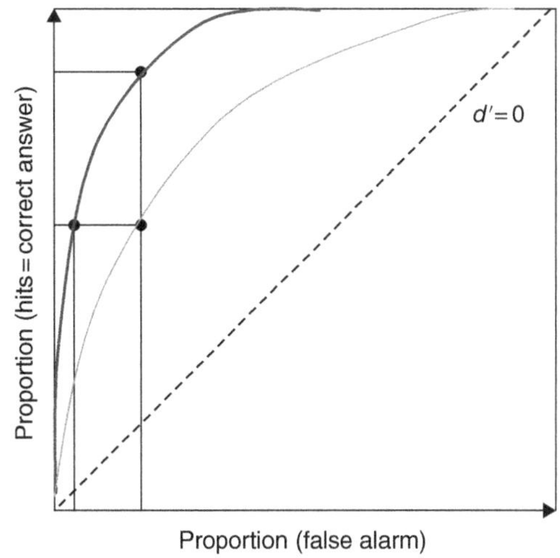


**Figure 5.6** The ROC curve or receiver operating characteristics. The curves in dark grey and light grey represent two observers with two different sensitivities. The dark grey observer achieves a higher proportion of hits at the same proportion of false alarms as the light grey observer. For the same proportion of hits, the light grey observer achieves more false alarms than the dark grey observer. The dotted line represents $d' = 0$.

be tabulated to calculate the number of "hits" (when the receiver correctly identified the signal), "miss" (when the receiver did not identify the signal) "false alarm" (when the receiver identified a signal when he only experienced a blank) and "correct rejection" (when the receiver correctly identified the blank). The receiver operating characteristics (ROC) can be plotted by reporting hit rates vs. false alarm. Figure 5.6 shows how two observers would behave in several experiments with the same levels of the stimulus and noise. As the receiver is put under different conditions, the ROC curve is plotted with the same condition of signal and noise. The level of discrimination is proportional to the area under the ROC curve and is related to $d'$. If the hit rate and false alarm rates were equal, there is no discrimination of the two levels, and $d'$ is zero. This is shown by the dotted diagonal line in Figure 5.6.

In Thurstone's law of comparative judgment the panelist compares two similar products, A and B, over several trials and we record the number of times A is judged stronger than B. Thurstone proposed that the sensory events produced by A and B would be normally distributed. Thurstone called these perceptual distributions "discriminal dispersions." The proportion of times A is judged stronger than B comes from a process of mental comparison of the difference between the two stimuli. Evaluating a difference is similar to a process of subtraction. Sometimes the difference will be positive (A stronger than B) and sometimes the difference will be negative (B stronger than A) since the two items are confusable. Figure 5.7 shows how the "discriminal dispersions" can be transformed into a sampling distribution.

Statistical analysis can help to relate the different sampling distributions (Figure 5.2) to the sensory event depicted in the upper distribution and relate it to $d'$.

#### 5.3.3.2 Application of Thurstone Theory to Sensory Forced Choice Tests

***Application to Detect Fully Match Reference*** The extension of this type of scaling to sensory discrimination tests has been performed by different authors on different types of tests (Frijters et al. 1980; Ennis 1993; Bi 2005).

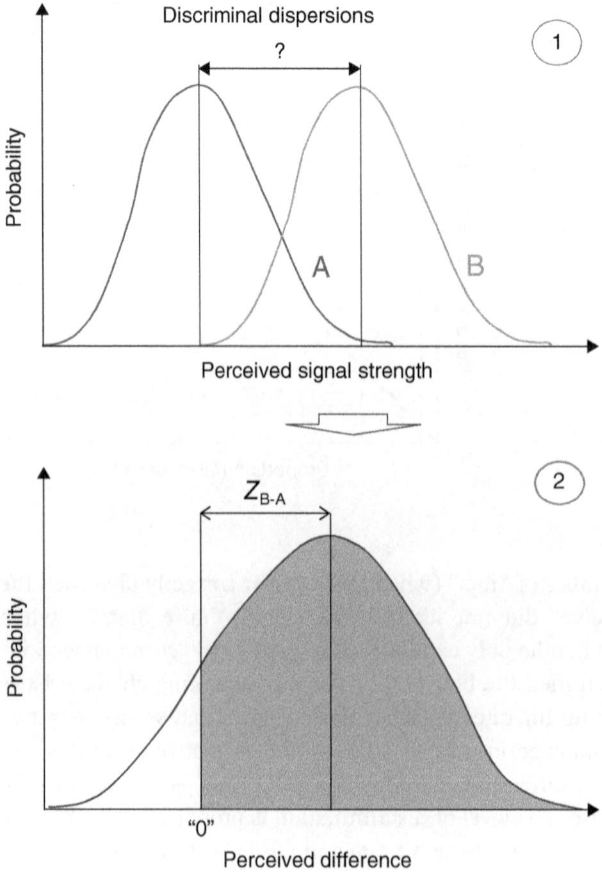


**Figure 5.7** Representation of how the Thurstone model proposes that the proportion of times one stimulus is judged greater than another is predicted by a difference sampling distribution, which in turn arises from the sensory variability and degree of overlap in the original stimuli. The colored area under the curve represents the probability of B > A in a paired comparison or forced choice method. Where A and B are the two products tested and $Z_{B-A}$ represents the proportion of times where A is judged stronger than B (*z*-score).

For a given level of scale difference, a lower level of proportion of correct answers is expected in a triangle test as opposed to the 3-AFC. This is notoriously known as the "paradox of discriminatory non-discriminators" where it is possible for many panelists to answer incorrectly in a triangle test but still accurately choose the strongest sample when it was the odd one (Byer and Abrams 1953). Frijters (1979) was able to be resolve the paradox by showing that the data actually represented the same $d'$. This effect is illustrated in Figure 5.8. This figure shows how product A, its duplicate and product B can be differently perceived. In this representation the smallest difference is not A vs. its duplicate, this would therefore lead to incorrect answers, whereas in a 3-AFC, B would be correctly found as the most intense.

The overall conclusion one can take from reviewing this work is that common unspecified discrimination tests such as triangle or duo-trio are not very sensitive tests compared

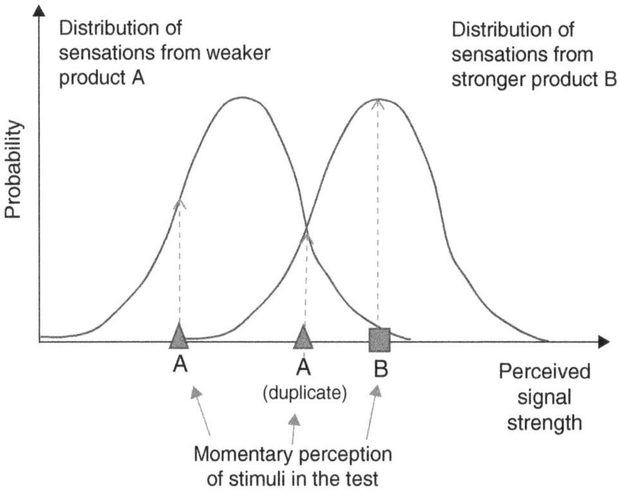


**Figure 5.8** Representation of an example of the paradox of discriminatory non-discriminators. In this example, in a triangle test, A and its duplicate is not the smallest comparison, hence would be an incorrect answer. In a 3-AFC, B would be correctly found as the strongest stimulus.

to specific discrimination tests like 3-AFC where a specific attribute needs to be selected to carry out the assessment (for example which sample is sweeter?). That is, for a given $d'$ value, a much larger panel size is needed to be sure that the difference is detected by the test. Put a different way, for a given panel size, the triangle test could easily miss a difference that the 3-AFC test would detect. Unfortunately, changes in complex food matrices can be difficult to predict, hence the sensory professional often has no other choice than using this less sensitive type of test.

The emergence of tetrad testing has provided an alternative to the triangle test that can provide more power and better discrimination with smaller panel sizes (Ennis and Rousseau 2012). Care needs to be taken to ensure that the addition of a fourth sample will not lead to panel fatigue which could be counterproductive. This will depend primarily on the product type.

Recent work by Castura (2019) looked at both the discriminal distance ($\delta$) and the response bias for responding the same ($\tau$) using polyurethane foam as a model by using same-different and tetrad testing. In this test, respondents were asked to assess the foam by pressing on it with their fingers. Both $\delta$ and $\tau$ estimates were large compared to those found in the evaluation of food.

The $\delta_{sd}$ (for same-different testing) and $\delta_{tet}$ (for tetrad testing) calculated for different paired tests respectively, do not obey a simple additive model. When $\hat{\delta}$ is large, the slope of the psychometric function is flat, leading to an imprecise estimation. High signal-to-noise ratios for the stimuli may lead to imprecise estimates that partially explain the fact that discriminal distances are not additive.

Results indicate that assessor performance on the tetrad test method was best when samples were presented in an XXOO pattern, and worst if presented in an XOOX pattern. This result is consistent with previously reported results from a web-based tetrad test (Castura et al. 2019).

Results also showed that $\delta_{sd} > \delta_{tet}$ as is consistent with previous studies (Ennis and Christensen 2015). The implication is that due to precision of measurement the tetrad test method will have higher statistical power and can sustain an increase of up to 50% in additional perceptual noise due to the evaluation of the additional two samples.

It was expected that memory effects and sensory fatigue would be relatively modest for these by-hand evaluations. Yet the differences in $\delta_{sd}$ and $\delta_{tet}$ indicate that perceptual variance associated with the tetrad task (involving four samples) was approximately 50% higher than the perceptual variance associated with the same-different task (involving two samples). Hence, $\delta_{sd} > \delta_{tet}$ may indicate higher levels of task understanding in the same-different test method than in the tetrad test method (Castura et al. 2019).

It is suggested that $\hat{\delta}$ estimates from same-different test data can be considered a psychological distance on a latent space with standard deviation units related to same-different categorization, whereas the $\hat{\delta}$ estimates from tetrad test data can be considered a psychological distance on a latent space with standard deviation units related to discrimination. If this were the case, these distances might lie on (possibly correlated) psychological continua that align approximately with what is considered by assessors to be "relevant" and "perceived," respectively.

**Application When Samples Do Not Fully Match a Reference**

Whilst many methods have been developed to answer the question "Are my test samples the same or different to my reference?," in some cases, samples to be tested do not exactly match the reference. This can be the case with samples having natural variability, for example cheese, or when a specific mixing ratio is altered. In this case the sensory professional is asking "How do my test samples relate to my references?" Forced grouping has been developed by Nijhof et al. (2021).

In this situation, two references are chosen (A and B). Panelists are asked to first taste them and describe them. Then the test products are presented to them monadically and they are asked to indicate for each of the tested products if it is more similar to reference A or reference B.

Data analysis is performed using the binomial test and Thurstonian model. Panel performance tests are also described to assess individual performances providing data on detectability, discrimination, reproducibility and agreement. The use of this method with mixtures of skimmed yoghurt and full fat quark showed good results and more application has shown its applicability in different cases from reformulation to quality control.

## 5.4 Example of Similarity Test Application in Claim and Recipe Change

### 5.4.1 Parity and Equivalence Claims

Sensory claims have been used by food and personal care manufacturers quite extensively. In Europe, more regulation has been put in place in order to help further protect consumers from misleading claims. It is therefore important to understand the industry standard framework and latest advances in this field when considering making a parity or equivalence claim.

First, the sensory professional needs to understand if marketing wants to deliver an "Equality claim" or a "unsurpassed' claim (Figure 5.9). Both types of claims can be based both on sensory perception or hedonic response, but equality claims are delivered against a given standard (e.g. "As good as brand X") whilst an unsurpassed claim assumes non superiority of other references (e.g. "No other yoghurt is creamier than ours").

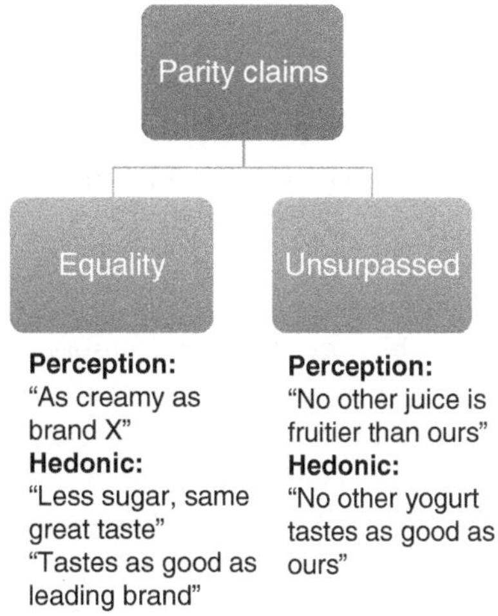


**Figure 5.9** Different types of parity sensory claims.

Typically, the ASTM guidance for a parity claim with a 2-AFC method is to fall within the 45%:55% preference split (ASTM 2020). Where paired tests have been used (typically the case when comparing two samples), for equality and unsurpassed types of claim, the ASTM provides separate tables with the minimum choice counts required depending on the number of consumer respondents used.

As an illustration, let's consider a biscuit manufacturer who would like to re-formulate its product with less sugar and be able to claim, "less sugar, same great taste."

To do so, a paired preference test is conducted with a panel of 800 consumers, 390 consumers chose the new reduced sugar formulation as their preferred product. To identify if this result substantiates an equality claim, we need to consider the null hypothesis that the choice probability is $\leq 0.45$ and $\geq 0.55$. Ennis (2006) provides a table to identify the critical counts. At 95% confidence level for a panel of 800 participants, the proportion of respondents to claim equality must be between [384;400]. You can thus substantiate your claim.

Note that comparative parity claims (two-sided test) are harder to support than superiority claims (one-sided test): they require more consumer respondents to protect the competitor, but also to provide an advantage to the advertiser. Examples of parity and unsurpassed claims are given in Table 5.1.

Unsurpassed claims are easier to achieve since they allow a superiority test (preferred or greater or less on some attribute) using only one of the two limits that are used to define an equality claim. For an unsurpassed preference claim, the preference probability may fall above 45% in favor of the manufacturer's product. Hence sample size for unsurpassed claims are generally smaller than for parity claims (Ennis 2006).

**Table 5.1** Examples of parity and unsurpassed claims (Google search).

| | |
|---|---|
| Parity claims | Less salt, same great taste |
| | Same great custard style (yoghurt) |
| Unsurpassed claims | Nothing tastes better than the unique taste of Brand X |
| | Best ever recipe |

## 5.4.2 Recipe Changes

### 5.4.2.1 Consumer Relevant Size of Difference

So-called "silent" recipe changes are very frequent in the food industry and require new formulations to pass similarity testing. Knowing when products are close enough to be equivalent is essential to the sensory professional but often difficult to know.

Rousseau and Ennis (2013) proposed a method to identify consumer relevant thresholds for difference perception.

In this case study, the sensory professional is interested in quantifying the size of the difference and if consumers would consider a change of that size acceptable so that ultimately an action standard could be developed within the mayonnaise category. Relating expert panel discrimination test results to consumer sensory preference requires a set of samples with well spread differences which is not always available. The proposed method uses the same-different task as it contains information about the maximal difference that will still elicit, on average, a "same" response from a selected group of consumers.

Using Thurstonian modelling with the same-different method can not only calculate the size of the underlying sensory difference, $\delta$; it also allows for the estimation of an additional parameter called $\tau$, the decision criterion. Like $\delta$, $\tau$ is measured in terms of the standard deviation of the products' underlying perceptual distributions. Figure 5.10 illustrates how the distance between the two products and $\tau$ relate to the consumer answering "same" or "different." In the first test, the distance between the two products is smaller than $\tau$ therefore the subject answers "same," in the second test, the difference between the two products is larger than $\tau$, leading the subject to answer "different."

In order to investigate the size of $\tau$ for the consumer, a study was planned with 300 consumers and 4 samples with different degrees of difference vs. the reference (10 pairs in total). For each pair the estimated sensory difference $\delta$ (called $d'$) and the decision criterion ($\tau$) has been calculated. The $\tau$ calculation showed that it was very constant across all tests and was on average 0.8. This value can therefore be used as the maximum threshold level.

In order to be able to routinely use this criterion with the in-house expert panel, this value needs to be related to the expert panel increased sensitivity compared to the general consumer population. Results from an earlier investigation had shown that the expert

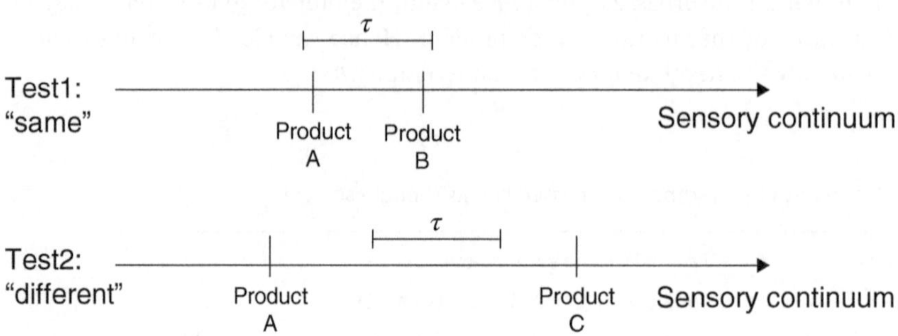


**Figure 5.10** Illustration of the use of $\tau$ as consumer decision criteria in same-different test with two products (adapted from Rousseau 2013).

panel was 40% more sensitive than consumers, this translated to a similarity criteria of 1.12 with the expert panel.

#### 5.4.2.2 Impact of Product Heterogeneity

Some products are heterogeneous in nature and can change from batch to batch due to natural variability. Common discrimination tests might not be suitable to determine if differences are detectable without modification.

Degree of difference (DoD) test, also called difference from control, has been highlighted as a method that can take into account inherent product variability by incorporating control batch variability (Aust et al. 1985). Whilst this method has been well used for product categories such as soups, cookies, rolls or multicomponent snack mixes, it requires a choice of a control batch that can be problematic.

The DoD approach uses a scaled difference rating (from "Not at all different" to "Extremely different") to measure the difference between the test and the control product and compare it to a baseline difference between control batches (both within batch variability and between batch variability). The selection of the control batch can affect the outcome of the test and may not detect the situation where the test batch is outside of normal control batch variability. A proposed method including control batch variability has been introduced by Pecore et al. (2006). In this method, the within batch difference for one control is eliminated and replaced with a comparison of the test batch to two controls.

Another type of heterogeneity that the sensory professional needs to deal with is between production facilities. An approach has been described by Rousseau and Ennis (2002), where a company is proposing an ingredient change to their vanilla yogurt. The study is complicated by the fact that the company has two production lines for the yogurt that produces slightly different products. An extension of the duo-trio method (Torgerson's method) is proposed to assess the current product produced on both lines vs. the new product produced also on both lines.

Using a unidimensional Thurstonian model, analysis of data provides the six possible $d'$ values as well as their variances as illustrated in Figure 5.11. The discriminal dispersions for current recipes are represented by $A_1$ and $A_2$, whilst the new recipe dispersions are represented by $B_1$ and $B_2$ for each of the lines. The estimated sensory difference calculated can be plotted showing the distance between line 1 current and new recipe, line 2 current and new recipe and the distance between the two lines' current recipes.

In this example, it is possible that all four products are different or that there is no difference in product lines but the only difference is made by the recipe modification. It is also possible that the modification had no effect. In order to understand if the distances between products are significant; various models can be used to fit the data using the method of

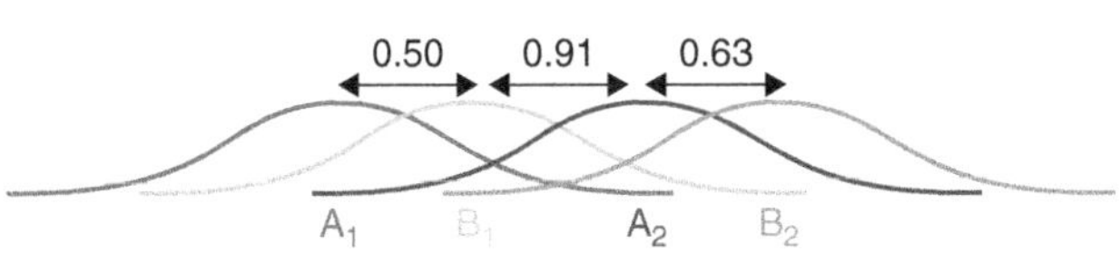


**Figure 5.11** Representation of product similarities using the Thurstonian model for current and new formulations produced on lines 1 and 2 (adapted from Rousseau 2002).

maximum likelihood. This simulation models the distribution using different numbers of distributions in a hierarchical way. It starts with the more complex model containing the total number of triads (here, this number is four) and will test the model against a simpler model (down to the simplest model at the end which is the pure guessing model). The model simulation showed that the model with four distributions was the best fit. The distance between current and new formulation is 0.5 which is significantly different from zero whilst the distance between the two production lines is 1.47. This means that whilst the new ingredient significantly altered the vanilla yogurt, the extent of the change is smaller than the difference between the two production lines. Hence, since consumers are already used to a broad variability, the new formulation might be appropriate.

## 5.5 Conclusion

In this chapter, we have looked at practical and common-sense approaches to similarity testing, then dived into the evolution of statistical approaches for testing for similarity and finally looked at how similarity testing is used in claims and recipe changes.

In science, proving the absence of difference is more complex than proving a difference exists. In the sensory field, testing for similarity has important implications for research and businesses, both in maintaining product quality and marketing communication. Good panel and protocol management practices are essential to measure equivalence between samples. With the acceleration of innovation processes, there is a need to develop ever quicker, cheaper and more accurate methods. The use of tetrads has been a very effective way to reduce panel size and increase discrimination power for some product categories. New techniques are required to assess complex products such as heterogeneous products (e.g. multicomponent snack foods) or products that change during consumption time such as chewing gum.

Knowing how to relate expert panel measures to a real-life consumer population is key in order to answer the research questions in the right context. Discrimination methods have often been criticized for being "lab experiments"; not representing real life as consumers rarely experience products in a non-monadic way. Yet, a stricter method is often looked for as maintaining sensory quality and for some, their distinctiveness year on year is a real issue that Fast-Moving Consumer Good manufacturers need to be able to effectively manage whilst being able to provide their companies with flexibility in formulation. To resolve this dilemma, more research is required to relate expert panel and discrimination methods to "real life." Linking discrimination to hedonic scores and different consumer populations (heavy vs. light users for example) would help to better understand how to use each tool and guide decision making. Similarly, more work in the area of sensory memory, sensory appreciation and preference over time would enable better understanding of key processes of memorization of sensory characteristics.

## References

ASTM (2020). Standard Guide for Sensory Claim Substantiation. E1958-20.

Aust, L.B., JR., M.C.G., BEARD, S.A. and II, R.W.W. (1985). Degree of difference test method in sensory evaluation of heterogeneous product types. *J. Food Sci.* 50: 511–513.

Bi, J. (2005). Similarity testing in sensory and consumer research. *Food Qual. Prefer.* 16: 139–149.

Bryant, K. (2021). 13 times food brands changed iconic recipes—and shocked the industry. RD. com (3 August 2021). https://www.rd.com/list/times-food-brands-changed-iconic-recipes/. Accessed date: 27 August 2022

Byer, A.J. and Abrams, D. (1953). A comparison of triangle and two-sample taste methods. *Food Technol.* 7: 183–187.

Castura, J.C., King, S.K., and Findlay, C.J. (2019). Does the τ estimate from same-different test data represent a relevant sensory effect size for determining sensory equivalency? *Food Qual. Prefer.* 75: 230–238.

Chambault, M. and Thomas, S. (2020). *Sensory Claim Substanciation: An Overview*. Campden BRI.

Dinnella, C. et al. (2022). Remote testing: sensory test during Covid-19 pandemic and beyond. *Food Qual. Prefer.* 96: 104437.

Ennis, D.M. (1993). The power of sensory discrimination methods. *J. Sens. Stud.* 8: 353–370.

Ennis, J.M. and Ennis, D.M. (2018). Consumer relevant confidence in sensory difference tests. *IFPress* 21: 2.

Ennis, J.M., Rousseau, B., and Ennis, D.M. (2012). Proper task insruction and the two-out-of-five test. *IFPress* 15 (4): 4–5.

Ennis, J.M. and Rousseau, B. (2012). Reducing costs with tetrad testing. *IFPress* 15: 1.

Ennis, M.E. (2006). Parity claims. *IFPress* 9: 4.

Finney, D.J. (1971). *Probit Analysis*, 3e. New York: Cambridge University.

Frijters, J.E.R., Kooistra, A., and Vereijken, P.F.G. (1980). Tables of *d′* for the triangular method and the 3-AFC signal detection procedure. *Percept. Psychophys.* 27 (2): 176–178.

Frijters, J.E.R. (1979). The paradox of discriminatory nondiscriminators resolved. *Chem. Senses* 4: 355–358.

ISO 20784 (2021). Sensory analysis – guidance on substantiation for sensory and consumer product claims (March–December 2021). https://www.iso.org/standard/69080.html. Accessed date: 23 August 2022

ISO. Sensory claim substantiation. https://www.iso.org/standard/69080.html.Accessed date: 27 August 2022

Jargon, J. (2014). Less salt, same taste? Food companies quietly change recipes. Wall Street Journal (24 June 2014–3 August 2021). https://www.wsj.com/articles/food-companies-quietly-cut-salt-fat-from-recipes-1403566403.Accessed date: 27 August 2022

John M. Ennis and Rune Christensen, A. (2015). Thurstonian comparison of the Tetrad and Degree of Difference tests, Food Qual. Prefer. Vol. 40, Part B, 263–269.

Lawless, H.T. and Hildegard, H. (2010). *Sensory Evaluation of Food: Principles and Practices*, 2e. Springer.

Meilgaard, M., Civille, G.V., and Carr, B.T. (2006). *Sensory Evaluation Techniques*, 4e. Boca Raton: CRC.

Moskowitz, H. (1998). Consumers versus experts in the light of psychophysics: a reply to Hough. *J. Sens. Stud.* 13.

Nijhof, A. et al. (2021). Forced grouping: a new methodology to measure the similarity between products and predefined references. *Food Qual. Prefer.* 88: 104103.

Pecore, S. et al. (2006). Degree of difference testing: a new approach incorporating control lot variability. *Food Qual. Prefer.* 17 (7–8): 552–555.

Pecore, S. and Kellen, L. (2002). A consumer-focused QC/sensory program in the food industry. *Food Qual. Prefer.* 13: 6.

Rogers, L. (2017). *Discrimination Testing in Sensory Science: A Practical Handbook*. Woodhead Publishing Series in Food Science, Technology and Nutrition.

Rousseau, B. and Ennis, D.M. (2013). When are two products close enough to be equivalent? *IFPress* 16: 1.

Rousseau, B. and Ennis, D.M. (2001). How retasting can improve the power of product testing. *IFPress* 4 (2): 2–3.

Rousseau, B. and Ennis, D.M. (2017). Preference without a difference. *IFPress* 20: 1.

Rousseau, B. and Ennis, D.M. (2002). Discrimination testing with multiple samples. *IFPress* 5: 1.

Rousseau, B. (2021). *Development in Discrimination Testing for Sensory Equivalence*, vol. 43, 122–125. IFP.

Rousseau, B. (2004). Relating expert and consumer sensitivities. *IFPress* 7 (2): 2–3.

Thurstone, L.L. (1927). A law of comparative judgment. *Psychol. Rev.* 34: 273–286.

WQPMag (2016). Company launches drinking water sensory kit. WQPMag (28 April 2016–6 February 2022). https://www.wqpmag.com/company-launches-drinking-water-sensory-kit. Accessed date: 27 August 2022

# 6

# Thurstonian Modeling and Signal Detection Theory

*Michael J. Hautus*

*School of Psychology, Faculty of Science, The University of Auckland, Auckland, New Zealand*

## 6.1 Signal Detection Theory and Thurstonian Modeling

Sensory difference tests are frequently used to obtain objective data on which business decisions can be made. Consequently, cultivating a full understanding of difference tests is an essential part of assessing the relevance and accuracy of the measurements they provide, and hence in making good business decisions. However, evidence shows that difference tests are not as simple to interpret as once thought. Not only does one need to be aware of response bias and its consequences, but also the fact that not all difference tests are equal in terms of sensitivity, or the equivalence of common measurements they produce. Furthermore, judges can perform commonly used difference tests using a variety of different decision strategies, each often leading to different levels of performance. A solid framework is required to tackle these issues. Two convergent approaches to sensory difference testing, Thurstonian analysis and signal detection theory (SDT), have sufficient similarity to allow integration into a unified framework. Both approaches have provided major advancements in the understanding of how judges behave in the context of a difference test, and how this affects their performance. The origins, similarities, and differences between these two approaches are briefly considered, as Thurstonian analysis and SDT have distinct historical roots. Thurstonian analysis, as the name suggests, can be traced back to the work of Louis Leon Thurstone (e.g., Thurstone 1927), whereas SDT has relatively modern origins in engineering, first appearing in the pioneering work of Peterson et al. (1954). The applicability of SDT to problems in psychophysics was recognized early on (e.g., Tanner and Swets 1954; Swets 2001).

Thurstone popularized the notion that the magnitude of the effect of a stimulus varies from moment to moment; this distribution of effects was termed a *discriminal dispersion*. This concept is practically identical to the construct in SDT of sensory evidence. The representation of this concept takes a familiar form in both lines of thought. They are both typically represented by a normal distribution, and two stimuli can be represented by a pair of distributions which may, or may not, differ in variance.

*Discrimination Testing in Sensory Evaluation*, First Edition. Edited by Lauren Rogers, Joanne Hort, Sarah E. Kemp and Tracey Hollowood.

The almost identical pictorial depictions in the literature of the models proposed by Thurstone, and those used in SDT, may lead one to consider them as alternative renditions of the same idea.

However, there are fundamental differences in the approaches as originally proposed.

Thurstone's models made use of the classical notion of a threshold – a level of sensory magnitude above which the stimulus is detected (not just said to be detected), and below which it is not. A major contribution of SDT was to replace the concept of threshold with that of *decision criterion* – a level of sensory magnitude above which the judge says the stimulus is present and below which they say it is not. Note that in some fields (e.g., medicine and epidemiology) the decision criterion is still referred to as a threshold; however, this modern usage bears no resemblance to the original usage by Thurstone and other classical psychophysicists.

Thurstone also considered the variation of the threshold. The threshold is set at a magnitude that changes subtly over time, just as the strength of the evidence from a stimulus varies. Thurstone encapsulated this variability of the threshold in many of his models. However, SDT normally does not consider criterion variability. This is because the developers of SDT explicitly subsumed this variability into that of the sensory evidence. This added variability reduces estimated sensitivity in the SDT context (Tanner and Swets 1954).

The null stimulus is important when considering detection tasks. While the idea that the magnitude of the experience of a stimulus varies over time can be traced back to Thurstone (and further back to Fechner), the idea that the magnitude of experience varies even when no stimulus is presented is probably original to SDT.

During the ensuing decades, several informative comparisons of SDT and Thurstonian analysis were published (e.g., Corso 1963; Lee 1969; Luce 1994). Distinctions between the two approaches became fairly blurred as concepts from both lines of scholarship were incorporated into later practical and theoretical developments. Thurstonian analysis blossomed in the field of sensory evaluation, but was almost forgotten within psychology, at least within the study of discrimination tests. Within the last 40 years, the two approaches, with very little left to distinguish between them, have made major contributions to the development, and approaches to analysis, of sensory difference tests within sensory evaluation. To avoid continuous alternation between these two approaches, the following description of the application of modern approaches within sensory evaluation to the A-Not A and 2AFC[1] difference tests will refer to SDT alone.

## 6.2 SDT and Two Stimulus Alternatives

Detection theory capitalizes on the notion that sensory noise is always present, positing a distribution of sensory evidence whether or not a stimulus is actually present. This means that the null (or blank) stimulus, as used in detection tasks, also produces a perceptual distribution and the same detection theory can be used with regard to both detection and discrimination tasks. As explained earlier, this differs from the approach of Thurstone, which could not account for the null stimulus.

1 The 2AFC test referred to in this chapter is not the same panellist procedure as the 2-AFC test described in ISO 5495:2005: Sensory analysis —Methodology — Paired comparison test.

For two stimulus alternatives – $S_1$ and $S_2$ in a discrimination task, or $S_0$ and $S_1$ in a detection task – standard SDT proposes two perceptual distributions that are normally distributed with identical standard deviations, but usually different means. This model can be named the binormal equal-variance model (BEVM). It suggests that if $S_2$ is greater than $S_1$ in some physical characteristic, then the mean of the $S_2$ distribution will be larger than that of the $S_1$ distribution on the dimension that sensorily represents that physical characteristic. This follows only if the sensory system can discriminate the difference in the physical characteristics. If not, the perceptual distributions will have the same mean. It is these distributions, which stem from the stimuli, that are employed as evidence upon which a decision is made by the judge.

Figure 6.1 shows these distributions in a format sometimes referred to as the *evidence space*. There is, in this case, considerable overlap between the two distributions. The left distribution corresponds to evidence available after the presentation of $S_1$, and the right distribution to the evidence from $S_2$. The axis, labelled the *decision variable*, can be loosely considered as the perceptual strength arising from a stimulus presentation. For example, if the two stimuli differ in sucrose content only, then the decision variable may be loosely considered as indicating strength of sweetness. Continuing with the example, it follows that if $S_1$ and $S_2$ are presented to the judge, sometimes $S_1$ will produce evidence that falls to the right of that for $S_2$. In this case, $S_1$ would be perceived as sweeter than $S_2$. Note that all other elements of Figure 6.1 are not included in the evidence space, just the two distributions and the decision variable.

The evidence space, by itself, tells nothing of how a response is generated by the judge. It simply conveys the nature of the information, or evidence, that the judge has available with which to make a decision. It also represents, in the form of the decision variable, the range of sensory experiences that the judge may encounter. It seems only natural that the range of sensory experiences needs to be mapped onto a response. This is accomplished with a *decision rule*, more colloquially referred to as a *decision strategy*. A reasonable decision strategy for this example is to place a boundary, usually referred to as a *criterion*, at a location somewhere along the decision variable axis. A very natural place for this criterion to be located is half-way between the two distributions, as illustrated in Figure 6.1. This criterion facilitates a response. The addition of a criterion to the evidence space engenders the

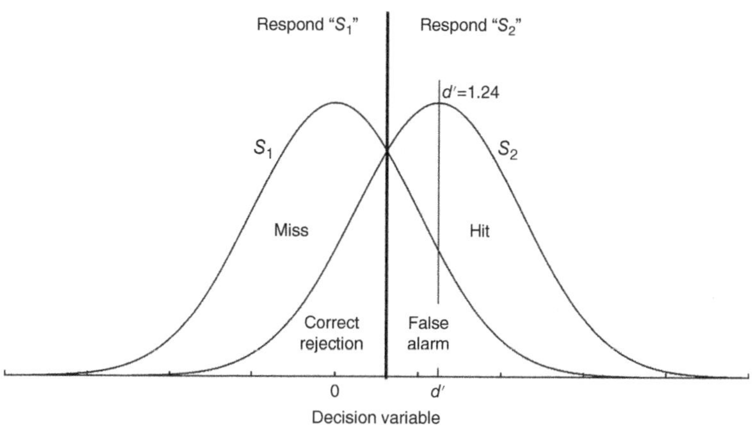


**Figure 6.1** The evidence space and the decision space for the case of two stimulus alternatives. This is also an example of the binormal equal-variance model for the A-Not A task.

*decision space*. In essence, the decision space illustrates how the evidence space is partitioned to produce a judgment. Once a stimulus is presented the judge can compare the sensory evidence that arises from it to the criterion. If the evidence were stronger than the criterion, the judge would indicate the stimulus was $S_2$, if it were weaker than the criterion, the judge would respond $S_1$. In the sucrose example, any evidence sweeter than the criterion level of sweetness would be attributed to having come from $S_2$, evidence less sweet would be ascribed to $S_1$.

## 6.3 The A-Not A Task

The preceding description of the evidence space and decision rule for the case of two stimulus alternatives maps to the most common sensory difference test; the A-Not A task. In this task the judge is presented with one of two possible stimuli ($S_1$ = A or $S_2$ = Not A) and must indicate which stimulus was presented. At the end of the assessment, the researcher forms a contingency table that contains the frequency with which each of the four outcomes occurs: hits, misses, false alarms, and correct rejections.

Consider an assessment of the ability of Judge 1 to discriminate between two versions of a product: the standard version is $S_1$ and the other is $S_2$. The researcher conducts 60 tests split over three sessions of 20 tests each. For each test, the researcher randomly presents a sample of $S_1$ or $S_2$ to the judge, who must then indicate whether the sample presented was $S_1$ or $S_2$. At the completion of the 60 tests, the researcher constructs the data in top section of Table 6.1.

### 6.3.1 Proportion of Correct Responses

The top section of Table 6.1 shows that hits and correct rejections occur when the judge gives a correct response, and false alarms and misses when the judge gives an incorrect response. Therefore, one measure of the judge's performance is the proportion of correct responses, generally denoted $p(c)$, which is equal to the sum of the frequencies of hits and correct rejections (44) divided by the total number of tests (60). Hence, in this case, $p(c) = 44/60 = 0.73$.

**Table 6.1** Frequencies for each outcome in a A-Not A task for Judge 1, together with the resulting hit (*H*), miss (*M*), false-alarm (*F*), and correct-rejection (*C*) rates.

| | Response | | |
|---|---|---|---|
| **Stimulus** | **$S_2$** | **$S_1$** | **Total** |
| $S_2$ | Hits = 22 | Misses = 8 | 30 |
| $S_1$ | False alarms = 8 | Correct rejections = 22 | 30 |
| $S_2$ | $H = 0.73$ | $M = 0.27$ | 1 |
| $S_1$ | $F = 0.27$ | $C = 0.73$ | 1 |

To relate this information back to the SDT case of two stimulus alternatives, it is necessary to realize that frequencies are uninformative on their own. Figure 6.1 deals in conditional probabilities. Thus, it is necessary to convert the frequencies in Table 6.1 into probabilities. Taking each frequency in Table 6.1 and dividing by the corresponding row total yields the hit ($H$), miss ($M$), false-alarm ($F$), and correct rejection ($C$) probabilities (usually referred to as rates) given in the lower half of Table 6.1. Figure 6.1 shows two normal distributions, with each distribution divided into two parts by the criterion, with the sum of the two parts of each distribution equaling one. In the same way, $H + M = 1$ because the corresponding frequencies have been divided by their sum. For the same reason, $F + C = 1$. This suggests that the four regions in Figure 6.1 correspond to the four rates for a particular level of difference between the means and a particular criterion. Working backward, one can determine the level of difference between the means and the criterion location that would give the observed rates. The normalized difference between the means is a measure of how well the sensory system can separate or discriminate between $S_1$ and $S_2$. It is typically referred to as $d'$ and can be calculated using

$$d' = z(H) - z(F) \tag{6.1}$$

where $z(.)$ is the inverse-normal transform, sometimes called the $z$-score. For Judge 1 in the example in Table 6.1, $d' = z(0.73) - z(0.27) = 0.62 - (-0.62) = 1.24$.

### 6.3.2 Bias as a Removable Problem

The placement of the criterion by a judge is a crucial aspect of decision-making. If the criterion is located midway between the two distributions, then the number of each type of response ("$S_1$" or "$S_2$") will be equal, as for Judge 1. If a judge places the criterion farther to the left of the decision space then they will make a greater number of "$S_2$" responses. This is a *liberal bias*. The opposite case, in which a judge places the criterion farther to the right of center in the decision space leads to a *conservative bias* (more "$S_1$" responses).

Consider Judge 2, who in this task gives two false alarms and 12 hits. For this judge $F = 0.067$ and $H = 0.400$ yielding $d' = 1.25$, which is almost the same as the value of $d'$ for Judge 1. Judge 2 demonstrates a conservative response bias, yet their sensitivity is the same as that of Judge 1. Calculation of $p(c)$ for Judge 2 shows an important weakness of this measure. Because the judge had two false alarms, and there were 30 $S_1$ trials, they produced 28 correct rejections. Therefore, $p(c) = (28 + 12)/60 = 0.67$. Thus, even though the ability of the two judges to separate out the evidence from the two perceptual distributions is the same, as assessed by $d'$, the commonly used measure of performance $p(c)$ yields different values, 0.73 and 0.67; $p(c)$ for Judge 2 is about 9% below that for Judge 1. Proportion correct as a measure of performance is prone to the effects of response bias. This means that proportion correct changes as a function of both the difference between the perceptual distributions – a characteristic desirable of a measure of sensitivity – and the location of the criterion, or response bias – a confounding effect to avoid. On the other hand, $d'$ is independent of response bias.

The slight difference in $d'$ (1.24 vs. 1.25) is because we are considering a finite and relatively small number of trials. The change of a single response in a row of the contingency table leads to a step change in the underlying probabilities, hence a step change in any values calculated from them, such as $d'$.

### 6.3.3 The Receiver Operating Characteristic Curve

A different perspective on the decision space portrayed in Figure 6.1 can be obtained by plotting the information it contains in another format. A pair of perceptual distributions and a single criterion give rise to a pair of independent values, $F$ and $H$. There are two ways that the decision space can change. The first is to hold the perceptual distributions constant but to adjust the criterion. Setting the criterion at a different location yields a new $F, H$ pair. A graph can be constructed that shows all of the possible $F, H$ pairs for a particular pair of distributions. By tradition, $F$ is plotted on the $x$-axis and $H$ on the $y$ axis. As $F$ and $H$ are probabilities, each axis extends from 0 to 1, and the domain of allowable values falls within a unit square. The graph constructed in this manner is called an isosensitivity curve (curve for same sensitivity), or more commonly a receiver operating characteristic (ROC) curve. The ROC curve for $d' = 1.24$ is illustrated in Figure 6.2. The solid symbol indicates the datum for Judge 1, and the hollow symbol that for Judge 2. The curve represents all possible response biases for a given level of sensitivity. Different locations on the curve, such as the two points displayed, represent different levels of response bias. The solid symbol lies on the minor diagonal of the unit square; a line going from (0, 1) to (1, 0). This location on the ROC curve represents a neutral bias, that is, a bias that shows no favoritism over the two responses. Any point on the ROC curve to the left of the minor diagonal, such as that of Judge 2, represents a conservative bias, and any point to the right, a liberal bias.

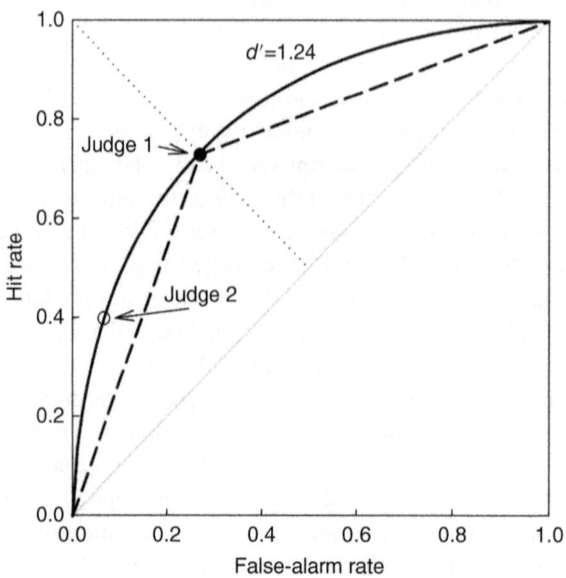


**Figure 6.2** Receiver operating characteristic curve for two judges with the same sensitivity ($d' \approx 1.24$) but different response biases. The area under the entire curve is $A_z$ whereas the area under the dashed lines is equal to the proportion correct, $p(c)$, obtained by Judge 1.

Bias can be quantified rather than simply described. While there is a considerable literature on this issue, the most frequently recommended measure of response bias is

$$c = -\frac{1}{2}\left(z(H) + z(F)\right) \tag{6.2}$$

which is the distance, in units of standard deviation, of the criterion from the crossover point of the two perceptual distributions. For Judge 1

$c = -1/2(z(0.73) + z(0.27)) = -1/2(0.62 + (-0.62)) = 0$, indicating that the criterion lies on the crossover point of the two distributions. For Judge 2, $c = 0.88$, indicating the criterion lies to the right of the crossover point.

For work in sensory evaluation, quantifying bias is usually not important (there are exceptions), however removing the contaminating effect of bias from measures of discriminative performance, or sensitivity, is a major benefit of using the measure $d'$. This raises the issue of how $d'$ is represented on the ROC curve. On the decision space, $d'$ is the distance between the two normal distributions. If this distance, and hence $d'$, is increased, given a fixed criterion, or value of $c$, $F$ must decrease and $H$ must increase. This is true for any value of the criterion, and so if $d'$ is increased, the ROC curve will lie above that for a lesser value of $d'$, except at the endpoints. This means that it is the location of the ROC curve in the unit square that is related to sensitivity, and $d'$.

### 6.3.4 Area Measures of Sensitivity

Quantifying the location of an ROC curve in the unit square can be tricky. For example, claiming that a curve lying above another curve represents greater sensitivity is not strictly correct. This is because all ROC curves are identical at two points in the unit square: (0, 0) and (1, 1). Various measures have been proposed, but that most consistent with SDT is the area measure, $A_z$, which is the entire area beneath the ROC curve. $A_z$ can range from zero to one. Values reliably below 0.5 represent performance of a judge who has performed the task "backwards"; for example, answers "$S_2$" when they meant to answer "$S_1$," and vice versa. This usually does not happen often, but permits a check to ensure judges are following the procedure correctly. Values reliably around 0.5 stem from judges who are unable to discriminate between the two stimuli. Some authors refer to performance in this region as guessing, but this is technically incorrect. $A_z$ of 0.5 corresponds to an ROC curve that is, rather oddly, a straight line that lies along the major diagonal. This means that $F = H$ for all points on the "curve," and occurs when the two perceptual distributions lie atop each other, hence $d' = 0$. In this case, the judge still has a criterion, and still makes reasoned perceptual judgments. However, their perceptual system cannot separate the two distributions of evidence, so $F$ and $H$ will be equal. The actual value of $F$ and $H$ for a particular judge depends on $c$, their criterion. They are not guessing, which implies a random allocation of responses, but performing using the same logical process they would follow given more discriminable stimuli.

However, considering performance from the judge's perspective, they may feel like they were guessing; after all, their performance was at the "chance level" that would be attained if responses had been randomly allocated.

Values of $A_z$ reliably above 0.5 indicate an ability of the judge to discriminate between the two stimuli. Perfect performance, where $F = 0$ and $H = 1$, leads to an ROC curve that goes from (0, 0) to (0, 1) and across to (1, 1). $A_z$ is therefore 1. Calculation of $d'$ for this case, gives $d' = z(1) - z(0) = \infty - (-\infty) = \infty$. This is not a very intuitive value, and indeed, getting a concept for what a particular value of $d'$ means in terms of performance, in general, is not that easy. On the other hand, $A_z = 1$ is easier to comprehend. $A_z$ and $d'$ are directly related by

$$A_z = \Phi\left(\frac{d'}{\sqrt{2}}\right) \tag{6.3}$$

where $\Phi(\cdot)$ is the cumulative normal distribution function.

Another benefit of $A_z$ is that there is a theorem that relates performance in the A-Not A task to performance in the two-alternative forced-choice (2AFC) task. This theorem states that $A_z$ obtained in the A-Not A task will be equal to the proportion correct, $p(c)$, for a neutral-biased judge in the 2AFC task. Thus, a judge in A-Not A who obtains $A_z = 1$ will obtain $p(c) = 1$ in 2AFC, and a judge who gets $A_z = 0.5$ in A-Not A will obtain $p(c) = 0.5$ in 2AFC. This theorem provides a natural meaning to values of $A_z$ in the A-Not A task.

Given that there are several benefits to the use of $A_z$ as an index of sensitivity, what reason could there be for the continued use of $d'$? The answer lies in the meaning of "sensitivity"; how well the sensory system can distinguish between two stimuli. A good index of sensitivity should be independent of the method used to measure it. As an example, for a particular pair of stimuli, a measure of sensitivity should be the same if it is acquired using the A-Not A task in one circumstance and using the 2AFC task in another. As the normalized difference between the perceptual distributions of $S_1$ and $S_2$, $d'$ is constant irrespective of the sequence in which stimuli are presented. After all, the functioning of the perceptual system is not greatly influenced by the ordering of stimuli. To a reasonable approximation, their perceptual distributions remain constant. As we will discover, a relatively invariant value of $d'$ can be estimated from any difference test, given the same judge and stimuli. In contrast, $A_z$ is highly dependent on the difference test used.

Another commonly used area measure of sensitivity is the R-index. This measure is rarely (if ever) applied to binary-response data; but considering this usage can be instructive. The R-index is equal to the area under the ROC curve constructed by connecting (0, 0) to the data point, and then connecting the data point to (1, 1), all with straight lines, as illustrated by the dashed lines in Figure 6.2. Using geometry, calculation of this area in terms of $F$ and $H$ indicates that the R-Index $= 1/2(H + (1 - F))$. For Judge 1, the case illustrated in Figure 6.2, the R-Index $= 1/2(0.73 + (1 - 0.27)) = 0.73$. Given that $C = 1 - F$, the R-index in the binary case is equal to $p(c)$. This is, in itself, a good reason not to use the R-index in the case of binary judgments. In addition, it indicates that $p(c)$ is also an area measure of sensitivity; albeit a poor one. Further discussion of the R-index will occur later in the context of confidence-rating procedures, a context in which the R-index has greater utility.

Another area measure of sensitivity, that holds both instructional utility as well as being applied in some contexts, is $p(c)_{max}$, the maximum value of $p(c)$ that could have been obtained given an observed level of sensitivity. Consider Judges 1 and 2 who both obtained

$d' \approx 1.25$, yet obtained proportions correct of 0.73 and 0.67, respectively. For these judges, $p(c)_{max}$ is the maximum value of $p(c)$ that could be attained given that $d' = 1.25$. Given the geometry of $p(c)$ in the ROC square, the maximum area occurs when the data point lies on the minor diagonal. The data point for Judge 1 already lies on this diagonal so his $p(c)_{max}$ requires no additional computation. It is equal to the obtained value of $p(c)$; $p(c)_{max} = 0.73$. For Judge 2, $p(c)_{max}$ equals the value of $p(c)$ that would have been attained if the data point were on the same ROC curve, but at the minor diagonal. This does not require computation as that point coincides with that for Judge 1; hence $p(c)_{max} = 0.73$. This demonstrates that using $p(c)_{max}$ as a measure of sensitivity effectively eliminates the effect of response bias. In general, for the A-Not A task,

$$p(c)_{max} = \Phi\left(\frac{d'}{2}\right) \tag{6.4}$$

(Hautus et al. 2021, Eq. 6.2). Thus, to calculate $p(c)_{max}$, $d'$ is required, or at least the values of $F$ and $H$ from which $d'$ is calculated; $p(c)_{max}$ cannot be calculated from $p(c)$ without this additional information. However, it is a bias-independent measure of sensitivity, and it therefore redeems $p(c)$ from its most undesirable property.

### 6.3.5 Alternative Distribution Assumptions

The BEVM is a specific case of the *binormal model* (BM), which allows the variance of the perceptual distributions of $S_1$ and $S_2$ to be different. The BM is that most frequently reported in the literature. Ignoring the criterial parameters, the BEVM has one additional parameter, $d'$, whereas the BM has two, usually referred to as the *intercept* and *slope* parameters. The BEVM is the special case of the BM in which the slope is one. For this case, the intercept is equal to $d'$. However, if the slope is not equal to one, the intercept is no longer equal to $d'$. The sensitivity measure $d'$ is only valid for the BEVM, that is, $d'$ is defined as the difference between the means of the perceptual distributions for $S_1$ and $S_2$, divided by their common standard deviation.

Thus, for the BM, another sensitivity measure is required. It should ideally take into account the fact that the variances are not "common." In the BM, the slope parameter is equal to the ratio of the standard deviation of the $S_1$ distribution to that of the $S_2$ distribution (Green and Swets 1966; O'Mahony and Hautus 2008). Thus, when the slope parameter is one, the two standard deviations are equal, and we have the BEVM. While there are many candidate measures of sensitivity for the BM, the best regarded is

$$d_a = \sqrt{\frac{2}{1+s^2}}\left(z(H) - sz(F)\right) \tag{6.5}$$

where $s$ is the value of the slope parameter (Simpson and Fitter 1973). While not obvious from Eq. (6.5), $d_a$ is the distance between the means of the $S_1$ and $S_2$ perceptual distributions divided by their root-mean-square standard deviation. When $s = 1$, Eq. (6.5) yields $d'$. Thus, just as the BM is a generalization of the BEVM, $d_a$ is a generalization of $d'$. Another example of this generalization comes from the area measure of sensitivity, $A_z$. For the BM, $A_z$ can be calculated by substituting $d_a$ for $d'$ in Eq. (6.3).

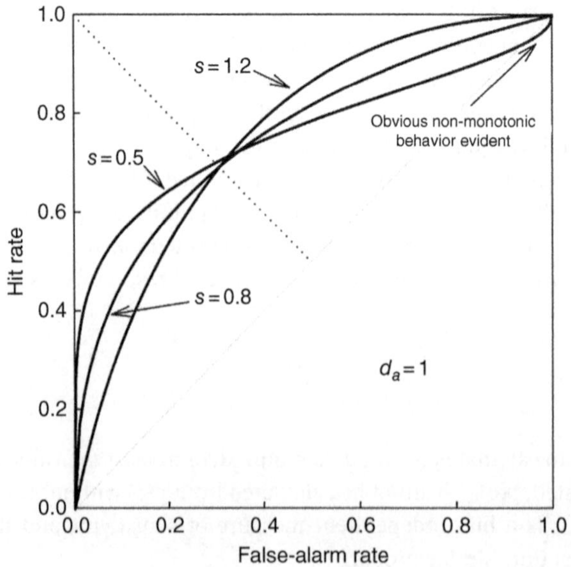


**Figure 6.3** Receiver operating characteristic (ROC) curves for three cases of the binormal model: $s = 0.8, 1.2$, and $0.5$. Sensitivity, as measured by $d_a$, is identical for all cases. Non-monotonic behavior is clearly evident in the third curve, but is present in all three.

ROC curves for the BM (when $s \neq 1$) are asymmetrical about the minor diagonal of the unit square. Figure 6.3 illustrates example ROC curves for three values of $s$: 0.8, 1.2, and 0.5, all with $d_a = 1$. The first two of these curves demonstrate slight departures from equal variance, while the third demonstrates a larger departure (the standard deviation of $S_2$ is twice that of $S_1$) that highlights an inherent problem with the BM. ROC curves for the BM do not have monotonically decreasing slope, whereas those for the BEVM do. This poses some theoretical problems for the BM (Egan 1975), which will not be elaborated here.

There is mounting evidence that the BEVM provides an adequate account of data acquired from discrimination experiments, at least for gustatory stimuli (e.g., O'Mahony 1972; Laming 1987; Irwin et al. 1992; Hautus and Irwin 1995). The majority of SDT models for other difference tests are based on equal variance normal perceptual distributions. Research into the efficacy of these other models has necessarily included investigation of the efficacy of the BEVM (e.g., Hautus et al. 2011; van Hout et al. 2011). If it could be shown that the BEVM does a poor job at accounting for the data – and hence the BM does so better – then the models developed for other difference tests would be cast into doubt.

### 6.3.6 Confidence-Rating Response Procedure

So far, discussion has been limited to the binary-response alternative. After the presentation of the test stimulus, the judge makes a binary response, indicating whether $S_1$ or $S_2$ had been presented. Confidence-rating responses (Green and Swets 1966) allow for more response alternatives on each test. A typical confidence-rating scale may have six response alternatives. For example: 1 = definitely $S_1$; 2 = probably $S_1$; 3 = possibly $S_1$; 4 = possibly

$S_2$; 5 = probably $S_2$; 6 = definitely $S_2$. However, there are many possible arrangements. A common alternative is to have the judge indicate whether $S_1$ or $S_2$ was presented, and then to indicate whether they are sure or unsure of that judgment. This amounts to a rating scale with four response alternatives. While some researchers prefer scales with an even number of alternative responses, which precludes a central "I don't know" response, others do employ that central response option (e.g., Gescheider 1997, p. 149). Thus, there seems no "correct" rule for the number of response alternatives.

The primary benefit of using confidence rating response procedures is that they provide more data points without having to increase the number of tests. These data points form an *empirical ROC*, that is, a discrete ROC curve based on the empirical data. As already seen, binary responses provide only one point on the ROC curve. Confidence ratings provide $n-1$ points on the ROC curve, where $n$ is the number of response alternatives. Any ROC curve, for example, based on the BEVM or the BM, can pass through a single point. Thus, if there is only one point, there is no way to determine how representative the model chosen for the ROC curve is of the data. This determination requires more data points. Conversely, a benefit of a single point is that $d'$ can be calculated directly using Eq. (6.1). However, $d_a$ cannot be calculated from a single point at all.

An empirical ROC consisting of several points allows the model of interest to be fitted to those data, and an indication can be obtained as to whether that fit is acceptable. Thus, greater certainty about the validity of the chosen measure of sensitivity is acquired. The downside is that $d'$ (or $d_a$) cannot be calculated by applying an equation. A model-fitting procedure is required, leading to estimates of the parameters of the fitted model, their standard errors, and a goodness-of-fit (GOF) statistic.

Table 6.2 demonstrates the calculation of false-alarm and hit rates from data collected using confidence-ratings in an A-Not A task. The same general approach is applicable for other tasks. The top rows of Table 6.2 present the frequency of use of each rating by Judge

**Table 6.2** Rating data for Judge 4 in the A-Not A task. Ratings range from "certain $S_1$" (1) to "certain $S_2$" (6). The shaded values are present for computational convenience. See text for details of calculations.

| | 1 | 2 | 3 | 4 | 5 | 6 | |
|---|---|---|---|---|---|---|---|
| | | | | **A-Not A data** | | | |
| $<S_2>$ | 0 | 14 | 27 | 73 | 33 | 3 | 150 |
| $<S_1>$ | 8 | 45 | 62 | 32 | 3 | 0 | 150 |
| $<S_2>$ | 0.00 | 0.09 | 0.18 | 0.49 | 0.22 | 0.02 | 1.00 |
| $<S_1>$ | 0.05 | 0.30 | 0.41 | 0.21 | 0.02 | 0.00 | 0.09 |
| $H$ | 1.00 | 1.00 | 0.91 | 0.73 | 0.24 | 0.02 | 0.00 |
| $F$ | 0.99 | 0.94 | 0.64 | 0.23 | 0.02 | 0.00 | 0.00 |
| | | | | **R-index** | | | |
| Mean $H$ | 1.00 | 0.96 | 0.82 | 0.49 | 0.13 | 0.01 | |
| Diff. $F$ | 0.05 | 0.30 | 0.41 | 0.21 | 0.02 | 0.00 | |
| Areas | 0.05 | 0.29 | 0.34 | 0.10 | 0.00 | 0.00 | 0.78 |

4 (from Hautus et al. 2011). The judge was discriminating cordials that differed subtly in flavor. The total at the end of these rows indicate that there were 150 tests in which $S_2$ was presented, with the same number of tests with $S_1$. The middle rows are the result of a conversion of the frequencies to proportions, accomplished by dividing each frequency by the row total in which that frequency lies. The row totals for the proportions will necessarily be equal to one, however the total for the $S_1$ row is 0.99 in this example, because of rounding error. The final rows contain false-alarm, hit rate pairs. These are obtained by cumulating each of the proportions (middle rows) systematically from right to left. The left-most cells (grey) will necessarily be equal to one (except for rounding error). They therefore contain no useful information. Recall that all ROC curves terminate at (1, 1). This leaves five useful data points that are the empirical ROC. Figure 6.6 illustrates the empirical ROC for these data.

This process makes apparent the fact that the data points on the empirical ROC collected with confidence ratings are not independent. It is still an experimental question as to whether estimates of sensitivity obtained from confidence ratings are more or less precise than those collected using the binary response format. It is clear, however, that confidence ratings, as the name accidentally suggests, can give the investigator greater confidence in the validity of the adopted measure of sensitivity. There is another issue to consider. Perhaps making confidence ratings is more cognitively taxing on judges than making binary responses. Fortunately, there is evidence from data collected in a few sensory modalities that binary or confidence rating procedures yield equivalent estimates of sensitivity (e.g., Gescheider et al. 1971 (vibrotactile); Markowitz and Swets 1967 (auditory)).

### 6.3.7 Fitting ROC Curves

An SDT model defines a family of ROC curves. What a particular model is appropriate for can be very specific. For example, the BEVM defines ROC curves that are symmetrical about the minor diagonal of the ROC square and provides appropriate estimates of sensitivity and bias when the empirical ROC was collected using an A-Not A task in combination with a confidence-rating response procedure. This model is generally not suited to fitting ROC curves to empirical ROCs collected from other difference tests (e.g., same-different or 2AFC). Each difference test requires a different model, more of which later.

There are many approaches to fitting SDT models to empirical ROCs. The most widely used, which will be used here, is maximum-likelihood estimation (e.g., Dorfman and Alf 1969), but various approaches based on standard statistical techniques are also gaining prominence. Generalized-linear models in particular seem promising, as they are capable of taking into account the effect of covariates (DeCarlo 1998; Pepe 2003; Brockhoff and Christensen 2010).

### 6.3.8 BEVM vs. BM Revisited

Hautus et al. (2011) asked judges to participate in three difference tests with cordials that differed subtly in flavor. Table 6.2 shows the rating responses for Judge 4, for the A-Not A task. Consideration of the adequacy of the BEVM to account for the A-Not A data is

**Table 6.3** Best-fitting parameter estimates and GOF statistics for the A-Not A data of Judge 4 provided in Table 6.2. The BM and BEVM are fitted to the same data.

| | Intercept | Slope (SE) | Sensitivity (SE) | $G^2$ | *df* | *p* | LL |
|---|---|---|---|---|---|---|---|
| BM | 1.15 | 0.86 (0.10) | $d_a = 1.23$ (0.14) | 4.36 | 3 | 0.225 | −389.5 |
| BEVM | 1.22 | 1 (fixed) | $d' = 1.22$ (0.13) | 6.02 | 4 | 0.198 | −390.4 |
| | | | | | | | $-2\ \Delta\text{LL} = 1.8$ |

instructive of the general approach to ROC analysis. Table 6.3 provides relevant parameter estimates and statistics obtained by fitting the BEVM and BM separately to the A-Not A data using the software, *SDT Assistant* (Hautus 2012).

Of primary interest is whether the BM provides a significantly better fit to the data than the BEVM. The important word here is "significantly" because the BM will always provide a fit as good as that for the BEVM, because the former model is a generalization of the latter. Formally, the BM has one more degree of freedom than the BEVM. The practical effect of this is that the GOF statistics, $\chi^2$ or $G^2$, will never be larger for the BM than for the BEVM. This is true even for data generated by simulation from a BEVM model; this is an example of over-fitting (Myung 2000).

Asymptotically, the GOF statistics are distributed as chi-square. The probability associated with the upper tail of the chi-square distribution with the appropriate degrees of freedom has a special interpretation. It is the probability that the data could have arisen from the model given that the model is correct. This probability presents an opportunity to determine if the model is appropriate. If the probability is very small then it is unlikely the data could arise from the model. It may be tempting to use a probability of 0.05 as a rejection criterion. However, this is very conservative. A value closer to 0.001 is more realistic in practice (Press et al. 1992). Note that the probability values for Judge 4 are 0.225 (BM) and 0.198 (BEVM); both acceptable fits. Based on these probability values, the best-fitting model could be interpreted as that which has the greatest probability of giving rise to the data (the BM in this case). However, this does not answer the question of whether the better-fitting model provides a significantly better fit.

One approach to this issue is to consider the value of the slope estimate for the BM in combination with its asymptotic standard error. The slope plus/minus two standard errors provides a rough 95% confidence interval for the slope. For Judge 4 the confidence interval is [0.66–1.06]. If unity falls within this interval, as it does marginally in this case, there is little evidence that allowing the slope to vary will significantly improve the ability of the model to account for the data.

A more theoretically grounded approach is to consider the BEVM model as nested within the BM. Tests of the significance of differences in the ability of nested models to account for the data are available (e.g., Kendall and Stewart 1979). In this context, minus twice the difference between the Log Likelihood (LL) of the data for each model ($-2\ \Delta$LL) is determined. $-2\ \Delta$LL is distributed as chi-square with one degree of freedom, because the BM has one more degree of freedom than the BEVM. The upper-tail probability of $-2\ \Delta$LL can be used to assess whether the additional parameter, the slope, provides a significantly better fit to the data. For Judge 4, $p\left(\chi_1^2 > 1.8\right) = 0.180$. Given that this probability is relatively large, there is little evidence that allowing the slope to vary provides a better fit to the data. Consequently, the BEVM is adequate.

### 6.3.9 Data Analysis Considerations

It is often important to know whether two sensitivity estimates are significantly different from one another, or whether a particular measure is significantly different from a fixed value, such as an action standard. For sensitivity measures obtained from model fitting, the process also leads to estimates of standard error. This is the information required to conduct z-tests. To compare two estimates of $d'$, for example, one calculates

$$z = \frac{d_2' - d_1'}{\sqrt{SE_{d_2'}^2 + SE_{d_1'}^2}} \tag{6.6}$$

where the two standard errors, $SE$, are those associated with each estimate of $d'$, and compares the value of $z$ with the percentiles of the normal distribution to obtain a probability. Similarly, to compare $d'$ to a fixed value, $k$,

$$z = \frac{d' - k}{SE_{d'}}. \tag{6.7}$$

For example, comparison of the sensitivity of Judge 4 to $d' = 1$, a level of sensitivity often used to define "threshold," yields $z = (1.22 - 1)/0.13 = 1.69$. A two-tailed test results in $p = 0.091$, leading to the conclusion that $d'$ for Judge 4 is not significantly different from one at the 0.05 level of significance.

For cases where there are more than two estimates of sensitivity being compared, an omnibus chi-square test can be undertaken prior to pairwise comparisons. For full details on this test, as well as a detailed description of the $z$-tests described above, see Bi et al. (1997). This paper also details the calculation of standard errors for estimates of sensitivity obtained using binary responses.

In sensory evaluation, it is often the case that each judge completes a relatively small number of tests. As a comparison, in visual and auditory research it is common for judges to undertake 600 tests, whereas there are few gustatory or olfactory studies where the test count reaches even 60, and in consumer research, that number may be as low as one. Pooling data from different judges introduces statistical bias into the results, and unbiased results are obtained by calculating sensitivity for each judge and then averaging those sensitivity estimates (Macmillan and Kaplan 1985). However, there is a practical problem in applying this unbiased approach. For any number of tests there is a finite probability that $F$, $H$, or both will be either zero or one. This probability increases as the number of tests decreases. As noted earlier, $z$-scores are infinite for these "extreme" probabilities; thus, $d'$ is also infinite. Calculating the average of several sensitivities, one or more of which is infinite, does not lead to a sensible result.

Resolution of this problem involves the application of a "correction" to the data before calculating $F$ and $H$. One approach is only to apply a correction if an extreme proportion (0 or 1) occurs. The standard correction is to adjust values of zero up, and values of one down, by the quantity $1/(2N)$ where $N$ is equal to the number of tests on which the probability is based (Snodgrass and Corwin 1988). Another approach is to correct all data, irrespective of the presence of extreme proportions. This approach, sometimes called the log-linear correction because of its link to this area of statistical analysis, adjusts upwards by ½ each value in the contingency table while adjusting the row totals upward by one (Fienberg 1980). The log-linear correction leads to more predictable, and usually less, statistical bias than the $1/(2N)$ correction (Hautus 1995).

Returning to the issue of summarizing data from a group of judges, the effect of statistical biases just described must be taken into consideration. While averaging $d'$ values leads to a group average that is unbiased, it is not always possible to determine $d'$ without introducing a correction, which in turn introduces bias. Thus, the approach of averaging estimates of sensitivity will, in the long run, still be biased. A second approach is to pool the data before estimating sensitivity. One problem with this approach is that estimates of sensitivity will be biased downwards. For a detailed explanation of why this is the case, see Hautus et al. (2021, pp. 315–321). In addition, pooling data suffers from the same biases due to the application of a correction as does the averaging method. There is no ideal solution therefore to determining group sensitivity. Hautus (1997) investigated the tradeoff between the different sources of bias. In general, the crossover point – in terms of number of tests – below which pooled $d'$ becomes less biased than average $d'$, is dependent on the true sensitivity of the group, the average bias of the group, and the variation in biases exhibited by group members. However, a general pattern of practical use emerged. Calculate group sensitivity using both methods (averaging and pooling) in conjunction with the log-linear correction. The larger of the two estimates is the best estimate of group sensitivity.

### 6.3.10 The R-Index

The R-index, as used in sensory science (see Lee and van Hout 2009, for a review), has appeared in at least three different guises: as the R-index (Brown 1974), as $p(A)$ (Green and Swets 1966), and as $A_g$ (Pollack and Hsieh 1969). Earlier discussion of the R-index, with regard to the binary-response option, demonstrated that this measure is equal to $p(c)$ in the binary case. Note again, that proponents of the R-index do not use it in this context. The benefits of this measure come about when used in combination with confidence-rating responses. Simply stated, the R-index is the area under the linear ROC constructed by connecting (0, 0) to the leftmost empirical ROC point, connecting that point to the next leftmost point, repeating this until the last remaining point is reached, and then connecting this point to (1, 1). This area can be calculated as the sum of the areas of a sequence of trapezia; a computationally simple task.

The lower section of Table 6.2 illustrates the manual calculation of the R-index for the A-Not A data of Judge 4. Each value in the row labeled *Mean H* is the mean of the hit rates for that rating and the next higher rating, and each value in the row labeled *Diff. H* is the difference between the false-alarm rates for that rating and the next higher rating. In the case of Rating 6, both *Mean H* and *Diff. H* are based on values of zero for the next highest rating. These values correspond to the lower-left corner of the unit-square, which is the starting point of the linear-segment ROC. Note that the values of *Diff. H* are the same as the un-cumulated proportion of responses to <$S_1$>, shown several rows above. Values in the row labelled Areas are the product of Mean $H$ and Diff. $H$ for the same rating. Each value is the area of one trapezium, and hence the sum of these areas (in bold at the right) is the R-index; in this case 0.78. Thus, both the empirical ROC and the R-index can be conveniently determined from a simple table structure.

The R-index, based on more than a single point, is a generalization of $p(c)$. While it is sometimes stated that the R-index is a bias-free measure (e.g., Bi 2006) it would be more precise to say that the effect of response bias on the R-index generally decreases with the number of empirical data points, and is additionally influenced by the spacing of those points in the ROC space. For example, for three points on an ROC curve that are closely spaced near the negative diagonal, the R-index will be higher than for a similar grouping of points away from this diagonal. This is true no matter what the shape of the "true" underlying ROC curve. Usually though, ROC points tend to be reasonably spaced out. With more points added to the empirical ROC – assuming good spacing of those points – the R-index will increase, and will become a better approximation of the area beneath the "true" underlying ROC curve (excluding the effects of sampling variability).

One benefit of the R-index is that it can be calculated without having to make any "corrections" to the data. Recall that corrections are required in the calculation of $d'$ because probabilities ($F$ or $H$) of zero or one may occur, and taking the $z$-score of these values is problematic. There is no such difficulty with the R-index. Direct calculation of the area beneath the linear ROC is always possible.

In addition, the R-index is non-parametric. In fact, this measure is related to the Mann–Whitney $U$ statistic (Bamber 1975). Thus, it does not make any assumptions about the underlying nature of the ROC curve. As the number of empirical ROC points increases, the R-index will get closer to the true area beneath that underlying ROC curve, whatever its form. This implies that the R-index is a conservative estimate of that true area. On average, the R-index will be less than the true area. This consistent underestimation is a strength of the measure, as uncertainty about the direction of error is reduced.

Given the relationship to the Mann–Whitney $U$ statistic, there is a considerable knowledge base about the mathematics of the R-index, including, importantly, how to calculate the standard error of the measure. Bi (2006) provides a thorough account of the calculation of the standard error of the R-index, and how this can be used to conduct tests of statistical significance.

## 6.4 The 2AFC (Paired Comparison) Task

The strength of SDT becomes more apparent when tasks with more than one observation interval are considered. In 2AFC, the quintessential two-interval task, the judge receives two different stimuli on a test. Consequently, there are two stimulus sequences: <$S_1$, $S_2$> and <$S_2$, $S_1$>. In essence – and there are many different instruction alternatives to accomplish this – the judge is required to indicate the stimulus sequence they received. To create a representation of the evidence space for this task one needs to consider the perceptual variability associated with each observation interval, and the fact that the judge makes a decision considering the information from both observation intervals. The approach taken by SDT is to represent the information from each interval as a separate dimension, hence yielding bivariate distributions of sensory evidence for each stimulus sequence, from which the judge can make a decision.

Evidence from the sequence <$S_1$, $S_2$> will have a bivariate normal distribution with a mean of zero on the first dimension, corresponding to $S_1$, and a mean of $d'$ on the second dimension, corresponding to $S_2$. This is because, as for the A-Not A task, the means of the perceptual distributions of $S_1$ and $S_2$ are defined to be zero and $d'$ respectively. Figure 6.4 shows the evidence space for the 2AFC task, with the distribution for <$S_1$, $S_2$> illustrated top-left. The distribution on the bottom-right is that for the sequence <$S_2$, $S_1$>. The distributions are represented by concentric circles, each of which represents a specific likelihood for the bivariate distribution it represents. A point represents the centroid of the distributions. As these equal-likelihood contours move further from the centroid, likelihood reduces and becomes zero when the contours have infinite radius.

These two distributions represent the evidence available to the judge in the 2AFC task.

There are two primary ways that a judge can employ this evidence to make decisions, that is, there are two decision strategies for 2AFC. The first is for the judge to decide, based on their knowledge of the stimuli, which stimulus the evidence on a trial was most likely to have come from. This decision strategy is based on a constant likelihood ratio decision criterion, where the likelihood ratio, $\beta$, is the ratio of the likelihoods for the <$S_1$, $S_2$> distribution to the <$S_2$, $S_1$> distribution. A neutral-biased judge would base their decisions on a criterion for which $\beta = 1$. This judge would be equally likely to judge $S_2$ to be in the second interval as the first interval. Therefore, the allocation of responses will mirror the probability of presentation of each sequence. This criterion is depicted on Figure 6.4 as a diagonal

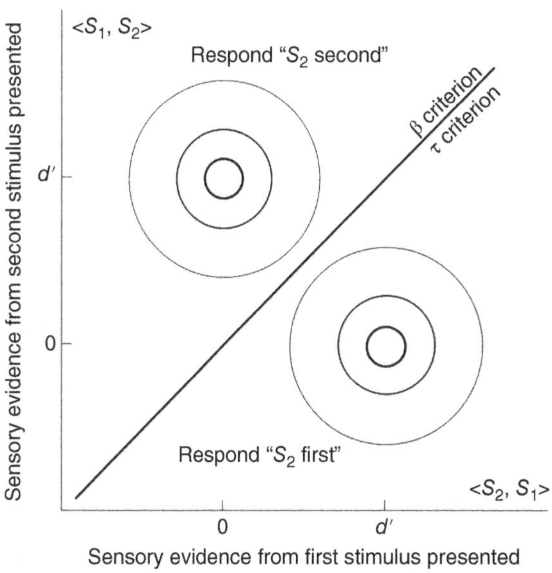


**Figure 6.4** Evidence space for the 2AFC task. Circles represent equal-likelihood contours for bivariate-normal distributions, with smaller line thicknesses indicating lower likelihood. These circles continue expanding infinitely with decreasing likelihood.

line. If a different value of $\beta$ were adopted for the criterion, the diagonal line would have the same slope (one) but have a different intercept. Wherever the criterion is located by a judge, a point, representing the evidence from both observation intervals, that falls above the criterion would be given the response "$S_2$ second" and a point that falls below the criterion would receive the response "$S_2$ first."

Another approach to decision making in 2AFC is the $\tau$ (or differencing) decision strategy. For the $\tau$-strategy judges assess the difference between the evidence from each stimulus. For example, they may subtract the magnitude of evidence from the first stimulus presented, A (which may be $S_1$ or $S_2$) from the evidence from the second stimulus, B. If this difference is greater than a difference criterion, the judge will respond "$S_2$ second" otherwise "$S_2$ first." If the judge has a neutral bias then the difference criterion will be zero, and the decision strategy amounts to selecting the observation interval with the greatest magnitude of evidence as containing $S_2$. It turns out that the criterion for the $\tau$-strategy takes on the same form (a line of slope one) as that for the $\beta$-strategy, as illustrated in Figure 6.4. Consequently, the response produced by either decision strategy will always be the same.

A hit can be defined as a response of "$S_2$" when the sequence $<S_1, S_2>$ was presented, and a false alarm as a response of "$S_2$" when the sequence was $<S_2, S_1>$. Given a value of $F$ and of $H$ collected from a 2AFC task (employing binary responses), sensitivity is (Green and Swets 1966).

$$d' = \frac{1}{\sqrt{2}}\left(z(H) - z(F)\right). \tag{6.8}$$

The interpretation of equations relating $F$ and $H$ to $d'$ for different tasks can be difficult. Comparison of Eqs. (6.1) (A-Not A) and 6.8 (2AFC) shows a $\sqrt{2}$ difference in this relationship for these particular tasks. This means that the quantity $z(H)-z(F)$ is $\sqrt{2}$ larger in 2AFC than in A-Not A, given the same level of sensitivity, $d'$. It also means that for a given value of $F$ and $H$, the estimate of $d'$ obtained for 2AFC will be smaller than that obtained for A-Not A by a factor of $\sqrt{2}$.

Returning to Figure 6.4, the false-alarm and hit rates are given respectively by the volumes of the $<S_2, S_1>$ and $<S_1, S_2>$ distributions that lie above the criterion (where the response is "$S_2$ second"). Consequently, an ROC curve for the 2AFC task can be constructed by plotting false-alarm and hit rates for all possible locations of the decision criterion, and the same ROC curve is formed for both decision strategies. Figure 6.5 illustrates two 2AFC ROC curve (the upper curves) for which $d' = 1.10$ and 1.22. Previous discussion about the representation of bias on the ROC curve in the A-Not A task applies to the ROC for the 2AFC task, as does that about area measures of performance. It is common practice to assume that judges in the 2AFC task have neutral bias. The existence of an ROC curve for 2AFC implies that, in theory, this may not be the case; there is also mounting empirical evidence that judges in the 2AFC task exhibit bias (e.g., Yeshurun et al. 2008; García-Pérez and Alcalá-Quintana 2010).

Thus, for binary responses, it is important to consider the implications of using $p(c)$ as a measure of sensitivity in 2AFC. These implications are clear from the discussion of the A-Not A task. For a biased judge, in the long run $p(c)$ will always be less than for a judge with neutral bias. This suggests that $p(c)_{max}$ should always be preferred over $p(c)$, as it controls/corrects for response bias. Use of $p(c)_{max}$ as a measure of sensitivity requires more

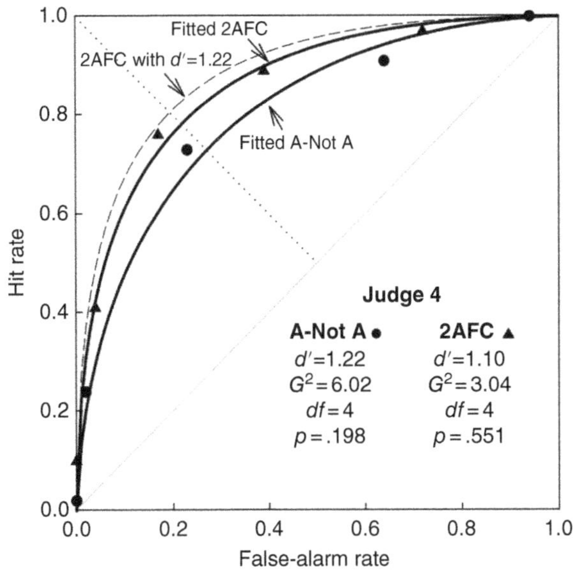


**Figure 6.5** Empirical receiver operating characteristic (ROC) curves (i.e., data) obtained from Judge 4 in the A-Not A and 2AFC task using the same stimuli. The solid ROC curves are the best fits to these data. The dashed ROC curve is where the best fitting 2AFC curve should have been located had the value of $d'$ on the 2AFC task been identical to that on A-Not A.

extensive recording of the data. One must record both the sequence of stimuli and the response on every test.

Also in Figure 6.5 is an ROC curve for the A-Not A task with $d' = 1.22$ (bottom curve). The two curves with $d' = 1.22$ (top and bottom curves), are clearly separated from each other. The curve for A-Not A lies below that for 2AFC with the same value of $d'$. Considering the relationship between $p(c)$ and the ROC curve, it is clear that $p(c)_{max}$ for 2AFC is larger than that for A-Not A for a given level of sensitivity, $d'$. This suggests why 2AFC can quantify smaller stimulus differences than A-Not A; A-Not A suffers from floor effects before 2AFC does as sensitivity decreases. With similar reasoning, as stimulus differences get larger and sensitivity increases, 2AFC suffers from ceiling effects before A-Not A does.

The separation of these two ROC curves for the same value of $d'$ also indicates that the calculation of $p(c)_{max}$ for the 2AFC task is different from that for $p(c)_{max}$ for the A-Not A task. Indeed, for 2AFC

$$p(c)_{max} = \left(\frac{d'}{\sqrt{2}}\right). \tag{6.9}$$

The right-hand side of Eq. (6.9) is identical to the right-hand side of Eq. (6.3), which gives $A_z$ in the A-Not A task. Recall that $A_z$ in the A-Not A task is identical to the proportion of correct responses obtained by a judge with neutral bias in the 2AFC task, given identical conditions. Given these relationships it will not be surprising that for the 2AFC task $A_z = \Phi(d')$.

**Table 6.4** Rating data for Judge 4 in a 2AFC task using the same stimuli as for the data reported in Table 6.2 for the A-Not A task. In this case, the ratings range from "certain $S_2$ was first" (1) to "certain $S_2$ was second" (6).

| | 2AFC | | | | | | |
|---|---|---|---|---|---|---|---|
| | **1** | **2** | **3** | **4** | **5** | **6** | **Total** |
| $<S_1, S_2>$ | 4 | 13 | 19 | 53 | 46 | 15 | 150 |
| $<S_2, S_1>$ | 42 | 50 | 33 | 19 | 6 | 0 | 150 |

The confidence-rating response procedure can be applied to the 2AFC task (or any other task). For example, ratings may range from (1) "certain $S_2$ was first" to (6) "certain $S_2$ was second." If 2AFC is used as an attribute specific test (e.g., van Hout et al. 2011) then an example would be a scale from (1) "certain the sweetest sample was first" to (6) "certain the sweetest sample was second." If used as an overall difference test (e.g., van Hout et al. 2011) ratings may go from (1) "certain $S_2$ first" to (6) "certain $S_2$ second." When applied to the 2AFC task, the confidence-rating procedure yields an empirical ROC to which can be fitted an ROC based on the 2AFC SDT model. Table 6.4 provides rating data for Judge 4 from Hautus et al. (2011). Fitting these data using SDT Assistant (Hautus 2012) yields, $d' = 1.10$; $SE = 0.10$, with about a 55% chance that the data arose from the model $\left(p\left(\chi_4^2 > G^2 = 3.04\right) = 0.551\right)$.

For Judge 4 there were two sets of rating responses, collected with two difference tests, using the same stimuli. Fitting the appropriate SDT model to the A-Not A data gave $d' = 1.22$; $SE = 0.13$. Fitting the appropriate SDT model to the 2AFC data gave $d' = 1.10$; $SE = 0.10$. Substitution of these values into Eq. (6.6) yields

$$z = \frac{1.22 - 1.10}{\sqrt{0.13^2 + 0.10^2}} = 0.73 \tag{6.10}$$

and a two-tailed test gives $p = 0.535$, leading to the conclusion that the sensitivity of Judge 4 is the same, as measured by these two difference tests. These data are therefore in accordance with the basic tenets of SDT. Figure 6.5 illustrates both the 2AFC and A-Not A data for Judge 4, together with the best-fitting ROC curves. To be clear, the two curves are separated; hence, the area under each curve, $A_z$, is different. Indeed, all area measures of sensitivity, $A_z$, $p(c)$, R-index, $p(c)_{max}$, differ between the two tasks. Only $d'$ is expected to be the same. Note that had the $d'$ estimate for the 2AFC data been the same as that for A-Not A ($d' = 1.22$) then the ROC curve would be located on the dashed curve in Figure 6.5.

## 6.5 Test Sensitivity

For Judge 4, performance on the A-Not A and 2AFC tasks was not significantly different, as measured by $d'$. Consideration of all ten judges in Hautus et al. (2011, table 2) yields average values of $d'$ of 1.28 ($SE = 0.08$) for A-Not A and 0.95 ($SE = 0.04$) for 2AFC. A paired

$t$-test for these data indicates a significant difference ($t_9 = 0.968$; $p = 0.003$). In addition, $d'$ obtained on 2AFC was lower than that for A-Not A for every judge; although not always significantly lower (e.g., Judge 4).

One interpretation of this finding is that SDT has flaws. However, there are many factors known to affect performance in difference tests, with some of these being more prominent for the chemical senses. Some of these factors are sensory fatigue, sensory adaptation, and stimulus carry-over. Factors that are more generic include memory load and inattention. SDT does not yet model these factors directly, although attempts have been made in this direction (e.g., Ennis and O'Mahony 1995; Lee and O'Mahony 2007). Therefore, it may be reasonable to take the position that differences in sensitivity across different tasks, as estimated by $d'$, may reflect the differential effect of these other factors. Thus, SDT provides a mechanism to control for variability and bias that stems from response tendencies, decision strategies, the effect of different tasks, as well as encouraging a standardized approach to test administration. By this means it makes possible a more direct study and analysis of factors, such as fatigue or carry-over, uncontaminated by these other major sources of variation. This may lead, in time, to the incorporation of the effects of carry-over, for example, into SDT models.

The effect of these other factors on estimates of sensitivity from a particular task will necessarily be downward. That is, the estimated value of $d'$ will be lower than an unaffected estimate. Different tasks can be compared with respect to their susceptibility to these effects. The lower the magnitude of these effects, the greater will be the sensitivity of the task to stimulus differences. In this regard, the A-Not A task appears to have better test sensitivity than 2AFC because $d'$ obtained from 2AFC is less than that obtained for A-Not A.

Test sensitivity is different from the power of a particular task. Power relates to the likelihood that a task will indicate a significant difference when a difference of at least a minimum prescribed size exists. Power is also dependent on the number of tests, the decision strategy adopted by the judge, and the true value of $d'$. There are many treatments of the power of difference tests, and tables are published to help with the calculations (e.g., Bi and Ennis 2001; Bi and O'Mahony 2013). These typically present the standard error of an estimate of $d'$ for a given task in a given context. They may also indicate the relative efficiency of the various tasks. The 2AFC task has greater statistical power (and hence efficiency) than the A-Not A task. A practical consequence of this is that fewer tests are needed in the 2AFC task than in A-Not A to achieve the same-sized confidence limit on the estimate of $d'$ (or the same sized standard error). Alternatively, fewer tests are required in 2AFC than in A-Not A to demonstrate a statistical difference (e.g., of $d'$ from a fixed value, or between two values of $d'$) when there is a difference.

The bold curves in Figure 6.6 illustrate the standard errors of estimates of $d'$ obtained from the A-Not A (solid) and 2AFC (dashed) tasks for a judge with neutral bias. Standard errors were determined using direct computation (e.g., Miller 1996; Hautus and Lee 1998, 2006) and the log-linear correction was applied. All curves are based on 40 tests, split equally between the two possible stimuli (A-Not A) or two possible stimulus sequences (2AFC). The lighter curves represent the same scenario, but for judges with a slight bias, equivalent to $c = 0.5$. The main feature of these curves is the smaller standard error for the 2AFC task compared to A-Not A. However, even though greater certainty of an estimate of

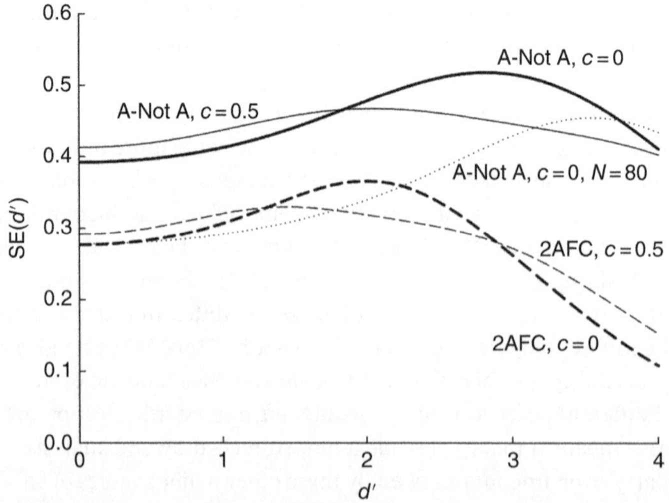


**Figure 6.6** The standard error (SE) of *d′* for different values of *d′* in the A-Not A (solid lines) and 2AFC (dashed lines) tasks when *d′* is estimated on the basis of 40 tests. The dotted line provides the case for 80 tests in the A-Not A task, which involves the same number of stimulus presentations (80) as 40 tests in the 2AFC task.

*d′* is provided by the 2AFC task for a fixed number of tests, it must be remembered that there are twice as many stimulus presentations in 2AFC than in A-Not A. If A-Not A is run with 80 tests, giving the same number of stimulus presentations as 2AFC with 40 tests, the dotted line in Figure 6.6 illustrates the standard error, which either is the same as, or marginally better than that for the 2AFC task with 40 tests.

Combining the knowledge of test sensitivity and power is not an easy task. There is little in the literature on this issue. While the computed power of a task in a particular context can give us important information for deciding the best task to use, and for statistically testing estimates of sensitivity that the task produces, there is a problem. The computations used to determine power assume a certain structure for perceptual information and decision-making (e.g., the type of distributions involved, their variance, and their relation to one another and the criterion). However, the fact that different tasks vary in test sensitivity (because of the differential effects of fatigue, carry-over, etc.) strongly suggests the addition of more variability into the perceptual and decisional processes, which will reduce performance below its theoretical level. This additional variability will inevitably increase the standard errors of any estimate of sensitivity. Furthermore, different levels of test sensitivity yield different levels of constant error (or statistical bias) in the estimates of sensitivity than are produced. This bias will most likely be in a downward direction. Sensitivity cannot become better than optimal (unless you modify the sensory-perceptual system in some way), but it can get worse. Hence, it is unlikely that estimates of sensitivity will systematically overestimate true sensitivity, but they can easily underestimate it. Until more information becomes available, it is important to be cognizant of the potential biases that can result when making decisions using data derived from difference testing, irrespective of whether or not SDT is applied.

## 6.6 Closing Comments

This chapter has provided an account of the A-Not A and 2AFC tasks from the SDT and Thurstonian perspective. The capability of these frameworks to deliver a unified approach to discrimination testing is evidenced by their ability to provide a common metric of sensory difference across tasks. Furthermore, the effects of response bias and decision strategies can be encapsulated within models, allowing for control and measurement of these sources of variability. Finally, these frameworks encapsulate a complete sensory-perceptual measurement system, allowing both estimation of sensory difference and statistical testing of the obtained measures.

The level of detail provided here is sufficient to enable practitioners and researchers to make evidence-based decisions about which difference tests to employ in a particular context, and how to interpret the ensuing results. The issues discussed in detail for the A-Not A and 2AFC tasks are generalizable to all difference tests. Whatever difference test is under consideration, the same issues need to be pursued. The broader literature has relevant details about most difference tests of practical interest.

## References

Bamber, D. (1975). Area above ordinal dominance graph and area below receiver operating characteristic graph. *J. Math. Psychol.* *12* (4): 387415.

Bi, J. (2006). Statistical analyses for R-index. *J. Sens. Stud.* *21* (6): 584–600. https://doi.org/10.1111/j.1745-459X.2006.00084.x.

Bi, J. and Ennis, D.M. (2001). The power of the "A"–"Not A" method. *J. Sens. Stud.* *16* (4): 343–359.

Bi, J. and O'Mahony, M. (2013). Variance of d' for the tetrad test and comparisons with other forced-choice methods. *J. Sens. Stud.* *28* (2): 91–101. https://doi.org/10.1111/joss.12004.

Bi, J., Ennis, D.M., and O'Mahony, M. (1997). How to estimate and use the variance of *d′* from difference tests. *J. Sens. Stud.* *12* (2): 87–104.

Brockhoff, P.B. and Christensen, R.H.B. (2010). Thurstonian models for sensory discrimination tests as generalized linear models. *Food Qual. Prefer.* *21* (3): 330–338. https://doi.org/10.1016/j.foodqual.2009.04.003.

Brown, J. (1974). Recognition assessed by rating and ranking. *Br. J. Psychol.* *65* (FEB): 13–22.

Corso, J.F. (1963). A theoretico-historical review of the threshold concept. *Psychol. Bull.* *60* (4): 356–370. https://doi.org/10.1037/h0040633.

DeCarlo, L.T. (1998). Signal detection theory and generalized linear models. *Psychol. Methods* *3* (2): 186–205. https://doi.org/10.1037//1082-989x.3.2.186.

Dorfman, D.D. and Alf, E. (1969). Maximum-likelihood estimation of parameters of signal-detection theory and determination of confidence intervals: rating-method data. *J. Math. Psychol.* *6* (3): 487–496.

Egan, J.P. (1975). *Signal Detection Theory and ROC Analysis*. New York: Academic Press.

Ennis, D.M. and O'Mahony, M. (1995). Probabilistic models for sequential taste effects in triadic choice. *J. Exp. Psychol. Hum. Percept. Perform.* *21* (5): 1088–1097.

Fienberg, S.A. (1980). *The Analysis of Cross-classified Categorical Data*, 2e. Cambridge, MA: MIT Press.

García-Pérez, M.A. and Alcalá-Quintana, R. (2010). The difference model with guessing explains interval bias in two-alternative forced-choice detection procedures. *J. Sens. Stud. 25*(6): 876–898. https://doi.org/10.1111/j.1745-459X.2010.00310.x.

Gescheider, G.A. (1997). *Psychophysics: The Fundamentals*, 3e. Mahwah, NJ: Lawrence Erlbaum Associates.

Gescheider, G.A., Wright, J.H., and Polak, J.W. (1971). Detection of vibrotactile signals differing in probability of occurrence. *Aust. J. Psychol. 78* (2): 253260.

Green, D.M. and Swets, J.A. (1966). *Signal Detection Theory and Psychophysics*. New York: Wiley.

Hautus, M.J. (1995). Corrections for extreme proportions and their biasing effects on estimated values of d′. *Behav. Res. Methods Instrum. Comput. 27* (1): 46–51.

Hautus, M.J. (1997). Calculating estimates of sensitivity from group data: pooled versus averaged estimators. *Behav. Res. Methods Instrum. Comput. 29* (4): 556–562.

Hautus, M. J. (2012). SDT assistant. http://hautus.org (accessed 2 March 2013).

Hautus, M.J. and Irwin, R.J. (1995). Two models for estimating the discriminability of foods and beverages. *J. Sens. Stud. 10* (2): 203–215.

Hautus, M.J. and Lee, A.J. (1998). The dispersions of estimates of sensitivity obtained from four psychophysical procedures: implications for experimental design. *Percept. Psychophys. 60* (4): 638–649.

Hautus, M.J., Macmillan, N.A., and Creelman, C.D. (2021). *Detection Theory: A User's Guide*, 3e. New York: Routledge.

Hautus, M.J. and Lee, A.J. (2006). Estimating sensitivity and bias in a yes/no task. *Br. J. Math. Stat. Psychol. 59*: 257–273. https://doi.org/10.1348/000711005x65753.

Hautus, M.J., Shepherd, D., and Peng, M. (2011). Decision strategies for the A Not-A, 2AFC and 2AFC-reminder tasks: empirical tests. *Food Qual. Prefer.* 22 (5): 433–442. https://doi.org/10.1016/j.foodqual.2011.02.008.

Irwin, R.J., Hautus, M.J., and Stillman, J.A. (1992). Use of the receiver operating characteristic in the study of taste perception. *J. Sens. Stud. 7* (4): 291–314.

Kendall, M.G. and Stewart, A. (1979). *The Advanced Theory of Statistics*, 4e. New York: Macmillan Publishing Inc.

Laming, D. (1987). The discrimination of smell and taste compared with other senses. *Chem. Ind. 1*: 12–18.

Lee, W. (1969). Relationships between Thurstone category scaling and signal detection theory. *Psychol. Bull. 71* (2): 101–107. https://doi.org/10.1037/h0026853.

Lee, H.S. and O'Mahony, M. (2007). Difference test sensitivity: cognitive contrast effects. *J. Sens. Stud. 22* (1): 17–33.

Lee, H.S. and van Hout, D. (2009). Quantification of sensory and food quality: the R-Index analysis. *J. Food Sci. 74* (6): R57–R64. https://doi.org/10.1111/j.17503841.2009.01204.x.

Luce, R.D. (1994). Thurstone and sensory scaling: Then and now. *Psychol. Rev. 101* (2): 271–277.

Macmillan, N.A. and Kaplan, H.L. (1985). Detection theory analysis of group data – estimating sensitivity from average hit and false-alarm rates. *Psychol. Bull. 98* (1): 185–199.

Markowitz, J. and Swets, J.A. (1967). Factors affecting slope of empirical ROC curves – comparison of binary and rating responses. *Percept. Psychophys. 2* (3): 91–100. https://doi.org/10.3758/bf03210301.

Miller, J. (1996). The sampling distribution of d'. *Percept. Psychophys. 58* (1): https://doi.org/10.3758/bf03205476.

Myung, I.J. (2000). The importance of complexity in model selection. *J. Math. Psychol. 44* (1): 190–204. https://doi.org/10.1006/jmps.1999.1283.

O'Mahony, M. (1972). Salt taste sensitivity: a signal detection approach. *Perception 1* (4): 459–464.

O'Mahony, M. and Hautus, M.J. (2008). The signal detection theory ROC curve: some applications in food sensory science. *J. Sens. Stud. 23* (2): 186–204.

Pepe, M.S. (2003). *The Statistical Evaluation of Medical Tests for Classification and Prediction.* New York: Oxford University Press.

Peterson, W.W., Birdsall, T.G., and Fox, W.C. (1954). The theory of signal dectectability. *IRE Trans. Inf. Theory 4*: 171–212.

Pollack, I. and Hsieh, R. (1969). Sampling variability of area under ROC-curve and of $d'$e. *Psychol. Bull. 71* (3): 161. https://doi.org/10.1037/h0026862.

Press, W.H., Flannery, B.P., Teukolsky, S.A., and Vetterling, W.T. (1992). *Numerical Recipes in C: The Art of Scientific Computing*, 2e. New York: Cambridge University Press.

Simpson, A.J. and Fitter, M.J. (1973). What is the best index of detectability. *Psychol. Bull. 80*(6): 481–488. https://doi.org/10.1037/h0035203.

Snodgrass, J.G. and Corwin, J. (1988). Pragmatics of measuring recognition memory – applications to dementia and amnesia. *J. Exp. Psychol. Gen. 117* (1): 34–50.

Swets, J.A. (2001). History of Signal Detection Theory. In: *International Encyclopedia of the Social & Behavioral Sciences* (ed. N.J. Smelser and P.B. Baltes), 14078–14082. Oxford: Elsevier.

Tanner, W.P. and Swets, J.A. (1954). A decision-making theory of visual detection. *Psychol. Rev. 61* (6): 401–409. https://doi.org/10.1037/h0058700.

Thurstone, L.L. (1927). A law of comparative judgment. *Psychol. Rev. 34*: 273–287.

Van Hout, D., Hautus, M.J., and Lee, H.-S. (2011). Investigation of test performance over repeated sessions using signal detection theory: comparison of three nonattribute-specified difference tests 2-AFCR, A-Not A and 2-AFC. *J. Sens. Stud. 26* (5): 311–321. https://doi.org/10.1111/j.1745-459X.2011.00346.x.

Yeshurun, Y., Carrasco, M., and Maloney, L.T. (2008). Bias and sensitivity in two interval forced choice procedures: Tests of the difference model. *Vis. Res. 48* (17): 1837–1851. https://doi.org/10.1016/j.visres.2008.05.008.

# 7

# Sureness Judgements and R-Index Calculations and Their Applications

*Danielle van Hout*[1] *and Hye-Seong Lee*[2]

[1] *Aigora, Richmond, VA, USA*
[2] *Department of Food Science and Biotechnology, College of Engineering, Ewha Womans University, Seoul, South Korea*

## 7.1 Quantifications of Sensory Differences

In sensory and consumer research, the aim is to study how much products differ, either in individual sensory characteristics or holistic perceptual dimensions determining consumer acceptance or specific benefits, or both. It is therefore important to have a measurement approach that can be used to effectively quantify the size of such sensory differences. Often this is measured by using direct scales, examples are the 9-point hedonic scale in consumer research, unstructured line scales, or category intensity scales that are used in sensory descriptive analysis.

When products exhibit subtle differences and are easily confused, direct scales may not provide reliable results, as the distinctions between products may be smaller than the scale's units. In such cases, discrimination tests can be employed to quantify the differences between these similar products. The signal detection parameter, $d'$ (dee-prime), estimates the perceived differences between products and can be interpreted as the proportion of test subjects who can detect these differences, thereby serving as an effect size measure.

For $d'$ to be employed reliably in quantifying differences, the data must adhere to specific assumptions, such as a normal distribution of responses and equal variances between products. If it is uncertain whether the data meets these assumptions, the R-index can be calculated as an alternative (Kim et al. 2017). The R-index is a non-parametric alternative to $d'$, a measure of the size of the difference that does not require these assumptions. It can be calculated for identification and discrimination tasks such

*Discrimination Testing in Sensory Evaluation*, First Edition. Edited by Lauren Rogers, Joanne Hort, Sarah E. Kemp and Tracey Hollowood.

as variants of A-Not A tests, categorization tasks such as same-different tests, and ranking tasks. The R-index can be either expressed in percentages or proportions and we will use percentages in this chapter. In its original form, the R-index is the probability of correctly identifying a target product (the signal + noise: S + N) when presented in a study with another product (the noise: N). If a subject cannot discriminate between the two products, the subject will have to guess and the chances of correctly identifying/discriminating "S + N" are 50%, so for identical products, the R-index will be around 50%. If subjects can discriminate perfectly between the two products, the R-index will be 100%.

- For A-Not A and same-different tasks, the R-index is a value between 50 and 100%, where 50% is no difference and 100% is completely different.
- For ranking tasks, in which products need to be ranked on an attribute (e.g., sweet, liked, different), an R-index can be calculated between each product pair and one product in the pair should be appointed as the reference. For ranking tests, R-index results can be between 0 and 100%. A value of 50% means there is no difference detected, a value above 50% means the other product is perceived as higher on the attribute than the appointed reference, and a value below 50% means that the other product is perceived as lower on the attribute than the appointed reference.

In this chapter, we will discuss considerations for sensory and consumer study designs, describe the computational strategies for R-index calculations and its various practical applications, and demonstrate a case study of an application of R-index analysis on preference ranking data.

## 7.2 R-Index

The R-index was developed by John Brown in 1974 (Brown 1974) as a non-parametric statistic and introduced to the field of sensory and consumer science by Michael O'Mahony as a short-cut signal detection method providing an alternative to $d'$. (O'Mahony et al. 1980; Green and Swets 1988; O'Mahony 1992; Lee and O'Mahony 2004; MacMillan and Creelman 2005). R-indices can be computed on sensory and consumer research data from identification, discrimination, categorization, and ranking methods.

Since its introduction, R-index has been widely used in sensory and consumer research data analyses because it produces numerical values indicating sizes of differences or effect sizes, which is different from traditional discrimination testing that produces *p*-values indicating the significance of differences.

The R-index calculation has been used in sensory and consumer research for various objectives such as sensitivity measurements, product reformulation, quality control, and preference testing (Lee and van Hout 2009). It was first applied for sensory difference testing in Research and Development (R&D), where it has been used for quality control, shelf-life studies, and to support product development in general (Miettinen et al. 2002; Brajkovich et al. 2005; Harker et al. 2005). Due to its flexibility, it has been applied more broadly, for example, as an alternative to hedonic scaling, and in marketing for the measurement of consumer concepts (Lee and O'Mahony 2007).

## 7.3 Sureness Judgements

### 7.3.1 Response Bias and Criterion Location

When subjects are evaluating confusable products, reporting a difference will not only depend on their sensory sensitivity and ability to detect the difference but also on a highly variable criterion of how big a difference must be to respond that it is different (Green and Swets 1988; Lee and O'Mahony 2004; MacMillan and Creelman 2005). This is called response bias, the subject's tendency to respond in a certain way. For example, one person might have a strict criterion and only respond "A" or "same" when very sure, while another might have a less strict criterion and respond "A" or "same" only when in doubt. In Chapter 10, an analysis to identify such criterion locations is introduced, as part of a series of signal detection-based data analysis tools for assessing and monitoring the quality of discrimination tests.

To address response bias in tests, two approaches can be employed: forced-choice tasks and criterion-free tasks. Traditional discrimination test methods often utilize forced-choice tasks, such as the triangle test, where participants must identify the odd product among two identical and one distinct item, or the 2-AFC test, where they must choose the sweeter product among two options. These tasks establish a set criterion between samples, ensuring participants report any perceived differences.

On the other hand, criterion-free tasks, also known as "response bias" methods, allow participants to apply their personal criteria when reporting differences. By asking them to indicate their confidence in their answers, researchers can analyze the data to separate the participants' actual sensory sensitivity from the response bias.

### 7.3.2 Number of Sureness Response Categories

The A-Not A and Same-Different tests are criterion-free discrimination tests and typically include a sureness judgment based on which an indirect scale can be constructed. Often, four or six response options are used: "S-sure," "S-unsure," "S-don't know but I guess," "N-don't know but I guess," "N-unsure," and "N-sure". The more response options, the more accurate the estimation of the size of the difference. Yet, if the middle categories are not being used, this advantage disappears. Also, increasing the response categories to more than six does not seem very useful because it adds too much complexity. In most studies published in the literature, either four or six response options were used (O'Mahony 1974, 1983, 1988, 1992; Lee et al. 2006, 2022; Pirc et al. 2022).

The R-index analysis can also be applied to ranking data, without incorporating confidence judgments. In this case, the product rankings serve as a scale. How to use R-index analysis for ranking data will be explained in the analysis section of this chapter.

## 7.4 Designing an R-Index Study, Using an A-Not A Task

### 7.4.1 Study Purpose

The design of an effective R-index study utilizing the A-Not A test method is heavily dependent on the objective of the study. The study's purpose, whether it is to quantify the differences between products using a trained analytical panel or to measure consumer sensitivity to these differences, significantly impacts the approach and design of the study. In this section, we will delve into the nuances of these objectives and how they require different study designs.

### 7.4.2 Familiarizing Sensory Panelists with the Reference Product

The subject's task in the A-Not A test is to determine whether a product is the reference "A" or Not. This requires subjects to be familiar with the reference product so that they can recognize it in the test.

As trained sensory panels play a crucial role in quantifying sensory differences, it's essential to ensure their performance is stabilized before beginning testing. The purpose of this stabilization process is to ensure that the panelists are familiar with the sensory perception of the reference product. The amount of effort required for this familiarization process can vary based on the subjects' prior experience with the products.

To streamline the familiarization process, one approach is to simply have the subjects assess the reference product twice before starting the test. However, if this is not enough to adequately stabilize their familiarity, other techniques can be employed. For example, providing the subjects with a brand image or description of the reference product can help fix their memory of its sensory perception (Kim et al. 2011). Another approach is to have subjects perform categorization tasks before the test begins, using techniques such as a 2-AFC (Two-Alternative Forced Choice) or tetrad test. These tasks provide additional opportunities for subjects to engage with and internalize the sensory perception of the reference product, helping to solidify their familiarity with it (Kim et al. 2014; Shin et al. 2016; Choi et al. 2023a, b; Lee et al. 2023).

Ultimately, the goal of these alternative approaches is to ensure that the subjects have a clear and comprehensive understanding of the reference product before participating in the A-Not A test. By doing so, we can increase the accuracy and reliability of the results, providing valuable insights into the sensory differences between products.

### 7.4.3 Consumer Familiarity with the Reference Product

Consumer studies, on the other hand, aim to represent the target consumers as accurately as possible and cannot involve prior training. The objective of consumer discrimination studies is to quantify consumer sensitivity, and it is crucial to maintain the consumers' representativeness throughout the testing process. Providing prior training to consumers would alter their sensitivity and result in them becoming less representative of the target population.

Instead of familiarization, a reminder of the reference product is offered to the consumers before the start of the test. The consumers are informed that this is the specific product they have used before, and they should look out for it in the test. Providing the consumers with an example of the reference product instead of solely relying on their memory of it is important because the testing context may impact the perception of the product. In a real-life scenario, the consumer might use the product in a different environment, with other people, while being distracted, and in combination with other products.

To refresh their memory of the reference product and provide an example of how the product is perceived in the testing environment, a warm-up procedure before the start of the test can be used. This procedure, referred to as a cognitive warm-up, involves asking consumers about their memory of the user experience of the product. Research has shown that this approach is effective in stabilizing consumer criteria in discrimination tests and hedonic scaling, reducing product range effects and leading to more stable performance across multiple testing sessions (Kim et al. 2015, 2018, 2022).

### 7.4.4 Number of Products in a Test

While conducting an A-Not A test with a single product is possible, it is crucial to include both the test product and the blind reference product in the evaluation process to calculate the R-index. This requires each subject to evaluate both products, either in one session or on separate days if the product characteristics necessitate such as in the case of intense or lingering products such as spicy pepper or toothpaste. The R-index represents the percentage of correct responses in a hypothetical 2-AFC test, where each test sample would be compared to each reference sample. To achieve this, the same subjects must evaluate both the test product and the reference product in a randomized, sequential, monadic order (Lee and van Hout 2009).

Although the minimum number of products in an A-Not A study with R-index calculation is two products (the reference and the test product), there is no fixed maximum number of products. If there are more test products to be compared to the reference, the test design can be more economical by including the reference product only once. This means that with two test products compared to a reference, subjects need to evaluate three products (the reference and the two test products). With three test products compared to the reference, subjects need to evaluate four products (the reference and the three test products).

When evaluating multiple test products compared to a single reference product, it's important to consider the impact on the study results. A study design with an equal number of reference and test products may lead to more stable results, as the signal-to-noise ratio is balanced. Recent studies in the food industry have confirmed that this design results in higher discrimination sensitivity compared to designs with only one reference and multiple test products. This hypothesis, first proposed by Green and Swets in 1966, has been backed up by recent research findings by Choi et al. (2023) and Lee et al. (2023). Thus, it is worth considering the number of reference products in a study to optimize results and increase the sensitivity of discrimination.

The number of products that can be tested in a single session using the A-Not A test method depends greatly on the characteristics and type of the products being evaluated. For more intense products, it may be necessary to conduct the study over several separate sessions. To determine the appropriate number of products to include in a session, it is important to conduct a pilot study when testing a new product type.

When splitting an A-Not A test with multiple products across multiple sessions, it is imperative to include the blind reference product in every session. This helps to counteract the potential impact of session effects, which can result from performance fluctuations across different sessions. This approach guarantees the accuracy and consistency of the results, regardless of the number of sessions required to complete the study.

## 7.5 R-Index Data Analysis

### 7.5.1 R-Index Analysis of A-Not A Data

For manual R-index analysis of A-Not A test results, the data must first be entered in a response matrix, after which a formula can be used that calculates the R-index value (Bi 2006; Bi et al. 2013a, b).

Table 7.1 illustrates the data obtained from an A-Not A study, in which 10 reference products and 10 test products were evaluated by a single subject. The evaluation of the 10 references and 10 test products can result in a theoretical 100 paired comparisons (10×10). The R-index represents the proportion of correct responses among these hypothetical 100 pairs and serves as a measure of the likelihood of detecting differences between the samples.

The results are synthesized in the response matrix, depicted in Table 7.2. The same response matrix can be found in Table 7.3, but with letters replacing the numerical values, representing the locations in the calculation formula where the results should be entered to manually determine the R-index. By applying the formula, the R-index value calculated from the example data would be 86.5%. If more than one product is compared with the reference, the structure of Table 7.4 can be used, which shows the response matrix and formulas for an A-Not A study in which three test products are being tested versus a reference.

**Table 7.1** Example of data sheet obtained from a randomized blind presentation of 10 test products and 10 references.

| Presentation order | Stimulus | Subject's responses |
|---|---|---|
| 1 | Not-A | "Not-A-unsure" |
| 2 | A | "A-sure" |
| 3 | A | "A-unsure" |
| 4 | Not-A | "Not-A-sure" |
| 5 | A | "A-sure" |
| 6 | Not-A | "Not-A-unsure" |
| 7 | Not-A | "Not-A-unsure" |
| 8 | Not-A | "A-sure" |
| 9 | A | "A-unsure" |
| 10 | Not-A | "Not-A-sure" |
| 11 | A | "A-sure" |
| 12 | Not-A | "Not-A-sure" |
| 13 | A | "A-unsure" |
| 14 | Not-A | "A-unsure" |
| 15 | A | "A-sure" |
| 16 | A | "Not-A-unsure" |
| 17 | Not-A | "Not-A-sure" |
| 18 | A | "A-sure" |
| 19 | A | "A-sure" |
| 20 | Not-A | "A-unsure" |

**Table 7.2** Response matrix, summarizing the responses of Table 7.1.

| | "Not A-sure" (Not A) | "Not A-unsure" (Not A?) | "A-unsure" (A?) | "A-sure" (A) | Total (*N*) |
|---|---|---|---|---|---|
| Test product (Not A) | 4 | 3 | 2 | 1 | 10 |
| Reference (A) | 0 | 1 | 3 | 6 | 10 |

**Table 7.3** Response matrix, with letters indicating how to enter the data in the R-index formulae.

| | "Not A-sure" (Not A) | "Not A-unsure" (Not A?) | "A-unsure" (A?) | "A-sure" (A) | Total (*N*) |
|---|---|---|---|---|---|
| Test product (Not A) | $a$ | $b$ | $c$ | $d$ | $N_{\text{NotA}}$ $(=a+b+c+d)$ |
| Reference (A) | $e$ | $f$ | $g$ | $h$ | $N_{\text{A}}$ $(=e+f+g+h)$ |

$$\text{R-Index }(\%)=\frac{a(f+g+h)+b(g+h)+ch+(1/2)(ae+bf+cg+dh)}{N_{\text{NotA}}N_{\text{A}}}\times 100$$

**Table 7.4** Response matrix and R-index formulas for an A-Not A test with multiple products.

| | "Not A-sure" (Not A) | "Not A-unsure" (Not A?) | "A-unsure" (A?) | "A-sure" (A) | Total (*N*) |
|---|---|---|---|---|---|
| Test product 1 (Not $A_1$) | $a$ | $b$ | $c$ | $d$ | $N_{\text{Not A1}}$ $(=a+b+c+d)$ |
| Test product 2 (Not $A_2$) | $e$ | $f$ | $g$ | $h$ | $N_{\text{Not A2}}$ $(=e+f+g+h)$ |
| Test product 3 (Not $A_3$) | $i$ | $j$ | $k$ | $l$ | $N_{\text{Not A3}}$ $(=i+j+k+l)$ |
| Reference (A) | $m$ | $n$ | $o$ | $p$ | $N_{\text{A}}$ $(=m+n+o+p)$ |

$$\text{R-Index }(\%)\text{ for Not }A_1=\frac{a(n+o+p)+b(o+p)+cp+1/2(am+bn+co+dp)}{N_{\text{NotA1}}N_{\text{A}}}\times 100$$

$$\text{R-Index }(\%)\text{ for Not }A_2=\frac{e(n+o+p)+f(o+p)+gp+1/2(em+fn+go+hp)}{N_{\text{Not A2}}N_{\text{A}}}\times 100$$

$$\text{R-Index }(\%)\text{ for Not }A_3=\frac{i(n+o+p)+j(o+p)+kp+1/2(im+jn+ko+lp)}{N_{\text{Not A3}}N_{\text{A}}}\times 100$$

**Table 7.5** Response matrix and formulas for R-index ranking.

| | Rank 1st | Rank 2nd | Rank 3rd | Rank 4th | Total |
|---|---|---|---|---|---|
| Test product 1 ($T_1$) | $a$ | $b$ | $c$ | $d$ | $N_{T1}$ $(=a+b+c+d)$ |
| Test product 2 ($T_2$) | $e$ | $f$ | $g$ | $h$ | $N_{T2}$ $(=e+f+g+h)$ |
| Test product 3 ($T_3$) | $i$ | $j$ | $k$ | $l$ | $N_{T3}$ $(=i+j+k+l)$ |
| Assigned reference (R) | $m$ | $n$ | $o$ | $p$ | $N_R$ $(=m+n+o+p)$ |

### 7.5.2 R-Index Analysis of Ranking Data

Table 7.5 displays the response sheet and formulae used in the data collection and analysis process for conducting an R-index analysis on ranking data. This method involves considering each product as a reference and comparing it to other products in the study. This allows for the calculation of R-index values for every product pairing, offering a thorough analysis of the differences among the products.

It is important to note that the formulae and data structure for the ranking R-index closely resembles those of the multiple A-Not A in Table 7.4, but with two key distinctions. First, the ranking serves as a scale, where the number of scale categories corresponds to the number of products. In contrast, the A-Not A approach uses the sureness levels as a scale. Second, in a ranking test, all products are directly compared to each other, allowing each product to be used as a reference and compared to the others. In the Multiple A-Not A method, products are only compared to the blind reference.

$$\text{R-Index }(\%)\text{ for R vs }T_1 = \frac{a(n+o+p)+b(o+p)+cp+1/2(am+bn+co+dp)}{N_{T1}N_R}\times 100$$

$$\text{R-Index }(\%)\text{ for R vs. }T_2 = \frac{e(n+o+p)+f(o+p)+gp+1/2(em+fn+go+hp)}{N_{T2}N_R}\times 100$$

$$\text{R-Index }(\%)\text{ for R vs. }T_3 = \frac{i(n+o+p)+j(o+p)+kp+1/2(im+jn+ko+lp)}{N_{T3}N_R}\times 100$$

### 7.5.3 Converting R-Index Into $d'$

The R-index and $d'$ ($d$-prime) are both measures of signal detection ability and are used in sensory analysis to determine the difference between products. While $d'$ is a parametric measure, the R-index is a non-parametric alternative. The relationship between the two measures is strong, as the R-index is directly related to the $p(A)$, which is the area under the ROC curve used for calculating $d'$ (O'Mahony and Hautus 2008). There are tables available for converting R-indices to $d'$ and vice-versa (Swets and Elliot 1964).

The R-index is commonly used in sensory analysis because it is easy to calculate and interpret. However, if results are to be compared across studies, it may be more useful to use $d'$. By converting the R-indices to $d'$, researchers can use a standardized measure to compare results across studies, even if different methods were used in each study.

This allows for a more accurate comparison of results and makes it easier to draw meaningful conclusions about the difference between products.

### 7.5.4 Sample Sizes

The sample size required to conduct an R-index analysis using the A-Not A ratings or same-different rating test is contingent on the magnitude of the difference being investigated. Similar to other statistical tests, enhancing the number of measurements bolsters the test's power and refines the accuracy of the outcomes. Bi et al. (2013) provides a comprehensive discussion about the statistical analysis of ROC curves for ratings given with the A-Not A and same-different methods, providing graded responses (ratings) for the degree of sureness of A-Not A or same-different judgments. As the R-index can be considered to be equivalent to the area under an ROC curve (AUC), the sample sizes can be computed using the same process presented in this paper. In Bi et al. (2013), for conservative estimations of powers and sample sizes, the maximum variance of AUC was used.

For difference testing using A-Not A rating test, according to their function, to detect a difference of $d' = 0.5$ (which can be converted to R-index = 64%) with a power of 0.8 and $\alpha = 0.05$, the sample size needed is 94.

R functions "blorocdpower" and "blorocdsamp" can be used for calculations of difference testing powers and sample sizes.

```
> blorocdpower(0.5,100,0.05,1)
A1: 0.64
[1] 0.8777315
> blorocdsamp(0.5,0.8,0.05,1)
A1: 0.64
[1] 94
```

For similarity testing using A-Not A rating test, according to their function, to test a difference of $d' = 0.5$ (which can be converted to R-index = 64%) with a power of 0.8 and $\alpha = 0.05$, the sample size needed is 99.

R functions "blorocspow" and "blorocssam" can be used for calculations of difference testing powers and sample sizes.

```
> blorocspow(1,0.5,100,0.05,1)
[1] 0.8597686
> blorocssam(1,0.5,0.8,0.05,1)
[1] 99
```

For both difference testing and similarity testing, the ratings of the A-Not A tests are more powerful than the ratings of the same-different tests, according to the power curves shown in Bi et al. (2013).

### 7.5.5 Significance of R-Index Values

The theoretical R-index value for products with no difference should be 50%. To establish statistical significance ($p < 0.05$) of the R-index being greater than 50, it must exceed a critical value ($cv$). The sample size influences the critical value for significance, with larger sample sizes yielding smaller critical values (Bi and O'Mahony 1995, 2007, 2020). If the R-index value is above $50 + cv$, it indicates that the product is significantly different from the reference. In ranking tests, the R-index may be significantly higher or lower than 50 ($50 \pm cv$) due to the directional differences present in ranking data, with the product being higher or lower in rank than the reference. The Appendix presents a table with critical values for determining the significance of R-index values, produced by Bi and O'Mahony (2007).

### 7.5.6 Significant Differences Between Two R-Index Values Using Confidence Intervals

Uncovering the significance between two R-indices is key to truly understanding the differences between products. The use of confidence intervals provides a powerful tool in this pursuit. When the intervals overlap, it means that there is no statistically significant difference between the two R-indices, but when they don't overlap, it signifies that there is indeed a difference between the two. This method is particularly useful in situations where both products are significantly different from the reference, as it confirms which product is closest to the reference.

For example, calculating a two-sided 90% confidence interval for an R-index of 64% (=0.64), performed with a sample size of 40 and with a Standard Error (SE) of 4.5% (=0.045), can be done as follows: (1) calculate the margin of error (ME) $= z^*$ SE $= 1.645\ 80.045 = 0.074$, (2) determine the upper and lower bounds of the 90% confidence interval: R-index $\pm$ ME $= 0.64 \pm 0.074 = (0.566–0.714)$, representing the range in which you can be 90% confident that the true R-index value lies. To express the interval in R-index percentages, multiply the lower and upper bounds by 100: (56.6%, 71.4%).

## 7.6 Applications

### 7.6.1 R-Index for A-Not A Tests, Comparing Multiple Products with a Reference

As R-index analysis provides sizes of difference between the reference and the test products, it can be a good output for knowledge management, for example on how product changes lead to perceived differences. When many test products are being compared to a reference, using the A-Not A test with R-index analysis is very efficient. The reference product can be treated as the noise, and the test products are treated as the signals. The subjects are instructed to compare each product to the reference product (Lee et al. 2006).

It is important to realize that when R-index results need to be compared across many experiments, the measurement method should be consistent, and the blind reference sample should be included at least once in each session. If it is going to be a big study with a large product set, it might also be worth including some anchor products. Anchor products are products that remain stable in terms of sensory properties and can be included in different sessions. The test results of such anchor products can be investigated to assess the fluctuation in subjects' performance on different days during the study (Meilgaard et al. 2006).

### 7.6.2 R-Index for Ranking Applied to Consumer Hedonic Measurements

R-Indices are primarily used for measuring discrimination in food research (O'Mahony 1992), but they can also serve as an alternative to hedonic scaling. They utilize preference ranking data or consumers' overall judgments on emotional attributes (Lee and O'Mahony 2007). Ranking tasks offer a comparative, behavior-based assessment rather than relying on numerical estimations on a scale.

Although hedonic scaling is the traditional method for measuring consumer liking, several studies (Parducci 1963, 1965; Lawless 1983; Lawless and Malone 1986) have revealed that untrained consumers tend to use a "relative" decision rule when using such scales. This means that consumers essentially rank products and use scores to indicate the perceived distances between them (Lee and O'Mahony 2007).

R-Index analysis of ranking data can yield relative spacings similar to scaling. While mean ranks measure distance within a specific experimental context, R-Indices derived from ranking provide a measure of spacing based on choice probabilities, making them less context-dependent (Lee and O'Mahony 2004; Feng and O'Mahony 2017). As a result, R-index analysis derived from ranking offers a straightforward alternative to scaling in consumer preference and emotion research.

The R-index for preference ranking represents the degree of preference rather than a degree of difference. More precisely, it indicates the probability of selecting a target food (the signal) when presented alongside a comparison food (the noise) in a paired preference test. In recent literature, hedonic scale data were subjected to conventional analysis of variance (ANOVA) and concurrently transformed into rank data for R-index computation. While both approaches yielded similar patterns, the R-index analysis identified significant differences between products that the traditional ANOVA failed to detect (Xia et al. 2023; Jara-Solis et al. 2023).

### 7.6.3 R-Index to Measure Consumers' Concepts or Emotions

Another application of the R-index is the measurement of consumer concepts, benefits, or emotions, related to conceptual attributes. Ishii and O'Mahony (1992) used this approach when comparing umami taste concepts held by a group of Japanese and a group of Americans. The "everyday" sensory concept of a target group of consumers could be measured for emotional sensations such as "refreshing," "thirst-quenching," "savory," "home-made," and so on, by having consumers rank the products on the given concept or emotional attributes as follows: "Most-like: the exemplar," "2nd most like," "3rd," "4th," and so on. Dacremont and Vickers (1994) measured the concept of cheddar cheese flavor in a group of 18 consumers, and Lee and O'Mahony (2004) measured an abstract consumer emotion, the refreshing appearance of toothpaste when spread on the toothbrush. The refreshing quality is an emotional attribute linked with the consumer benefit of the toothpaste. The ranking test with R-index analysis was a fast and straightforward way to identify which products were perceived to be most refreshing and therefore expected to deliver strongest to the benefit.

### 7.6.4 R-Index for Sensitivity Measurements

The R-index analysis can also be applied as an individual sensitivity measurement to basic tastes such as salt, sweet, and bitter (O'Mahony 1974; Matsuda and Doty 1995) and oral tectile sensitivity (Lee et al. 2022). Such sensitivity measurements were one of the earliest applications of the R-index in sensory, to measure the discrimination of salt and water stimuli (O'Mahony 1974). Matsuda and Doty (1995) used this method to investigate regional taste sensitivity to salt (NaCl) as a function of subject age, tongue locus, and area of stimulation.

For sensitivity measurement, the taste stimulus (the signal: S) is compared to a "blank" stimulus (the noise: N) (Robinson et al. 2005; O'Mahony 1974, 1992). For example, for measuring human sensitivity to NaCl, we can assume that 10 replicates of each taste stimulus, such as a low concentration "threshold" NaCl solution and a "blank" stimulus such as distilled water, need to be tasted in one session for a given subject. A total of 20 stimuli are presented one by one in random order and the subjects need to respond to each stimulus whether a taste could be detected, or whether it tasted like water. In addition, the subjects must indicate how sure they are of their decision. This procedure results in four response options: "S-sure," "S-unsure," "N-unsure," and "N-sure." The response frequencies for all four rating categories are counted for both the NaCl stimulus (S) and the distilled water stimulus (N) and summarized in a response matrix. Table 7.2, presented earlier in this chapter, contains the raw data for this example in a response matrix, and Table 7.3 contains the formula for computing an R-index from the response matrix. The R-index value calculated from the data is 86.5%. This is the degree to which the panelists can distinguish a NaCl taste stimulus from distilled water.

## 7.7 Future Developments

The R-index, which measures the agreement among sensory panelists or consumers in detecting differences between products, has several potential future developments and applications:

1) **Integration with machine learning and artificial intelligence**: By incorporating machine learning algorithms and artificial intelligence techniques, it may be possible to predict consumer preferences based on sensory data. This could aid companies in the development of new products or the modification of existing ones, aligning them more closely with consumer preferences and reducing the need for extensive consumer testing.
2) **Cross-cultural studies**: Implementing R-index calculations in cross-cultural sensory studies could provide insights into how preferences and sensory sensitivities vary among different consumer populations. Such information would enable companies to customize their products for specific markets, ensuring their offerings align with local tastes and preferences.
3) **Time series analysis and monitoring**: Employing R-index calculations for monitoring changes in consumer preferences and sensory panel performance over time could help companies identify trends, adjust action standards, and maintain the relevance and effectiveness of their sensory evaluations. By tracking these changes, businesses can better adapt to evolving consumer preferences and ensure that their products continue to meet the expectations of their target audience.

By exploring these future developments, the R-index can potentially become a more versatile and powerful tool for understanding consumer preferences and sensory panel performance in various contexts.

## Case Study: Preference Ranking with R-Index Analysis

A company wants to understand consumers' perception of different kinds of margarine to learn how much these products differ in preferences. Seven products were selected to be included in the study. The selected products varied in multiple sensory directions and reflected the range of products in the market. The seven kinds of margarine were spread on bread before serving to consumers.

In total, 105 consumers were included who use these types of margarine on bread at least three times a week. Consumers were given all products at the same time. Consumers were instructed to taste the products from left to right, after which they could freely re-taste as much as they needed to. The consumer task was to rank the products from most to least liked. Table 7.6 is the response matrix containing the consumer responses in this preference ranking study.

The results were analyzed with R-index analysis for ranking to quantify the percentages of consumers that preferred one product over the other, using the formulae in Table 7.5. The results of the R-index preference ranking analysis are summarized in Table 7.7.

**Table 7.6** Response matrix with consumer preference ranking data.

| | Most liked | | | | | | Least liked | |
|---|---|---|---|---|---|---|---|---|
| $P_1$ | 3 | 7 | 9 | 20 | 19 | 23 | 24 | 105 |
| $P_2$ | 20 | 23 | 17 | 13 | 14 | 12 | 6 | 105 |
| $P_3$ | 25 | 19 | 13 | 14 | 14 | 15 | 5 | 105 |
| $P_4$ | 11 | 17 | 15 | 21 | 15 | 14 | 12 | 105 |
| $P_5$ | 15 | 12 | 19 | 13 | 19 | 13 | 14 | 105 |
| $P_6$ | 14 | 14 | 14 | 8 | 9 | 18 | 28 | 105 |
| $P_7$ | 17 | 13 | 18 | 16 | 15 | 10 | 16 | 105 |

**Table 7.7** Results of the R-index analysis of consumer preference ranking data.

| Products | $P_1$ | $P_2$ | $P_3$ | $P_4$ | $P_5$ | $P_6$ |
|---|---|---|---|---|---|---|
| $P_1$ | | | | | | |
| $P_2$ | **73.9** | | | | | |
| $P_3$ | **73.3** | 50.3 | | | | |
| $P_4$ | **65.8** | **40.7** | **40.9** | | | |
| $P_5$ | **64.8** | **40.8** | **40.9** | 50.3 | | |
| $P_6$ | 56.1 | **35.7** | **35.5** | 43.2 | 43.7 | |
| $P_7$ | **66.0** | 42.6 | 42.6 | 51.4 | 51.6 | 57.4 |

**Table 7.8** Summary of results of the preference ranking study.

| Preference cluster | Products | Percentage of consumers that prefers the product over $P_1$ | Percentage of consumers that prefers the product over $P_2$ |
|---|---|---|---|
| Most liked | $P_2$ | 73.9 | — |
| | $P_3$ | 73.3 | — |
| In-between group | $P_7$ | 66.0 | — |
| | $P_4$ | 65.9 | 59.2 |
| | $P_5$ | 64.8 | 59.2 |
| | $P_6$ | — | 64.4 |
| Least liked | $P_1$ | — | 26.1 |

The results in the table should be interpreted as follows:

- The difference between $P_2$ (test product) and $P_1$ (reference) is 73.9, meaning that 73.9% of the consumers preferred $P_2$ over $P_1$.
- The difference between $P_6$ (test product) and $P_2$ (reference) is 35.6, meaning that only 35.6% of the consumers preferred $P_6$ over $P_2$.
- The critical value for significance is $50 \pm 7.71$. This is based on the table with critical values for R-index in the Appendix. The table only lists values for $N = 100$ and $N = 110$, so we take the mean of the difference (2-sided, $\alpha = 5\%$). This means that values between 42.29 and 57.71 are not significantly different. R-Indices in bold indicate the significant differences.

From the results in Table 7.7, it is difficult to see at one glance which products are most or least preferred. Therefore, for easier interpretation, the results are further summarized in Table 7.8 in a table in which we can see that two products ($P_2$ and $P_3$) are most preferred, one product ($P_1$) is least preferred, and the other four products are in the middle.

## Appendix

Table 7.9 of the Appendix contains a table with critical values for determining the significance of R-index values, Bi and O'Mahony (2007).

**Table 7.9** Critical values expressed in percentages of R-index-50%.

| One-sided | $\alpha$ = 0.200 | 0.100 | 0.050 | 0.025 | 0.010 | 0.005 | 0.001 |
|---|---|---|---|---|---|---|---|
| Two-sided | $\alpha$ = 0.400 | 0.200 | 0.100 | 0.050 | 0.020 | 0.010 | 0.002 |
| $n$ | % | % | % | % | % | % | % |
| 5 | 15.21 | 21.67 | 26.06 | 29.22 | 32.27 | 34.01 | 36.91 |
| 6 | 13.91 | 20.01 | 24.26 | 27.40 | 30.48 | 32.26 | 35.30 |
| 7 | 12.89 | 18.68 | 22.80 | 25.89 | 28.97 | 30.78 | 33.91 |
| 8 | 12.07 | 17.59 | 21.58 | 24.61 | 27.67 | 29.49 | 32.68 |
| 9 | 11.39 | 16.67 | 20.53 | 23.51 | 26.54 | 28.36 | 31.58 |
| 10 | 10.81 | 15.88 | 19.63 | 22.54 | 25.54 | 27.36 | 30.60 |
| 11 | 10.31 | 15.19 | 18.84 | 21.69 | 24.65 | 26.46 | 29.71 |
| 12 | 9.88 | 14.59 | 18.13 | 20.93 | 23.85 | 25.64 | 28.89 |
| 13 | 9.49 | 14.05 | 17.50 | 20.24 | 23.12 | 24.90 | 28.14 |
| 14 | 9.15 | 13.57 | 16.93 | 19.62 | 22.46 | 24.22 | 27.45 |
| 15 | 8.84 | 13.13 | 16.42 | 19.05 | 21.85 | 23.59 | 26.81 |
| 16 | 8.56 | 12.74 | 15.95 | 18.53 | 21.29 | 23.01 | 26.21 |
| 17 | 8.31 | 12.38 | 15.51 | 18.05 | 20.77 | 22.48 | 25.65 |
| 18 | 8.08 | 12.04 | 15.11 | 17.61 | 20.28 | 21.98 | 25.13 |
| 19 | 7.86 | 11.74 | 14.74 | 17.19 | 19.83 | 21.51 | 24.64 |
| 20 | 7.66 | 11.45 | 14.40 | 16.80 | 19.41 | 21.07 | 24.17 |
| 21 | 7.48 | 11.19 | 14.08 | 16.44 | 19.02 | 20.65 | 23.74 |
| 22 | 7.31 | 10.94 | 13.78 | 16.11 | 18.64 | 20.26 | 23.32 |
| 23 | 7.15 | 10.71 | 13.50 | 15.79 | 18.29 | 19.89 | 22.93 |
| 24 | 7.00 | 10.49 | 13.23 | 15.49 | 17.96 | 19.55 | 22.56 |
| 25 | 6.86 | 10.29 | 12.98 | 15.21 | 17.65 | 19.21 | 22.20 |
| 26 | 6.72 | 10.09 | 12.75 | 14.94 | 17.35 | 18.90 | 21.86 |
| 27 | 6.60 | 9.91 | 12.52 | 14.68 | 17.07 | 18.60 | 21.54 |
| 28 | 6.48 | 9.74 | 12.31 | 14.44 | 16.79 | 18.32 | 21.23 |
| 29 | 6.37 | 9.57 | 12.11 | 14.21 | 16.54 | 18.04 | 20.93 |
| 30 | 6.26 | 9.42 | 11.92 | 13.99 | 16.29 | 17.78 | 20.65 |
| 31 | 6.16 | 9.27 | 11.73 | 13.78 | 16.06 | 17.53 | 20.38 |
| 32 | 6.06 | 9.12 | 11.56 | 13.58 | 15.83 | 17.29 | 20.11 |
| 33 | 5.97 | 8.99 | 11.39 | 13.39 | 15.61 | 17.06 | 19.86 |
| 34 | 5.88 | 8.86 | 11.23 | 13.21 | 15.41 | 16.84 | 19.62 |
| 35 | 5.80 | 8.74 | 11.08 | 13.03 | 15.21 | 16.63 | 19.38 |
| 36 | 5.72 | 8.62 | 10.93 | 12.86 | 15.02 | 16.42 | 19.16 |

(*Continued*)

**Table 7.9** (Continued)

| One-sided | $\alpha$ = 0.200 | 0.100 | 0.050 | 0.025 | 0.010 | 0.005 | 0.001 |
|---|---|---|---|---|---|---|---|
| Two-sided | $\alpha$ = 0.400 | 0.200 | 0.100 | 0.050 | 0.020 | 0.010 | 0.002 |
| *n* | % | % | % | % | % | % | % |
| 37 | 5.64 | 8.50 | 10.79 | 12.70 | 14.83 | 16.22 | 18.94 |
| 38 | 5.57 | 8.39 | 10.65 | 12.54 | 14.65 | 16.03 | 18.73 |
| 39 | 5.49 | 8.29 | 10.52 | 12.39 | 14.48 | 15.85 | 18.52 |
| 40 | 5.43 | 8.19 | 10.39 | 12.24 | 14.32 | 15.67 | 18.33 |
| 45 | 5.12 | 7.72 | 9.82 | 11.58 | 13.57 | 14.87 | 17.43 |
| 50 | 4.85 | 7.33 | 9.33 | 11.02 | 12.92 | 14.18 | 16.65 |
| 55 | 4.63 | 7.00 | 8.91 | 10.53 | 12.36 | 13.57 | 15.96 |
| 60 | 4.43 | 6.70 | 8.55 | 10.10 | 11.87 | 13.04 | 15.36 |
| 65 | 4.26 | 6.44 | 8.22 | 9.72 | 11.43 | 12.56 | 14.82 |
| 70 | 4.10 | 6.21 | 7.93 | 9.38 | 11.04 | 12.14 | 14.33 |
| 75 | 3.96 | 6.01 | 7.66 | 9.08 | 10.68 | 11.75 | 13.89 |
| 80 | 3.84 | 5.82 | 7.42 | 8.80 | 10.36 | 11.40 | 13.49 |
| 85 | 3.72 | 5.65 | 7.21 | 8.54 | 10.06 | 11.08 | 13.12 |
| 90 | 3.62 | 5.49 | 7.01 | 8.31 | 9.79 | 10.79 | 12.78 |
| 95 | 3.52 | 5.34 | 6.82 | 8.10 | 9.54 | 10.51 | 12.46 |
| 100 | 3.43 | 5.21 | 6.65 | 7.89 | 9.31 | 10.26 | 12.17 |
| 110 | 3.27 | 4.97 | 6.35 | 7.53 | 8.89 | 9.80 | 11.64 |
| 120 | 3.13 | 4.76 | 6.08 | 7.22 | 8.53 | 9.40 | 11.17 |
| 130 | 3.01 | 4.57 | 5.85 | 6.94 | 8.21 | 9.05 | 10.76 |
| 140 | 2.91 | 4.41 | 5.64 | 6.70 | 7.92 | 8.73 | 10.39 |
| 150 | 2.81 | 4.26 | 5.45 | 6.47 | 7.65 | 8.45 | 10.05 |
| 160 | 2.72 | 4.12 | 5.28 | 6.27 | 7.41 | 8.19 | 9.75 |
| 170 | 2.64 | 4.00 | 5.12 | 6.09 | 7.20 | 7.95 | 9.47 |
| 180 | 2.56 | 3.89 | 4.98 | 5.92 | 7.00 | 7.73 | 9.21 |
| 190 | 2.49 | 3.79 | 4.85 | 5.76 | 6.82 | 7.53 | 8.98 |
| 200 | 2.43 | 3.69 | 4.73 | 5.62 | 6.65 | 7.34 | 8.76 |
| 210 | 2.37 | 3.60 | 4.61 | 5.49 | 6.49 | 7.17 | 8.56 |
| 220 | 2.32 | 3.52 | 4.51 | 5.36 | 6.34 | 7.01 | 8.37 |
| 230 | 2.27 | 3.44 | 4.41 | 5.24 | 6.21 | 6.86 | 8.19 |
| 240 | 2.22 | 3.37 | 4.32 | 5.14 | 6.08 | 6.72 | 8.02 |
| 250 | 2.17 | 3.30 | 4.23 | 5.03 | 5.96 | 6.58 | 7.86 |
| 260 | 2.13 | 3.24 | 4.15 | 4.94 | 5.84 | 6.46 | 7.71 |
| 270 | 2.09 | 3.18 | 4.07 | 4.84 | 5.74 | 6.34 | 7.57 |

Table 7.9 (Continued)

| One-sided | $\alpha$ = 0.200 | 0.100 | 0.050 | 0.025 | 0.010 | 0.005 | 0.001 |
|---|---|---|---|---|---|---|---|
| **Two-sided** | **$\alpha$ = 0.400** | **0.200** | **0.100** | **0.050** | **0.020** | **0.010** | **0.002** |
| *n* | **%** | **%** | **%** | **%** | **%** | **%** | **%** |
| 280 | 2.05 | 3.12 | 4.00 | 4.76 | 5.63 | 6.23 | 7.44 |
| 290 | 2.02 | 3.07 | 3.93 | 4.68 | 5.54 | 6.12 | 7.31 |
| 300 | 1.98 | 3.02 | 3.87 | 4.60 | 5.45 | 6.02 | 7.19 |
| 310 | 1.95 | 2.97 | 3.80 | 4.52 | 5.36 | 5.92 | 7.08 |
| 320 | 1.92 | 2.92 | 3.74 | 4.45 | 5.28 | 5.83 | 6.97 |
| 330 | 1.89 | 2.88 | 3.69 | 4.39 | 5.20 | 5.74 | 6.87 |
| 340 | 1.86 | 2.84 | 3.63 | 4.32 | 5.12 | 5.66 | 6.77 |
| 350 | 1.84 | 2.79 | 3.58 | 4.26 | 5.05 | 5.58 | 6.67 |
| 360 | 1.81 | 2.76 | 3.53 | 4.20 | 4.98 | 5.50 | 6.58 |
| 370 | 1.79 | 2.72 | 3.48 | 4.14 | 4.91 | 5.43 | 6.49 |
| 380 | 1.76 | 2.68 | 3.44 | 4.09 | 4.85 | 5.36 | 6.41 |
| 390 | 1.74 | 2.65 | 3.39 | 4.04 | 4.78 | 5.29 | 6.33 |
| 400 | 1.72 | 2.61 | 3.35 | 3.99 | 4.72 | 5.22 | 6.25 |
| 410 | 1.70 | 2.58 | 3.31 | 3.94 | 4.67 | 5.16 | 6.17 |
| 420 | 1.68 | 2.55 | 3.27 | 3.89 | 4.61 | 5.10 | 6.10 |
| 430 | 1.66 | 2.52 | 3.23 | 3.85 | 4.56 | 5.04 | 6.03 |
| 440 | 1.64 | 2.49 | 3.19 | 3.80 | 4.51 | 4.98 | 5.96 |
| 450 | 1.62 | 2.46 | 3.16 | 3.76 | 4.46 | 4.93 | 5.90 |
| 460 | 1.60 | 2.44 | 3.12 | 3.72 | 4.41 | 4.88 | 5.83 |
| 470 | 1.58 | 2.41 | 3.09 | 3.68 | 4.36 | 4.82 | 5.77 |
| 480 | 1.57 | 2.39 | 3.06 | 3.64 | 4.32 | 4.77 | 5.71 |
| 490 | 1.55 | 2.36 | 3.03 | 3.60 | 4.27 | 4.73 | 5.66 |
| 500 | 1.54 | 2.34 | 3.00 | 3.57 | 4.23 | 4.68 | 5.60 |
| 550 | 1.46 | 2.23 | 2.86 | 3.40 | 4.03 | 4.46 | 5.34 |
| 600 | 1.40 | 2.13 | 2.74 | 3.26 | 3.86 | 4.27 | 5.12 |
| 650 | 1.35 | 2.05 | 2.63 | 3.13 | 3.71 | 4.11 | 4.92 |
| 700 | 1.30 | 1.98 | 2.54 | 3.02 | 3.58 | 3.96 | 4.74 |
| 750 | 1.25 | 1.91 | 2.45 | 2.92 | 3.46 | 3.83 | 4.58 |
| 850 | 1.18 | 1.79 | 2.30 | 2.74 | 3.25 | 3.60 | 4.31 |
| 900 | 1.15 | 1.74 | 2.24 | 2.66 | 3.16 | 3.49 | 4.19 |
| 950 | 1.11 | 1.70 | 2.18 | 2.59 | 3.08 | 3.40 | 4.08 |
| 1000 | 1.09 | 1.65 | 2.12 | 2.53 | 3.00 | 3.31 | 3.97 |

Reproduced from Rogers (2017) with permission.

## References


Bi, J. (2006). Statistical analysis for R-index. *J. Sens. Stud.* 21: 584–600.

Bi, J. and O'Mahony, M. (2007). Updated and extended table for testing the significance of the R-index. *J. Sens. Stud.* 22: 713–720.

Bi, J. and O'Mahony, M. (2020). R-Index critical value. *J. Sens. Stud.* 35 (4): 1–3.

Bi, J. and O'Mahony, M. (1995). Tables for testing the significance of the R-index. *J. Sens. Stud.* 10: 341–347.

Bi, J., Lee, H.-S., and O'Mahony, M. (2013). Statistical analysis of receiver operating characteristic (roc) curves for the ratings of the A-Not A and the same-different methods. *J. Sens. Stud.* 28: 34–46.

Bi, J., O'Mahony, M., and Lee, H.-S. (2013a). Nonparametric estimation of $d'$ and its variance for the A-Not A with reminder. *J. Sens. Stud.* 28 (5): 381–386.

Bi, J., Lee, H.-S., and O'Mahony, M. (2013b). Statistical analysis of ROC curves. *J. Sens. Stud.* 28: 34–46.

Brajkovich, M., Tibbits, N., Peron, G. et al. (2005). Effect of screwcap and cork closures on $SO_2$ levels and aromas in a sauvignon blanc wine. *J. Agric. Food Chem.* 53: 10006–10011.

Brown, J. (1974). Recognition assessed by rating and ranking. *Brit. J. Psychol.* 65: 13–22.

Choi, E.S., Lee, H.L., Kwon, B.K. et al. (2023a). Sensory test A-Not A rating signal detection: panel's sensory learning and stability of decision criteria. *Food Res. Int.* 167: 112592.

Choi, E.-S., van Hout, D., and Lee, H.-S. (2023b). Improving the performance of A-Not A sensory discrimination ratings by modifying sample presentation probability. *Food Qual. Prefer.* 104: 104748.

Dacremont, C. and Vickers, Z. (1994). Classification of cheeses according to their closeness to the Cheddar cheese concept. *J. Sens. Stud.* 9: 237–246.

Feng, Y. and O'Mahony, M. (2017). Comparison between American and Chinese consumers in the use of verbal and numerical 9-point hedonic scales and R-index ranking for food and personal products. *Food Qual. Prefer.* 60: 138–144.

Green, D.M. and Swets, J.A. (1988). *Signal Detection Theory and Psychophysics*, 3e. Los Altos, CA: Peninsula Publishing.

Harker, F.R., Norquay, C., Amos, R. et al. (2005). The use and misuse of discrimination tests for assessing the sensory properties of fruit and vegetables. *Postharvest Biol. Technol.* 38: 195–201.

Ishii, R., Vié, A., and O'Mahony, M. (1992). Sensory difference testing: ranking R-indices are greater than rating R-indices. *J. Sens. Stud.* 1: 57–61.

Jara-Sokis, F., Araya-Quesada, Y., O'Mahony, M., and Cubero-Castillo, E. (2023). Is the R-index method for eliciting preference measures from the 9-point hedonic scale fit for purpose? *Food Qual. Prefer.* 103: 104710.

Kim, M.-A., Chae, J.-E., van Hout, D., and Lee, H.-S. (2011). Discriminations of the A-Not A difference test improved when "A" was familiarized using a brand image. *Food Qual. Prefer.* 23: 3–12.

Kim, M.A., Chae, J.E., van Hout, D., and Lee, H.S. (2014). Higher performance of constant-reference duo–trio test incorporating affective reference framing in comparison with triangle test. *Food Qual. Prefer.* 32: 113–125.

Kim, M.-A., Dessirier, J.-M., van Hout, D., and Lee, H.-S. (2015). Consumer context-specific sensory acceptance tests: effects of a cognitive warm-up on affective product discrimination. *Food Qual. Prefer.* 41: 163–171.

Kim, I.A., Hopkinson, A., van Hout, D., and Lee, H.S. (2017). A novel two-step rating-based 'double-faced applicability' test. Part 2: introducing a novel measure of affect magnitude ($d'$ A) for profiling consumers' product usage experience based on Signal Detection Theory. *Food Qual. Prefer.* 59: 141–149.

Kim, M.-A., van Hout, D., Dessirier, J.M., and Lee, H.-S. (2018). Degree of satisfaction-difference (DOSD) method for measuring consumer acceptance: a signal detection measurement with higher reliability than hedonic scaling. *Food Qual. Prefer.* 63: 28–37.

Kim, M.-A., Lee, Y.-J., Kim, M.-S., and Lee, H.-S. (2022). Reminder–preference test, affective difference-preference test using reference framing with a brand: 1. Sensitivity comparisons with the same-different difference-preference test. *Food Res. Int. 155*: 111065.

Lawless, H.T. (1983). Contextual effects in category ratings. *J. Test. Eval.* 11: 346–349.

Lawless, H.T. and Malone, G.J. (1986). A comparison of rating scales: sensitivity, replicates and relative measurement. *J. Sens. Stud.* 1: 155–174.

Lee, H.-S. and O'Mahony, M. (2004). Sensory evaluation and marketing: measurement of a consumer concept. *Food Qual. Prefer.* 16: 227–235.

Lee, H.-S. and O'Mahony, M. (2007). Some new approaches to consumer acceptance measurement as a guide to marketing. *Food Sci. Biotechnol.* 17: 863–867.

Lee, H.-S. and Van Hout, D. (2009). Quantification of sensory and food quality: the R-index analysis. *J. Food Sci.* 74: 57–64.

Lee, H.-S., Van Hout, D.H.A., and O'Mahony, M. (2006). Sensory difference tests for margarine: a comparison of R-indices derived from ranking and A-Not A methods considering response bias and cognitive strategies. *Food Qual. Prefer.* 18: 675–680.

Lee, J.-K., Russell, C.G., Mohebbi, M., and Keast, R. (2022). Grating orientation task: a screening tool for determination of oral tactile acuity in children. *Food Qual. Prefer.* 95: 104365.

Lee, H.L., van Hout, D., and Lee, H.S. (2023). Improving the performance of A-Not AR discrimination test using a sensory panel: effects of the test protocols on sensory data quality. *Food Qual. Prefer.* 104: 104740.

Macmillan, N.A. and Creelman, C.D. (2005). *Detection Theory: A User's Guide*, 2e. UK: Cambridge University Press.

Matsuda, T. and Doty, R.L. (1995). Regional taste sensitivity to NaCl - relationship to subject age, tongue locus and area of stimulation. *Chem. Senses* 20: 283–290.

Meilgaard, M.C., Civille, G.V., and Carr, B.T. (2006). *Sensory Evaluation Techniques*, 4e. CRC Press.

Miettinen, S.M., Piironen, V., Tuorila, H., and Hyvonen, L. (2002). Electronic and human nose in the detection of aroma differences between strawberry ice cream of varying fat content. *J. Food Sci.* 67: 425–430.

O'Mahony, M. (1983). Adapting short cut signal detection measures to the problem of multiple difference testing. In: *Sensory Quality in Foods and Beverages, Its Definition, Measurement and Control* (ed. A.A. Williams and R.K. Atkins), 69–81. Chichester, UK: Ellis Horwood.

O'Mahony, M. (1988). Sensory difference and preference testing: the use of signal detection measures. In: *Applied Sensory Analysis of Foods* (ed. H. Moskowitz), 145–175. Boca Raton, FL: CRC Press, Inc.

O'Mahony, M. (1992). Understanding discrimination tests: a user-friendly treatment of response bias, rating and ranking R-index tests and their relationship to signal detection. *J. Sens. Stud.* 7: 1–47.

O'Mahony, M. and Hautus, M.J. (2008). The signal detection roc curve: some applications in food science. *J. Sens. Stud.* 23: 186–204.

O'Mahony, M., Garske, S., and Klapman, K. (1980). Rating and ranking procedures for short-cut signal detection multiple difference tests. *J. Food Sci.* 45: 392–393.

O'Mahony, M. (1974). A simple flow system for controlled gustatory stimulation avoiding stimulus delay or mixing. *Med. Biol. Eng.* 12: 547–568.

Parducci, A. (1963). Range-frequency compromise in judgement. *Psychol. Monogr.* 77: 1–50.

Parducci, A. (1965). Category judgement – a range-frequency model. *Psychol. Rev.* 72: 407–418.

Pirc, M., Maas, P., De Graaf, K. et al. (2022). Humans possess the ability to discriminate food fat content solely based on retronasal olfaction. *Food Qual. Prefer.* 96: 104449.

Robinson, K.M., Klein, B.P., and Lee, S.-Y. (2005). Utilizing the R-index measure for threshold testing in model caffeine solutions. *Food Qual. Prefer.* 16: 283–289.

Rogers, L. (2017). *Discrimination Testing in Sensory Science: A Practical Handbook*. Woodhead Publishing.

Shin, H.-K., Hautus, M.J., and Lee, H.-S. (2016). Unspecified duo-trio tests can be as powerful as the specified 2-AFC: effects of instructions and familiarization procedures on cognitive decision strategies. *Food Res. Int.* 79: 114–125.

Swets, J.A. and Elliott, P. (1964). *Signal Detection and Recognition by Human Observers*. New York: Wiley. Page 683.

Xia, Y., Song, J., Zhong, F. et al. (2023). The 9-point hedonic scale: using R-index preference measurement to compute effect size and eliminate artifactual ties. *Food Res. Int.* 133: 109140.

# 8

# Replicated Discrimination Testing

*Michael Meyners*[1] *and Bernard Thomas Carr*[2,3]

[1] *Discovery & Innovation Platforms, Procter & Gamble Service GmbH, Schwalbach am Taunus, Germany*
[2] *Carr Consulting LLC, Evanston, IL, USA*
[3] *National Wine and Grape Industry Centre, Charles Sturt University, Wagga Wagga, New South Wales, Australia*

## 8.1 Introduction

In applications, the number of assessors or consumers available for a discrimination test might be insufficient to warrant reasonable statistical power of such a test. As an alternative, repeated assessments of the same pair of products by the same individuals might be considered. While this might appear as an easy solution to address a potential lack of assessors, replicated discrimination tests require careful design and dedicated statistical methods for data analysis. An important implication of using replications is that observations might no longer be independent from each other, such that sample size and power determinations are no longer straightforward and standard analyses based on the binomial law may not be appropriate anymore. In this chapter, we discuss various watchouts when using replicated discrimination testing, and we provide an overview of available methods for data analysis including their respective pros and cons. We also address the potential for additional insights to be gained from replicated tests which are not available from unreplicated tests; the potential of the related analyses and interpretation is largely ignored in the literature. We will focus on replicated discrimination testing for the most part of this chapter, but briefly discuss the relatively few specificities of replicated preference testing in Section 8.7.

In what follows, we will start with some general discussion on advantages and limitations of replicated discrimination tests (Section 8.2). After a brief introduction of the required notation in Section 8.3, subsequent sections will focus on the data analysis. We discuss various analysis options with an emphasis on statistical testing ("Are there any differences at all?", Section 8.4) as well as estimation ("How big a difference is it? What is the proportion of discriminators?", Section 8.5); we will actually distinguish more strictly than usual in the literature between these two purposes, as the analyses to be performed might be different, and the second aspect offers potential additional insights. Section 8.6 briefly addresses specific issues and limitations when the objective is to show equivalence (also referred to as similarity or parity at times), with the variations of the discussed methods

*Discrimination Testing in Sensory Evaluation*, First Edition. Edited by Lauren Rogers, Joanne Hort, Sarah E. Kemp and Tracey Hollowood.

required for replicated preference tests being discussed in Section 8.7. Two examples are given in Section 8.8, before we provide conclusions (Section 8.9) and a short summary of key recommendations (Section 8.10). Finally, Section 8.11 provides a glossary of terms and mathematical notation used in this chapter for ease of reference.

## 8.2 General Considerations and Study Design

### 8.2.1 Fatigue and Learning Effects

The most common rationale for using replications in discrimination tests is the lack of a sufficient number of independent assessors and the limited statistical power that comes with it. To increase the power, asking individuals twice or more often appears a viable remedy in many situations. Yet, it is obvious that there are some practical limitations to this, most notably related to learning or familiarization and fatigue. Exposing individual assessors multiple times to the same two products might sharpen their senses with regards to which product attributes to pay attention to in order to identify the correct sample, and this might result in an increase in the sensitivity of the evaluations. Publications on the topic of learning in replicated discrimination tests are sparse, but many would consider it likely to occur. Kim et al. (2012) show a positive effect on discrimination rates from familiarization, so it is worth bearing it in mind as a potential limitation. At the same time, repeated exposure over a short time period could easily lead to fatigue and hence decreased sensitivity. These two effects could occur in isolation or simultaneously, and it is impossible to say which of the two would outweigh the other, if either. Accordingly, all possible measures should be taken to avoid these effects in practice by, e.g., warranting sufficient breaks between assessments and avoiding over-exposure of individuals with the task. Still, any such effects can never be completely ruled out. Consequently, the results from a replicated discrimination test may not be directly comparable to those from a test without replications. The investigator should carefully think upfront which of such effects are likely to occur and note that this might affect interpretation. If, for example, the objective is to show product differences, some fatigue might be acceptable: it decreases the probability to detect the correct sample and hence would provide a conservative solution, i.e., if statistically significant differences can still be established (despite the potentially decreased true discrimination probability), it is safe to conclude the products are truly discriminable. In turn, if the learning effect is more pronounced, the results might no longer be representative of what the related unreplicated discrimination test might be representative of; more specifically, the difference might be overestimated and "proven," even though it is not relevant or consumer-meaningful. Learning can obviously occur only in the presence of perceptible differences, so in that sense, it is fair to conclude that products differ. However, in many projects, even if not explicitly stated, sample sizes (and hence power) are determined based on experience about observed differences in an unreplicated test that are consumer meaningful, hence for the interpretation we may want to bear this in mind to avoid overinterpretation of the observed discrimination rates. The opposite applies when the objective is to show equivalence (see also Chapter 5 of this book), i.e., that two products are not too different. Fatigue will increase the probability to unduly establish equivalence, as a potential difference will be underestimated. An effect from learning would not harm too much in this situation: if equivalence can still be established (when learning is the dominant bias), the result is reasonably reliable as the setup with replications is more sensitive due to the learning than the same setup using independent assessors throughout.

### 8.2.2 Randomization and Presentation Order

General considerations for designing discrimination tests should be adhered to (see Chapter 2 of this book) as well for replicated tests and are hence not discussed in further detail. It is worth emphasizing that the order of samples should ideally be randomized among all trials, including the replications (see Kunert and Meyners 1999). Any assessor must not (in a planned way) always receive the same order, but it should not be ruled out by the design that any assessor might receive the same order multiple times, even if in direct sequence. Though statistically not ideal, rotation schemes warranting that each possible order is used equally often might be used (as, e.g., suggested in ISO 4120; International Organization for Standardization 2021) if this avoids that the same order is given to all assessors, or that the same order is given to a single assessor in all their replications.

### 8.2.3 Replicated Discrimination Tests to Monitor Panelist Performance

There are a few situations where replications in discrimination testing have value by themselves, offering benefits beyond the gain in statistical power which could also (and usually more sustainably) be obtained by employing more assessors. One of these situations is when individual performance of assessors in discrimination testing should be evaluated. In this case, it might not be necessary to consider only replications in a strict sense, but it could include data from various tests using the same protocol and the same product category. In that situation, replications will allow estimation of the individual assessor's discrimination rate, and by comparing it to others, it might identify re-training opportunities and the like.

### 8.2.4 Benefits from Replications of Preference Tests

The other situation where replications can add insights is in preference tests. Without replications, it is impossible to decide whether the population of interest is homogeneous and has a slight overall preference for product A, say, or whether there are clearly distinct segments of possibly different sizes, one of which strongly prefers A while the other strongly prefers B. This information can be crucial in many projects but is impossible to obtain from an unreplicated preference test. We defer further discussions on this topic until Section 8.7.

### 8.2.5 Required Sample Sizes

Determination of appropriate sample sizes for replicated discrimination tests is not straightforward, unfortunately, as using replications imposes a more complex correlation structure onto the data. Some guidelines have been established in the past, typically based on worst-case scenarios (e.g., Meyners and Brockhoff 2003). We will provide further discussion in the subsequent statistical section, and only note here that replicated evaluations cannot be considered as independent for sample size determination purposes. This implies that in replicated discrimination tests, the total number of assessments needs to increase relative to an unreplicated test in order to achieve a comparable power. As an example, if the project requirements dictate using a triangle test to establish product differences at $\alpha$-level 5% and a power of 90% is required to detect a proportion of discriminators of 37.5% (Schlich 1993), without replications 35 assessors would be needed. However, it will not suffice to have 18 assessors performing two replications, or seven assessors performing five. Instead, from Table 1 in Meyners and Brockhoff (2003), we find that we might alternatively use 20 assessors with two replications or 11 assessors with five replications, resulting in 40 and 55 evaluations in total, respectively. In other words, if replications should be used, we need to counter-balance the implicit potential loss in power for a given total number of assessments due to potential assessor heterogeneity by increasing the total number of assessments. We again defer more details to Section 8.4.

### 8.2.6 Replications can Offer Additional Benefits and Insights

Though the focus of analyzing replicated discrimination test data is often on formal statistical significance testing, we do believe that this is too limiting and easily misleading. Instead, we believe that more useful insights can typically be obtained by powerful estimation procedures that will help to better understand the underlying phenomena, which includes the overall discrimination (or possibly preference) rate, but also the potential differences among assessors or even segmentation. These aspects are heavily underutilized in applications, and in Section 8.5, we will emphasize approaches that leverage the availability of replicated data beyond the pure addition of data to increase statistical power for significance testing.

## 8.3 Notation

### 8.3.1 Basic Notation: Number of Assessors and Replications

To allow for a concise yet exact description of the statistical methodology, we introduce the following general notation, which is largely adopted from Meyners and Brockhoff (2003). While all notation is explained in the following text, for ease of referencing Section 8.11 at the end of this chapter provides a glossary of key terminology as well as a table with explanations of the mathematical notation used.

Let $n$ be the number of assessors and $k_i$ the number of replications which assessor $i$ has performed. In many applications, the number of replications will be the same for all assessors (i.e., $k_i = k$, for all $i$), such that the index $i$ can be dropped. We refer to the case of $k = 1$ as the unreplicated test, while $k > 1$ is a scenario with replications. Note that $k$ refers to the total number of replicates, including the first assessment. As an example, if $k = 3$, we refer to it as three replicates, though it is 1 initial assessment and 2 (true) repeats of the task. We will refer to these as first, second, and third replication, respectively.

### 8.3.2 Correct and Incorrect Answers and Guessing Probability

We consider a forced-choice discrimination test and first consider the case in which an assessor's answer can be classified as "correct" or "incorrect." (We will come back momentarily to the situation of two-way paired comparisons like, for example and most notably, paired preference tests, in which no "correct" answer exists.) The true success probability across all evaluations from assessors randomly drawn from some overall population of interest is denoted by $p_c$, whereas $p_0$ denotes the probability to obtain a correct result by pure guessing. For example, in a triangle or tetrad test $p_0 = 1/3$, while in a duo-trio test we have $p_0 = 1/2$. We refer to $p_0$ as the "guessing probability" at times in the remainder of this chapter.

### 8.3.3 Success Probability and Probability of True Discrimination

The parameter $p_c$ refers to the overall success probability of the population of interest, whereas $p_d$ refers to the probability of true discrimination, i.e., the probability that the assessor selects the correct sample based on appropriate perception, as opposed to pure guessing. An individual assessor may have a success probability that differs from $p_c$, and we refer to it as $p_i$. These probabilities come from a general distribution $P$ on the interval $[p_0, 1]$, with the only general assumption being that the mean of that distribution is $p_c$, and the individual $p_i$'s are independent from each other, which is warranted by randomized sampling of assessors from the population of interest, and by warranting that they do not influence each other's assessments.

### 8.3.4 Number of Successes

Finally, to describe the observed data, $X_i$ denotes the number of successes of assessor $i$, and $X = \sum_{i=1}^{n} X_i$ the total number of successes across assessors. An implicit assumption being made is that all $X_i$ are binomially distributed with parameters $k_i$ and $p_i$, respectively, which implies that no effect of fatigue or learning is present.

### 8.3.5 Replicated Discrimination Tests as a Hierarchical Model

It may be helpful to think about the underlying model for replicated discrimination tests as a hierarchical one: On a higher level, the success rates follow a certain distribution $P$; a single draw $p_i$ is taken by selecting an assessor. On the lower level, the outcomes from the multiple tests performed by assessor $i$ are $k_i$ draws from a Bernoulli distribution[1] with success rate $p_i$, such that the number of successes from this assessor is binomially distributed with parameters $k_i$ and $p_i$. For illustration, consider the success rates across a population of assessors for a triangle test to be uniformly distributed between 1/3 (the lowest reasonable value, see Brockhoff 2003; Meyners 2007a) and 1. Once an assessor $i$ is selected, they will have their individual success rate, say $p_i = 0.6$, which is assumed to be constant for this assessor (a reasonable assumption provided proper design and randomization, and in absence of learning or fatigue effects). Performing the task $k_i = 5$-times will result in five random draws from a Bernoulli distribution with parameter 0.6, such that overall, the number of successes among the five tasks of assessor $i$ follows a binomial distribution with

1 The Bernoulli distribution is the same as the binomial distribution with a sample size of $n = 1$, i.e. only a single observation which takes either value 0 or value 1. The sum of $k_i$ independent Bernoulli distributions with the same probability parameter, here $p_i$, is binomially distributed with parameters $k_i$ and $p_i$.

parameters 5 and 0.6. Another assessor $j$, say, might have a different success probability, for example $p_j = 0.8$, and perform $k_j = 8$ replications. The number of successes obtained from this assessor is then binomially distributed with parameters 8 and 0.8. This illustrates the aforementioned hierarchy: at the higher level is the selection of assessors, which gives different individual success rates which are random draws from some unknown distribution on $[p_0, 1]$. At the lower level, replicates from one assessor, once selected and with success probability $p_i$, follow a binomial distribution with their number of replicates $k_i$ and success probability $p_i$.

### 8.3.6 Success Probabilities for Preference Tests

In a preference test, there is no correct answer. Instead, without loss of generality, we consider $p_c$ as the overall probability of selecting product A. This probability can then be considered in the same way as the probability for a correct answer in a discrimination test; for the sake of readability, we refrain from distinguishing between the two throughout this chapter except for the dedicated Section 8.7 on preference tests. $p_0$ then reflects the situation that an assessor truly has no preference or is not at all able to discriminate the samples. In paired preference tests, $p_0 = 1/2$, as it is in most other practical applications of preference testing. It is important to note that, unlike a discrimination test, in a preference test the probability for selecting product A can fall below $p_0$, namely when an assessor prefers B. Therefore, the support for the distribution $P$ of probabilities to select A ranges from 0 to 1 (as opposed to $[p_0, 1]$ for a discrimination test).

### 8.3.7 Assessor Heterogeneity

Along with this notation, it is essential to define the concept of assessor heterogeneity, which is important for discrimination tests using replications. It denotes the situation in which the individual success probabilities, $p_i$, vary across assessors, formally implying $p_i \neq p_j$ for at least one pair $(i, j)$. In the above example on the hierarchical model, we considered two assessors, one with $p_i = 0.6$ and the other with $p_j = 0.8$. With that, assessor $j$ has a higher success probability than assessor $i$;[2] as these rates are different, they constitute heterogeneity in assessors' success rates. If all assessors have the exact same true success rate,[3] there is no heterogeneity (sometimes referred to as assessor homogeneity instead). In other words, assessor heterogeneity in this context means that assessors vary in their sensitivity and hence in their true success probabilities for the (discrimination) task that is repeatedly presented to them.

---

2 Note that this does not necessarily imply that we truly observe more correct answers from assessor $j$ than from assessor $i$. Due to random variation within the individual binomial distributions, we may observe four out of five correct if the true success rate is $p_i = 0.6$, but only three out of five correct if it is $p_j = 0.8$. It is on the long run (or on average) that assessor $j$ will outperform assessor $i$, but not necessarily in every instance.

3 Again, note that this is about the true but unknown underlying success rates which refers to the higher level of the hierarchical model. It does by no means imply that observed success rates are the same, as at the lower level of this model, the binomial distribution should and will lead to variation in the observed rates. Observed success rates might even fall below 1/3 in a triangle test as will be discussed later (Section 8.4), while the underlying true success rates cannot.

A consequence of existing assessor heterogeneity is that at least some assessors necessarily have a success probability different from $p_0$ as not all can be equal to this value. Consequently, some assessors are able to truly discriminate the samples, and thus assessor heterogeneity generally implies product differences. In contrast, assessor homogeneity denotes the situation in which all assessors have the identical success probability, i.e., $p_i = p_j = p_c$ for all pairs $(i, j)$. In this situation, $p_c$ may be equal to $p_0$ but it does not have to. In fact, in practice we deem it very unlikely that all assessors have the exact same success probability unless it is $p_0$, as there will always be some level of differences in sensitivity among the assessors if the products perceptually differ at all. Note that the definition of assessor heterogeneity ignores the degree of differences among success rates. In theory, it is sufficient if a single assessor has a marginally different success probability than all others. In practice, that would not be meaningful and almost certainly not detectable anyway with reasonable sample sizes. Whether the degree of assessor heterogeneity is meaningful and (consumer-) relevant is to be determined based on project objectives. However, for the statistical tests in the next section, it is important to note that there is a clear formal distinction between assessor heterogeneity and assessor homogeneity per the above definition, and that this necessarily ignores the magnitude of potential heterogeneity. As an example, if all assessors have a true success rate between 0.59 and 0.61 (but not all equal), formally speaking assessors are heterogeneous. However, for all practical matters these success rates are sufficiently close to be considered the same, and a statistical test with reasonable sample sizes will not be able to prove assessor heterogeneity. In this case, though assessors are heterogeneous, the differences between the success rates are so small that we deem the magnitude of this heterogeneity small and thereby not meaningful. In contrast and as the most extreme scenario, consider half of the assessors to be unable to detect a difference and hence only guess, such that their success rate is equal to $p_0$ (i.e., equal to the guessing rate like, e.g., 1/3 for the triangle test). Assume the other half of the assessors always identifies the correct sample, leading to a success rate of 1. In that case, the magnitude of assessor heterogeneity is maximized, as some assessors always answer correctly, while the others only guess and succeed by chance alone, if they do. Such heterogeneity is more important to learn about in applications, and statistical tests will be much more likely to detect it. Real world scenarios are typically somewhere between these extremes, of course, but the examples illustrate that small differences in assessors' success rates (and hence small heterogeneity) are less relevant than large differences and thereby large heterogeneity.

### 8.3.8 Further Notation

Some additional notation is specific to certain models (most notably Beta-Binomial (BB) models, see Section 8.4.6) and will be introduced along with those to avoid unnecessary duplication; that notation is also included in the overview in Section 8.11.

## 8.4 Statistical Testing to Show Product Differences

### 8.4.1 Foundations and the Need for Proper Randomization

Statistical testing to establish significant product differences using replicated discrimination tests has been exhaustively addressed in the literature. We start by introducing the simplest analysis and move to more advanced methods in the sequence. Note that this section is about formal statistical testing for product differences; whether an observed difference is meaningful will be ignored for now but revisited in Section 8.5. Further, this section is only about showing differences, not showing parity, which again is the topic of a separate Section 8.6. In what follows, we limit the amount of technical details to a bare minimum; model details can be found in the respective literature and will be skipped as they are likely not of interest to most readers, given that the respective calculations can easily be performed using readily available software. We show how to use R and the contributed package `sensR` (Christensen and Brockhoff 2023) with examples in Section 8.8.

In the following, at times we emphasize the need for a proper randomization and execution of the study. It is easy to conceive compelling examples that render the binomial as well as any other model considered in this chapter (and in the literature) invalid from a statistical perspective if these conditions are not met, and the violations of model assumptions could lead to both over- as well as underdispersion (see Kunert 2001, for an example). All reasonable methods that have been (or could be) suggested for the analysis of replicated discrimination tests equally rely on these assumptions.

### 8.4.2 The Naïve Binomial Test

A naïve approach of statistical testing to show product differences is to use the binomial test based on all available observations, i.e., $nk$ observations in case the same number of replications is available from each assessor. In fact, Kunert and Meyners (1999) show that this is indeed a valid test if product differences are to be shown. Consequently, if the binomial test shows a statistically significant result, it is reasonable to conclude that the products statistically significantly differ. In a nutshell, this is because under the null hypothesis of no product differences, there is no way to discriminate the products and therefore assessor heterogeneity cannot occur (as all assessors have the same success probability of $p_0$, see end of Section 8.3). Provided proper randomization and execution of the study, this renders all observations independent, and hence using the binomial law is appropriate. It is important to note that this implies in particular that the binomial test is valid in this context, i.e., that it respects its nominal type I error rate $\alpha$. In other words, the probability to conclude for perceivable product differences when, in fact, there are none, does not exceed the pre-defined level $\alpha$. (A test would be considered invalid if the probability to conclude for product differences in case there are truly none exceeds the nominal level $\alpha$, e.g., when the test uses a 5% level, but the false positive rate is 10%; see Meyners and Hasted 2021, 2022, 2023, for examples in a remotely related context.) Unlike repeatedly yet falsely claimed at times (Bi and Ennis 1999b; Bi et al. 2000; Lee and O'Mahony 2006; Cubero-Castillo et al. 2019), alternative methods do not offer any additional or better protection against type I errors (except possibly via unnecessary and not quantifiable conservatism). For instance, as much as the critical values offered by, e.g., Bi and Ennis (1999b, 2001) differ from those based on the binomial law, they are unnecessarily conservative: They are stricter than those from the binomial distribution as exemplified in their concluding example, and as the binomial test respects the type I error rate, the approach by Bi and Ennis (1999b, 2001) is necessarily overly conservative.

### 8.4.3 Potential Power Loss in the Binomial Test and Limitations for Equivalence Problems

Kunert and Meyners (1999) use an extreme example to illustrate that using replications can have a large detrimental effect on the statistical power of the naïve binomial test in the presence of assessor heterogeneity, when the objective is to prove product differences.[4] This huge decrease in power is only observed in extreme cases, however, while in most more relevant scenarios for practical applications, the decrease is more moderate and might be acceptable, as will be discussed later. Brockhoff (2003) uses simulations to show that within the models considered, the drop in power of the binomial test is usually not very large; a reasonable strategy would hence be to use a not too large number of replications (maybe $k \leq 5$) along with a still reasonable number of assessors such that the total number of trials $nk$ increases moderately versus the recommended sample size for an unreplicated test. By that approach, the power loss arising from heterogeneity will be compensated by a slightly-to-moderately larger total number of evaluations. This is quantified by Meyners and Brockhoff (2003), who offer tables that allow to select $n$ and $k$ to obtain a comparable power by using $k$ replications relative to the unreplicated test. It is worth mentioning that there are (upper) limits regarding what power might be achieved with a certain number of independent assessors $n$, even if the number of replications $k$ tends to infinity, which is reasonable and in line with the extreme example discussed by Kunert and Meyners (1999). Brockhoff (2003) summarizes his findings by arguing that with a small number of independent assessors, the achievable power is limited regardless of the size of $k$. Hence it does not appear as good advice to heavily sacrifice on the number of assessors $n$ and attempt to counterbalance that with large numbers of replications. Yet, using a few replications, provided the products under investigation and the test setup allow this, can increase the power, and allow for a moderate decrease in the number of assessors employed in a study.

It is worth emphasizing that the rationale for the binomial test being a valid approach does not equally apply to equivalence problems (see, e.g., Meyners 2012, for an overview) or for evaluations of statistical power. For these scenarios, evaluations are for the case of product differences of a certain size, and those may or may not be (partly) due to assessor heterogeneity. In the presence of assessor heterogeneity, the binomial law is no longer applicable, and consequently power determinations or equivalence tests based on the binomial distribution are not reliable. In particular, this holds also for the discouraged (see Meyners 2012) but common practice of concluding on product similarity in case of absence of a statistically significant result, as this relies on the prior power determination again. We provide more details on the evaluation of equivalence and the entailed difficulties when using replications in Section 8.6.

### 8.4.4 Reduction of Sample Size in the Analysis to Account for Heterogeneity

Primarily for sake of completeness, we mention another simple model by Brockhoff and Schlich (1998) who suggest a correction of the sample size based on an estimate of the assessor heterogeneity. We do not provide details here, as Kunert and Meyners (1999) show

4 In their example, in presence of large assessor heterogeneity, using 100 assessors without replications, i.e. $k = 1$, might give a power of >99% as opposed to using one single assessor with $k = 100$ replications resulting in a power of 52.5% only due to the high likelihood that the single assessor chosen might be one of those insensitive ones that are not discriminating and hence only guessing.

the correction to be unnecessary and hence detrimental to power for the same reasons as mentioned above, namely the naïve binomial test being a valid test of the null hypothesis of no product differences.

### 8.4.5 A Confidence Interval-Based Approach

Kunert (2001) provides a (conservative, i.e., wider than necessary) confidence interval for the proportion of discriminators or the discrimination rate $p_d$, which could easily be adopted to a confidence interval for the proportion correct or success rate, $p_c$, if desired. Noting the duality between confidence intervals and statistical tests, such a confidence interval could be used to establish a test for differences: if the confidence interval for the success rate does not contain $p_0$, we can reject the null hypothesis and conclude that the products must be noticeably different. In turn, if $p_0$ is included in the confidence interval, we cannot reject that null hypothesis. The test could alternatively be considered for the discrimination rate, for which we would conclude on product differences if this interval does not contain the value 0. For a one-sided test like, e.g., the triangle test, a one-sided interval would be used. To the best of our knowledge, Kunert's (2001) confidence interval has not been widely adopted, though. This might be due the fact that many practitioners prefer to report and interpret p values (or use tables with critical values for statistical testing, as provided in, e.g., ASTM International (2020) rather than confidence intervals, despite the latter being more informative.

### 8.4.6 The Beta-Binomial Model

The more advanced models suggested for replicated discrimination tests use two parameters, one representing the average success probability across assessors (also referred to as location parameter), and the other representing assessor heterogeneity (variance). Among the most commonly used methods is the BB model, properties of which were discussed by Skellam (1948) and which was introduced for replicated triangle tests by Harries and Smith (1982) and later made more popular by Ennis and Bi (1998). A major reason for the popularity of the BB model is likely its simplicity from a mathematical perspective. Without going into much detail, the Beta distribution is the conjugate prior to the binomial distribution in a Bayesian context,[5] which means that the posterior distribution is a Beta

5 Using a conjugate prior implies that, within that model, the posterior distribution is from the same probability distribution family. In this context, using the Beta distribution as prior distribution implies that the posterior distribution can be analytically determined (rather than using computational-intensive simulations using Monte Carlo methods, see, e.g., Robert and Casella 2010) and is hence mathematically simple. The posterior distribution is again a Beta distribution, which aggregates the prior information ("What did we assume the distribution of assessors' success rates to be like?") with the observed data ("What did the data tell us about the distribution of the assessors' success rates?") into a combined, posterior information ("Combining our prior beliefs with the observed data, what do we now know about the distribution of assessors' success rates?"). The stronger the prior information and the less data, the more the posterior distribution will resemble the prior distribution. If a weak prior is used and more data are available, the posterior distribution will be mainly influenced by the data, illustrating the fine-tuning options in Bayesian approaches. Using a Beta prior distribution warrants that the resulting posterior distribution is again a Beta distribution, with all its pros (e.g. simplicity of notation in a closed form, simple derivation of its parameters) and cons (many relevant shapes cannot be represented by a Beta distribution, even if the data strongly suggests so, see Meyners 2007b).

distribution again and easily determined. This used to be a major advantage in times of limited computing power (the technically interested reader is referred to a textbook on Bayesian methods, e.g., Gelman et al. 2004; Marin and Robert 2014). The Beta distribution is relatively flexible with regard to the various shapes it can take, but to the best of our knowledge there is no general theory suggesting that the success rates in a discrimination test should be Beta-distributed.

The underlying model postulates that the distribution $P$ of success probabilities across assessors is a Beta distribution. While in its general form the Beta distribution uses parameters $a$ and $b$, in the context of replicated discrimination tests it is typically re-parameterized such that $\mu = a/(a+b)$ is the expected value and hence represents the average success probability $p_c$ across assessors (this is the location parameter), while $\gamma = 1/(a+b+1)$ represents the heterogeneity among assessors, taking values between 0 and 1 (variance parameter). In this model, assessor heterogeneity, i.e., when the $p_i$'s vary across assessors, is, strictly speaking, established whenever $\gamma > 0$, and is more pronounced for larger $\gamma$. In contrast, assessors would be considered (perfectly) homogeneous if they all have the exact same success probability. In that case, $\gamma = 0$. In practice, small values of $\gamma$ would still be considered as indicating reasonable homogeneity, while larger values, typically also statistically significantly different from 0, would be considered as evidence for noteworthy heterogeneity. We are not aware of any general guidelines regarding a potential threshold, but we note that presence or absence of a statistically significant difference from 0 should not be considered a sole criterion in applications, as statistical significance depends heavily on the sample size. In case of replicated discrimination tests, this is composed of both the number of assessors $n$ and the number of replications $k$. The estimation procedures in Section 8.5 may provide better guidance on how to interpret respective data.

### 8.4.7 The Chance-Corrected Beta-Binomial Model

In a BB model for a replicated discrimination test with no difference between products, $\mu$ would equal $p_0$, the guessing probability, and it could maximally take a value of 1. Regarding the minimum value $\mu$ might take, there has been quite some discussion in the literature. In the original proposition by Harries and Smith (1982) as well as in the work by Ennis and Bi (1998), $\mu$ was allowed to vary freely in [0, 1]. Yet, Kunert and Meyners (1999) argued that the success probability in a properly randomized and executed study using replicated discrimination tests cannot fall below $p_0$, and that hence values below the guessing probability are not reasonable. (Note that this is different for replicated preference tests, where $\mu$ can indeed take values in [0, 1], as will be discussed in Section 8.7.) An alternative approach suggested by Brockhoff (2003) and later also by Meyners and Brockhoff (2003) and Meyners (2007a) confines the range of the Beta distribution via a simple linear transformation to warrant $\mu$ only taking values in $[p_0, 1]$. In earlier work, this model was referred to as corrected Beta-Binomial model, but this terminology was mis-interpreted at times to suggest that the BB model was generally flawed and needed to be fixed. Instead, it is the application of the BB model in replicated discrimination tests that is flawed without a correction for the guessing probability, which is clarified by the terminology introduced by Meyners (2007a). He refers to this model as the chance-corrected Beta-Binomial (cBB) model, a term that we will use in this chapter. Note that in the cBB model, the parameter $\mu$ of the Beta distribution no longer represents the success rate, but the probability of success above the guessing rate. In other words, in the cBB model, $\mu$ represents the probability of true discrimination, $p_d$, while it represents the success probability $p_c$ in the standard BB model. No product differences generally occur if $p_d = 0$, which implies $\mu = 0$ in the cBB model, but $\mu = p_0$ in the BB model. This is an important difference to bear in mind when interpreting results from an analysis using such a model. Bi and Ennis (1998, 1999a) provide considerations on power within the BB model, based on assumptions made for both the location (i.e., average success rates) and the overdispersion parameter (i.e., variability between individual assessors' success rates). Studying both parameters simultaneously is important as, e.g., in an application of a triangle test to investigate product differences, it might be important to understand whether an average true discrimination rate of 50% represents the individual rates of consumers well, or whether some might never while others almost always discriminate. If the latter are the heavy users of the product and will notice the changes made for, e.g., cost-saving reasons, this might be a disaster for the product's success, while it might be less critical if all consumers only notice the difference at times even when products are presented back-to-back in a triangle test, and hence likely less often under normal usage conditions.

Some authors have argued against the cBB model as observed success rates might fall below $p_0$ (Liggett and Delwiche 2005). While it is true that observed rates might fall below this value, the theoretical ones will not (again, provided proper design and execution of the experiment). Of course, with a guessing probability of $p_0 = 1/2$ in a duo-trio test, say, we may nevertheless observe assessors always failing to select the correct sample; they are just guessing, and with a guessing probability of 1/2, there is a certain likelihood to observe no correct answer or, more generally, less than 50% of correct answers. Yet, the underlying theoretical success rates are no different from 1/2 if no difference is perceived (see also Meyners 2007a). Hence, the rationale for using the classical BB model is not compelling and we consider it conceptually flawed for replicated discrimination tests, therefore focusing on the cBB model instead. We will revisit the use of the BB model and some generalizations for replicated preference tests in Section 8.7.

### 8.4.8 Heterogeneity Implies Product Differences

For both the BB and the cBB model, Meyners (2007a) argued that in order to investigate product differences, both parameters, $\mu$ and $\gamma$, need to be considered. This is due to the reasoning mentioned earlier that assessor heterogeneity implies product differences, with the same reservation as before that this does not take relevance of the heterogeneity into account. Consequently, the overdispersion parameter cannot be ignored; if it were, it would imply that one might conclude that there are no significant perceivable differences (i.e., the overall success rate is not significantly different from $p_0$), while at the same time concluding (based on the overdispersion being significant) that consumers differ with regard to their success rates, i.e., they vary in their sensitivity with regard to the task. This would establish an obvious contradiction: if they are not all the same, some must be different from $p_0$, and hence at least some assessors must be able to perceive some product differences. For further details and a contrast to the interpretation by Ennis and Bi (1998) of their simulation results, we refer to Meyners (2007a). The same rationale applies to other models proposed in the literature (e.g., Brockhoff 2003). Consequently, in a BB or any other multiparameter model, to investigate product differences we should perform tests on both (all) parameters. Statistically, it would be possible to confine evaluation to one parameter (assuming only differences in location exist, i.e., no heterogeneity), but that would render the use of the more advanced model useless: we would implicitly assume that there is no heterogeneity, in which case we do not need to model it and the simple binomial test on all evaluations will be perfectly appropriate. However, if we take the second parameter into account, we need to consider the arising multiplicity problem, implying that we need to correct for the multiple tests of the same hypothesis (no product differences) by, e.g., applying a significance level of $\alpha/2$ for each hypothesis according to Bonferroni's correction. That, in turn, will decrease the power of the test, however (see Meyners 2007a, section 4 for more details). Overall, we do think that either both parameters should be tested in any such model, or a simpler model should be used. Some authors explicitly or implicitly suggest conducting a test for the overdispersion parameter first or applying a threshold for the estimated overdispersion, and continue with a binomial test if the test is not significant or the overdispersion does not exceed the threshold (e.g., Brockhoff and Schlich 1998; Brockhoff 2003). This is discouraged, however, as it might render the ultimate test liberal (i.e., invalid per the above-mentioned definition, as the test does not maintain the type I error rate) as the test procedure is chosen based on the observed data; this issue can easily be confirmed by simulations (not reported here) under the null hypothesis.

### 8.4.9 Generalized Linear Mixed Models

Hunter et al. (2000) suggest a generalized linear mixed model (GLMM), which Brockhoff (2003) considers along with its chance-corrected version (cGLMM), following the same rationale as described above for the BB model. It differs from the BB model in that the success rates of assessors are not assumed to follow a Beta distribution, but a normal distribution with mean equal to the true success rate $p_c$ and some variance $\sigma^2$. A GLMM-type model uses a normal distribution to model the assessors' success rates, which offers lower flexibility in terms of potential shapes relative to the BB model (the Beta distribution is more flexible in that regard than the normal distribution). In contrast, the GLMM models are more natural in the context of Thurstonian models (Thurstone 1927; Ennis 1993), which also postulate normal distributions, and hence is closer to the respective models broadly used to make discriminations tests better comparable across different protocols (refer to Chapter 6 in this book for more details on Thurstonian models). While Brockhoff (2003) concludes that the GLMM models estimate heterogeneity more extremely than the BB models do, the differences in performance (most notably power) seem rather marginal (see also Cochrane et al. 2005). Given the limited adoption of the GLMM models in applications, we will not go into further details here.

### 8.4.10 The Common Limit Model

Brockhoff (2003) identifies a so-called common limit model, which is the same as the model used by Kunert and Meyners (1999) for their extreme example regarding power. In that model, assessors will either always or never truly discriminate; the good assessors will hence always identify the correct sample, while the poor assessors will guess throughout and hence choose the correct sample randomly with probability $p_0$. This model has been determined by Brockhoff (2003) as the one that the BB and GLMM models converge to as overdispersion (assessor heterogeneity) tends to infinity (in the BB models, this means that $\gamma$ tends to 1).

### 8.4.11 The $\chi^2$ Test

Kunert and Meyners (1999) briefly discussed using the well-known $\chi^2$ test for replicated preference tests, an idea which was elaborated upon further by Meyners (2007b). It can be applied if the number of replications is the same for each assessor, namely $k$. In that case, the expected number of assessors with $0, 1, \ldots, k$ correct answers under the null hypothesis of no product differences can be easily determined from the binomial law, and then compared with the truly observed numbers via the well-known $\chi^2$ goodness-of-fit test (Cochran 1952; see also, e.g., Agresti 2002; Bilder and Loughin 2015, or any other suitable textbook). It should be noted that the $\chi^2$ goodness-of-fit test, as a rule of thumb, requires expected values of 5 or more per category, which is rarely fulfilled in the context of replicated discrimination tests, so using Monte Carlo simulations is recommended. However, even with that complication it is easily implemented, as will be shown in the example in Section 8.8. The approach is also shown to be powerful across a variety of distributions of the assessors' success rates, including shapes that none of the other models would allow for like, e.g., multi-modal mixture distributions that avoid huge weights on the extremes (i.e., success rates in a discrimination test of either $p_0$ or 1). In a simulation study, it was also found to be more powerful than, e.g., the BB model even if the data truly came out of a BB distribution (Meyners 2007b). It has not been adopted widely, though, so we will not go into further detail here. An important side result of the related simulation study is that, though inferior to the $\chi^2$ test in many situations, the naïve $nk$ binomial test was not inferior or even superior in power relative to the BB model. Meyners (2007b) attributes this to the fact that in the BB model, two parameters have to be considered simultaneously, which appears to rather reduce than enhance power. The limitation of this approach is that it is only easily available for balanced scenarios where all assessors performed the same number $k$ of replications. It could be generalized to other scenarios by aggregating the $\chi^2$ statistic for different tables (one for each number of replications occurring) but would require a more complex approach and noteworthy programming to execute.

### 8.4.12 Further Methods and Recommendations

For the sake of completeness, we also refer to the approach by Linander et al. (2019). It relies on multiple tests performed against the same control, and it uses a more advanced GLMM model to investigate assessor effects as well as assessor * product interactions, thereby generalizing the GLMM idea (Brockhoff 2003) to replicated discrimination testing in an advanced Thurstonian framework (Thurstone 1927; Ennis 1993), building on earlier work by Brockhoff and Christensen (2010) on unreplicated tests. As we focus on comparisons between two products here, we refer the interested reader to Linander et al. (2019) for details.

Overall, we conclude that the naïve binomial test is a valid (Kunert and Meyners 1999) approach. It was also found to be powerful (Brockhoff 2003; Meyners 2007b); Brockhoff (2003, p. 414) concludes that the "limit powers are so close to the independent case powers that there is little room for improvements (more correct power assessment) of major practical importance by applying any of the formal models." Consequently, it can be safely applied in most if not all applications. The $\chi^2$ test might be a useful and simple alternative, while more complex models do not seem to provide major advantages in statistical testing for product differences. We consider them more relevant in the context of describing the data well, as discussed in the next section.

## 8.5 Descriptive Analysis and Parameter Estimation

### 8.5.1 Descriptive Analyses Enhance Insights

As already mentioned in Section 8.2.6, an important aspect of any data analysis, beyond pure statistical testing for significant differences, is the description of the data in a useful way. It is our opinion that replicated discrimination tests are heavily underutilized in this regard, even though this enables noteworthy additional insights due to the use of replications that cannot be obtained from unreplicated tests. In this section, we discuss various approaches to analyze and interpret the data obtained.

### 8.5.2 Tabulation of Observed and Expected Success Rates and $\chi^2$ Test

For situations where all assessors provide the same number $k$ of assessments, we generally recommend, as a first step, tabulation of the data and amending it with the expected values under the null hypothesis of no product differences. In other words, we tabulate how many assessors have given a certain number or correct answers, for values between 0 and $k$, and we further determine via the binomial law how many correct answers for any such value we would expect under the null hypothesis of no product differences (as exemplified later by Table 8.2 in Section 8.8). This is also the first step to perform the $\chi^2$ test mentioned in Section 8.4.11 for statistical significance testing (Meyners 2007b). It provides a quick overview of whether the data is in line with the null hypothesis, or whether it has distinctly more assessors with many (or possibly few in case of a preference test) correct answers. Thereby, it also provides a quick assessment of potential overdispersion, which would manifest itself in fewer than expected intermediate success rates, and more extreme values than expected. To avoid over-interpretation, determining the corresponding test statistic and p value is deemed helpful. In the imbalanced situation where the number or replications varies across assessors as in Example 8.2 in Section 8.8, the tabulation is less useful or not even possible, and a formal test rather cumbersome requiring dedicated programming. However, a graphical display using (empirical) cumulative distribution functions of what success rates are expected under the null hypothesis (no product differences, implying also no overdispersion) together with the observed success rates can serve a similar purpose by contrasting the true data with what we would expect by chance under the null hypothesis. We show a respective plot in Example 8.2 along with its interpretation, for which the slopes of the cumulative distribution functions (i.e., the rates of change) are compared across the range of data. Either of these approaches will become more useful with increasing $k$. For small $k$, e.g. $k = 2$, assessors will have either 0, 1, or 2 correct answers, i.e., there are few categories, which will usually render interpretation of differences between observed and expected values difficult or meaningless.

### 8.5.3 Mean Discrimination Rate with Confidence Interval

Meyners and Brockhoff (2003) discuss the estimation of the mean discrimination rate based on the naïve binomial model. This estimate can serve as a good starting point, though it should be noted that, in the presence of overdispersion, it comes with an increased uncertainty and hence a wider confidence interval as $k$ increases while $nk$ remains constant. Meyners and Brockhoff (2003) provide tables for varying $k$ that show which combinations of $nk$ would provide similar variance and hence similar certainty for the overall estimate relative to an unreplicated test design. Note that these tables, though similar in their idea, will result in different combinations of $n$ and $k$ than the tables in the same paper related to power as discussed in Section 8.4.3. Consequently, the researcher should decide upfront whether estimation or formal testing is the primary objective, and hence plan accordingly. In case of doubt, the larger numbers for $n$ at a given $k$ (or vice versa) should be used. Alternatively, the (conservative) confidence interval for the proportion of discriminators or the success rate by Kunert (2001) might be used (see also Section 8.4.5).

### 8.5.4 Lower Bound for the Proportion of True Discriminations

Meyners (2002, 2005) considers a closed set of hypotheses under intersection and related multiple testing procedures to determine a (conservative) upper bound for the number of discriminators, i.e., assessors that at least occasionally perceive a difference rather than just guessing throughout. The resulting procedure is rather simple: The naïve binomial test is used on all available data. If the test gives a statistically significant result, the assessor with the highest number of correct answers is considered a discriminator and subsequently removed from the data set. The naïve binomial test is then repeated on this smaller data set, and here as well as in possible further iterations, in case of statistical significance again the assessor with the most correct answers is considered a discriminator and removed for the next iteration. This procedure continues until the naïve binomial test no longer gives a statistically significant result. The proportion of discriminators is then at least as large as the number of assessors identified as discriminators during this procedure divided by the total number of assessors. It is important to note that the procedure does not classify any individual as a discriminator or non-discriminator – consequently, it does not matter which of those assessors that might have the same number of correct answers is removed in any specific iteration. The theory leading to this approach is more involved; given the low adoption of this approach, we abstain from providing further theoretical details and rather refer the interested reader to Meyners (2005), but we will exemplify the approach in Example 8.1 in Section 8.8.

### 8.5.5 Model Fitting: Beta-Binomial, Generalized Linear Mixed Models, and Related Approaches

As a second step, it might be useful to fit more complex models like the (c)BB or (c)GLMM models to allow quantification of the overdispersion, if any, as also emphasized by Cubero-Castillo et al. (2019). This will typically be particularly helpful in identification and estimation of the size of potential consumer segments, and to visualize the (estimated) distribution of success rates across assessors for communication purposes. Yet, it is important to remember that these models are either unimodal, or they are bimodal with the modes observed at the extremes, i.e., at $p_0$ and 1. Ultimately, the (c)BB and the (c)GLMM models will only provide an indication of whether the distribution is rather narrow (low overdispersion), about uniform, or has peaks at the two extremes (only the BB models; this could not be found from the GLMM models). In particular with larger numbers of replications $k$, this becomes less realistic and often not a good fit to the data, as it implies that noteworthy proportions of assessors have a very high success rate, though in real life, most assessors will get at least one or two tests out of a dozen wrong, say, even if only due to batch-to-batch variability of products. Meyners and Duineveld (2008) illustrate the poor fit of the BB model for overdispersed data.

Along with the BB and the GLMM models, Cochrane et al. (2005) compare further models with a focus on replicated preference tests (though results might equally be adopted to replicated discrimination tests), including generalized linear models with either a Pearson adjustment for overdispersion or with a correlated binomial distribution, and two variants of the BB model. It needs to be mentioned that they only consider the location parameter (i.e., the average choice or success rate) of the various models. In that work, overdispersion is considered a nuisance parameter that is only used for modeling purposes to improve the fit to the data, but not considered of interest in its own right, different from our rationale in Section 8.4.8. In their simulations, these authors do not find major differences among the various models but note a slightly increased power at the price of showing a tendency for liberalism for the BB model (thereby as well contradicting assertions that the binomial test is invalid and the BB would help or even be needed to protect against an increase in type I errors as made by, e.g., Lee and O'Mahony 2006; Cubero-Castillo et al. 2019). We refer to Cochrane et al. (2005) for further details.

We conclude that a main benefit of the BB model is its inherent measure of overdispersion $\gamma$, while there is no evidence for the BB model to be superior to other methods with regard to type I or type II error rates.

### 8.5.6 Bayesian Methods

Bi (2003) was likely the first to suggest a Bayesian model for preference and discrimination tests, assuming constant success probabilities across tests and among consumers. His approach was motivated by arguing that the assumption of identical success probabilities from different consumers is usually violated and hence even questions the validity of the binomial test in unreplicated tests. However, if assessors are considered to be randomly drawn from some population, the assumption can actually be warranted to hold as argued by Kunert and Meyners (1999). This is due to the fact that the success rate is established through a combination of the probability of any assessor to be selected from the population, and their individual success rate. Assume we have one third of the population each with a true success probability of 1/3, 2/3 and 1, respectively. If we randomly sample an assessor from the population and realizing that we cannot know upfront in which of the three groups the assessor falls, the success rate of an individual sampled this way is 2/3.[6] Net, the success rate of an individual randomly sampled can safely be assumed to be identical across samples, as the process of random sampling is part of the design and warrants the use of the binomial law. Consequently, we do not see compelling evidence that the binomial test might be inappropriate for (replicated or unreplicated) discrimination and preference tests. The model by Bi (2003) as well as the approach in Bi (2011) were explicitly developed for unreplicated tests, so we do not consider these any further here.

Based on earlier work by Carbonell et al. (2007) for unreplicated tests, Meyners and Duineveld (2008) provide an alternative estimate of the distribution of success rates without relation to a parametric model like a BB or GLMM model based on a simple application of Bayes' rule. As the approach will generally result in discrete distributions, it is extended by Duineveld and Meyners (2008) to a full hierarchical Bayesian model using Monte Carlo Markov Chains (MCMC, see, e.g., Robert and Casella 2010) which overcomes the issue, i.e., it allows estimating a continuous distribution of success rates across assessors. In contrast to the parametric BB and GLMM models, the hierarchical Bayesian model allows thereby a more granular estimation of distributions without noteworthy restrictions to their shape. Furthermore, it allows estimating the posterior probability distribution of individual assessors, providing an indication of the range in which the individual success rate is likely to fall. Due to the Bayesian nature of the approach, it will borrow information for this estimate if only few evaluations are available for any particular assessor, while it will be increasingly dominated by the individual's success rate with increasing $k_i$. This can be particularly useful in monitoring and training sensory panels, where the approach will allow the panel leader to differentiate between well- and not-so-well-performing assessors. In a more generic way, it might even be considered across a series of tests with different products or even different product categories, i.e., not formally replicated tests in the sense of this chapter. Though we would not be able to estimate the success probability for any particular product, we might get a good understanding of the assessors' performance in general or in a specific product category. True replications would not be necessary in that case. The main drawback of the hierarchical Bayesian model is its complexity, as it requires dedicated programming of the approach and the choice of a prior distribution, even if non-informative (see Duineveld and Meyners 2008). The other drawback is that the model may not be overly robust with few replications per individual. Duineveld and Meyners (2008) suggest $k \geq 5$ to enable a somewhat robust estimation of the individual true discrimination rates; with fewer replications, the number of theoretically possible counts of successes $(0, 1, \ldots, k)$ per assessor is too limited to warrant estimation of a more granular distribution. In order to get a proper understanding of individual performance, a larger number of

---

6 More specifically, the rate is determined as the sum over the probability to fall into a certain group multiplied by the respective success rate of that group, which here gives $1/3*1/3+1/3*2/3+1/3*1 = 1/3(1/3+2/3+1) = 2/3$.

replications is desirable, which is in contrast to other considerations, where we rather try to consider as many independent assessors as possible and use only some replications to increase the power. An application of this approach if used, for instance, with consumers, would be to further investigate the serial discriminators. If these are the heavy users of the current product, that might lead to a different business decision relative to everyone having a success probability which is only slightly above the guessing rate, $p_0$.

Along similar lines and using Bayes' rule as in Carbonell et al. (2007) sequentially, Bayarri et al. (2008) model the individual success probabilities replicate by replicate. In this approach, the prior distribution used for a later replicate is obtained as the posterior distribution from the preceding replicate's data. In this model, a fully data-independent prior distribution is only needed for the first replicate, for which the authors chose the uniform distribution (equivalent to the Beta distribution with $a = b = 1$, see Bi 2011).

Another Bayesian approach to the analysis of replicated paired preference data is suggested by Dubnicka (2013). The analysis suggested stays within the framework of a BB and the Dirichlet-Multinomial (Ennis and Bi 1999) models and focuses on whether the success rates of individual assessors change over replicates. This approach is based on a Bayesian approach to McNemar's test (McNemar 1947) as suggested by Altham (1971). It allows determination of various probabilities that might be of interest in specific situations like, for instance, the probability that a consumer changes their preference from A to B over replicates rather than the other way around. This could be of interest if familiarity of the products might be important for consumers' acceptance. Overall, Dubnicka (2013) focuses on preference changes over time rather than preference differences among consumers.

### 8.5.7 Thurstonian Approaches

Thurstonian methods (Thurstone 1927) could be applied to the results from any of these analyses by using the respective transformations (see, e.g., Ennis 1993; Bi and Ennis 1998; Jesionka et al. 2014; Linander et al. 2019) to determine the Thurstonian measure $d'$, which is less dependent of the test protocol used and eases comparison across various protocols. Yet, this transformation is most useful only for the location parameter (proportion of discriminators), not any measure of heterogeneity, and it is unclear to us if and how that might be accomplished, and whether that would offer an easily interpretable measure of overdispersion. We also note that the underlying assumption of Thurstonian models, namely that the product characteristics follow two normal distributions with possibly different mean values but same variance, are not easily translated into the models we consider here, which imply that the distance between the mean values of the two normal distributions will vary among assessors.

### 8.5.8 Conclusions for Descriptive Analyses

To conclude, the naïve estimate given by the total number of successes divided by $\sum_i^n k_i$ ($=nk$ for balanced data) will typically provide a useful first approximation of the overall success rate, while cross-tabulation of success counts or visualization of respective rates might be useful to gain further insights into the data. More complex models (cBB or cGLMM) could be used to quantify heterogeneity. Provided a decent number of replications per assessor is available and interest is (also) in individual success rates, a hierarchical Bayesian approach will prove useful but is more cumbersome to execute.

## 8.6 Tests for Equivalence/Similarity

Formal equivalence or similarity testing is required to show that two products are reasonably similar to each other. Absence of a statistically significant result is insufficient to conclude that there is no meaningful difference (Hartung et al. 1983; Altman and Bland 1995). It is worth mentioning that a proof of exact equality is statistically impossible. Instead, some margin $\delta$, say, needs to be defined, such that products between which the difference does not exceed this margin are considered comparable for consumers. The choice of such a margin is a non-statistical one and requires taking consumer noticeability and relevance into account. Details are beyond the scope of this section; a review of equivalence tests in sensory and consumer tests is provided by Meyners (2012).

The common null hypothesis for an equivalence test in an unreplicated discrimination test is that the true average success rate $p_c$ does not exceed a certain value, $p_0 + \delta$, say (see also Chapter 5 in this book), i.e., $H_0$: $p_c \leq p_0 + \delta$.

The confidence interval suggested by Kunert (2001) which was mentioned earlier (Section 8.4.5) could also serve to provide an equivalence test: We would reject the null hypothesis if and only if the confidence interval does not include the equivalence margin $\delta$. For discrimination tests, a one-sided approach should be employed. The limitation is in the lack of critical values or p values in the original paper. In recent work, Meyners et al. (2023) use the test statistic from Kunert (2001) to close this gap. For studies with equal numbers of replications across subjects, it can be shown that the test statistic

$$T(X) = \frac{X - nkp_0 - nk(1-p_0)p_d}{\sqrt{nk(1-p_0)p_0(1-p_d) + nk^2(1-p_0)^2 p_d(1-p_d)}}$$

is asymptotically normally distributed with mean 0 and a variance that cannot exceed 1, such that comparisons with the standard normal distribution can be used to determine (usually conservative) p values. To do so, $p_d$ takes the value of the equivalence margin. If that margin is defined as a value for $p_c$, it needs to be transformed onto the scale of $p_d$ first, of course. Details and generalizations to the case of unequal number of replications are given by Meyners et al. (2023). They also provide tables with critical values and resulting power, as well as R-code for the respective calculations. Their tables indicate that, for reasonable numbers of replications and assessors, the critical values are not much smaller than those for an unreplicated test with the same number of total evaluations, and that the power does hence not decrease drastically either. Using a small number of replications while slightly increasing the total number of evaluations (along the lines of Section 8.2.5) could offer a viable way to test for equivalence using replicated discrimination test data. It is worth noting that this approach does not make any noteworthy (distributional) assumptions about the data that could not be warranted by proper design and execution of the study.

For the sake of completeness, note that unlike the null hypothesis in the test to show differences, the null hypothesis for an equivalence test does not rule out potential overdispersion even with proper randomization and execution of the study. As discussed earlier (Section 8.4.8), overdispersion implies product differences in its own right, and that implies that we cannot simply re-apply the same null hypothesis from the unreplicated case. A composite null hypothesis would be required instead, using margins for both the success rate and a suitable parameter for overdispersion like, e.g., $\gamma$ in the different BB models. Though two parameters are considered simultaneously, the usual complication arising from multiplicity (i.e., several tests at level 5%, see, e.g., Lehmann and Romano 2005, or Bretz et al. 2011) can safely be ignored in equivalence testing (Berger 1982). Yet, while defining an equivalence margin for the success rate seems doable in many situations, defining a similar margin for a parameter related to overdispersion is not straightforward. Meyners (2008) describes various scenarios that, in a BB model, would lead to a certain overdispersion. This provides a sense about what a given magnitude of overdispersion could mean in practice, but it remains unclear how reasonable margins could be derived from this, and even less how to drive broader agreement for any of such choices, so this approach has not been pursued. Bi (2007, 2015) suggests equivalence tests based on such models using a Beta- or a Dirichlet-distribution while ignoring the critical role of the second parameter, possibly explaining why these approaches have not been adopted by practitioners and could, if anything, only result in rather small power improvements at the cost of making strict and usually unjustifiable model assumptions. The interested reader is referred to Meyners et al. (2023) for further details and rationales that render, to us, the above-mentioned test based on $T(X)$ superior to those approaches, and to Bi (2007, 2015) for details about those models and the respective tests.

## 8.7 Replicated Preference Tests

### 8.7.1 Success Rates and Test Problems

So far, we have confined ourselves to replicated discrimination tests like the triangle, tetrad, or duo-trio test. All of these tests have in common that any answer can be classified as correct or incorrect (see Section 8.3.2), and that the success rate in any such test is bounded below by $p_0$, the guessing rate. In those tests, only one-sided tests are of interest. To show differences, we want to statistically prove the true success rate to be larger than $p_0$ (or, at times, any other value larger than $p_0$). In contrast, to show parity, we want to show that the true success rate is smaller than $p_0 + \delta$, with an equivalence margin $\delta$. Preference tests and similar two-way tests on specific attributes allowing for reasonable two-sided interpretation ("A is *sweeter* than B," or "B is *sweeter* than A"; see, e.g., Linander et al. 2019) are conceptually different from this: Given the choice of A or B, there is no "correct" answer; instead, we use the preference rate for A (without loss of generality – if we are interested in the preference for B, we just change notation; see also Section 8.3.6). Most of what has been described in previous sections applies all the same for this scenario if success rate is replaced by preference rate, so we will confine the discussion in this section to highlighting the important differences. By far the most common test in this context is the paired preference test, which is why we will focus the following discussion around this protocol.

### 8.7.2 When Preference Tests Must Not Be Replicated

An important consideration is related to the purpose and design of the test. Whereas in discrimination tests, products are typically blinded and potential differences in other modalities (e.g., in visual appearance) tried to be masked, this may or may not be true in preference tests. The question about preference is still very reasonable when product concepts, appearance and/or branding are included. If so, the products might be easily distinguishable, and assessors can be expected to realize that they are asked (at least about) the same question over and over again. This might have all sorts of consequences that are not under the control of the experimenter, e.g., assessors at most superficially re-evaluating the products as they remember their previous choice, a point made previously by Marchisano et al. (2003, see also Wilke et al. 2006). They might also get annoyed by the repetitive task and choose to give fake answers, say, or deliberately choose a different answer as they might assume their first answer was deemed wrong in some sense and they were therefore asked again. Neither these nor other scenarios will lead to robust data. In the best case, it is the same answer every time (in which case replications do not add any information and are hence useless); in the worst case, the data basically consists of random numbers only. We are convinced that replications in such a scenario are at best a waste of effort and money, and that they should not be used. The same applies if products are clearly discriminable and possibly even recognizable by the assessors in a blinded study. In contrast, replications in preference testing might be useful when the products do not obviously differ and the fact that the same products are presented repeatedly is not apparent. In that case, replications might be useful to allow investigation of potential consumer segments and to judge, e.g., whether some assessors have a strong preference while others might be rather undecided, which is information that cannot be retrieved from unreplicated tests.

Another important aspect when considering replications in preference tests is that the representativity of the (consumer) sample suffers from asking fewer consumers but repeatedly. The needs or even regulations on selection of a representative sample might vary and are beyond the scope of this section, but it is worth taking this limitation into account when deciding to run a replicated preference test. Consequently, we would expect the number of potentially useful and reliable replications in preference testing to typically be more strictly limited than in discrimination tests.

### 8.7.3 Statistical Tests and Models

From an analysis point of view, in a preference test, the lower bound for the preference rate is 0, not $p_0$. Consequently, for replicated paired preference data, the chance-corrected models (cBB and cGLMM) discussed above do not apply; instead, the standard versions of these models (BB and GLMM) should be used, if any. Still, they come with the limitations in the potential shape of the distribution of assessors' preference rates as illustrated by Meyners (2007b). For the detection of product differences, they have been reported to have little power advantage over the naïve binomial test using all available observations (Brockhoff 2003; Cochrane et al. 2005), even less as the number or replications would typically be at most moderate.

### 8.7.4 Segmentation in Consumer Preferences

In contrast, we consider it even more important to leverage estimation procedures as described in Section 8.5 for replicated preference tests. If replications are available, it would be a huge missed opportunity to not also study the distribution of preferences. The individual's distributions are of less relevance, as interest is not in a single consumer, but it might help to identify consumer segments. With that, it can be essential to know why a preference rate for A of 55% is observed, as it could be due to any of the following (or many other intermediate scenarios):

- All consumers tending to have a slight preference for A, but not consistently.
- 90% of consumers having no preference at all (and might not even discriminate), while 10% always prefer A.
- 55% of consumers strictly preferring A, while the remaining 45% prefer B.

Obviously, these scenarios might lead to very different business decisions, e.g., depending on the group into which the heavy consumers of a product fall.

### 8.7.5 Equivalence Tests

To show equivalence of two products with regard to preference, the null hypothesis to test is no longer one-sided as in Section 8.6, but two-sided. It could be stated as $H_0$: $p_c \leq p_0 - \delta$ or $p_c \geq p_0 + \delta$, while the alternative to be shown would be $p_0 - \delta < p_c < p_0 + \delta$.[7] As discussed by Meyners (2012), the two-one-sided-tests procedure (TOST, Schuirmann 1981) is a simple, valid and reasonably powerful approach to turn this test problem into two separate tests of one-sided hypotheses. These hypotheses can then be robustly addressed by the approach mentioned in Section 8.6 (see Meyners et al. 2023 for details).

### 8.7.6 The No Preference Option

Sometimes a no preference option is allowed for in preference testing (e.g., ASTM International 2022; an equivalent is rarely considered in discrimination tests like the triangle test). This renders the statistical analysis more difficult as the binomial distribution cannot be applied directly anymore. In guide E1958-22 (ASTM International 2022), the recommendation is to split the "no preference" data between the two products with the last "no preference" vote being assigned in disfavor of the advertiser in case their total number is odd. It should be noted that it affects the resulting distributions by introducing underdispersion as the data is pulled into the direction of a 50% preference for either product. This is undesirable for the advertiser as long as the true preference for their product is larger than 50% (otherwise, the power to successfully show at least non-inferiority is small with reasonably sized studies), and hence will render the usually considered tests for superiority or non-inferiority less powerful, which is a consequence of allowing for (and observing a

7 Note that in principle, different margins could be used for the upper and lower bounds, see Meyners (2012). That would be very uncommon and hardly justifiable in this context, so we use a single margin for the lower and the upper bound.

noteworthy number of) “no preference” votes. Further details are beyond the scope of this chapter, but we note that the approach could be equally applied in the presence of replications. Ennis and Bi (1999) generalize the BB model to the Dirichlet-multinomial model that addresses tests with more than two products and could also be used for scenarios where an assessor could give three answer like, e.g., when offering the “no preference” option in a paired preference test (see, e.g., Dubnicka 2013).

## 8.8 Examples

In what follows, we will analyze two data sets using some of the aforementioned methods. Calculations were done in R 4.2.1 (R Core Team 2022), and respective example code and results are provided along with the analysis details. The scope of this book does not allow a detailed introduction into R, but we encourage readers unfamiliar with this software to visit https://www.r-project.org/ or https://www.rstudio.com/ for an easy installation and some first steps. Alternatively, Brockhoff and Linander (2017) also provide a section about starting out in R. We make use of an add-on package, `sensR` (Christensen and Brockhoff 2023) that needs to be installed along; apart from that, the code should not require any further software or packages to be installed.

**Example 8.1 Same Number of Replications**

In the first example, we look at data representing $n = 30$ panelists performing $k = 20$ replications of (unspecified) tetrad tests for non-food products; project or product details cannot be disclosed for confidentiality reasons but are not limiting the illustration. For transparency, we note that the data does not come from a replicated discrimination test in the common sense (it is across various pairs of products); clean and useful data is hard to get hold of, in particular with a noteworthy number of replications. We will interpret the data as if it relates to the same product in what follows, and we believe that it serves well to illustrate the methods and interpretation, nevertheless.

The number of panelists with the respective number of correct answers is shown in Table 8.1. It is worth mentioning that the probability of observing two or fewer correct answers out of 20 tests is, with a guessing probability of 1/3 as in the tetrad test, only about 2%. Yet, this does by no means imply that the success probability of any assessor was truly lower than 1/3, as that cannot occur. In fact, given there are 30 assessors, it is reasonably likely (40%) that at least one of them has no more than two correct answers by pure guessing, and it is almost certain (>99%) that at least one of them has less than five correct answers (i.e., an observed success rate of only 25%). Apparently, these

**Table 8.1** Number of panelists out of $n = 30$ with a given number of correct answers out of $k = 20$ tetrad tests.

| # correct | 2 | 5 | 6 | 7 | 8 | 9 | 10 | 11 | 12 | 13 | 14 | 18 |
|---|---|---|---|---|---|---|---|---|---|---|---|---|
| # assessors | 1 | 1 | 2 | 6 | 2 | 4 | 4 | 2 | 4 | 1 | 2 | 1 |

numbers do not apply directly here, as it seems likely at first glance that at least some of the assessors were not just guessing, but it illustrates that observed success rates of 0.1 are not uncommon even if the true yet unknown success rate cannot fall below 1/3. It is also worth noting that the number of correct answers occurring most is 7, which is as close as possible to the expected value (6.67) in case of no product differences.

The data can be loaded into R for further processing as[8]

```
> exmpl1 <- data.frame(correct=rep(c(2,5:14,18),
+                               c(1,1,2,6,2,4,4,2,4,1,2,1)),
+                  k=20)
```

In a total of 282 out of 600 tetrad tests, the correct pairs were identified:

```
> colSums(exmpl1)
  correct     k
      282   600
```

The one-sided naïve binomial test based on all observations is obtained via

```
> binom.test(x=282, n=600, p=1/3, alternative= "greater")
number of successes = 282, number of trials = 600,
p-value = 2.956e-12

alternative hypothesis:
true probability of success is greater than 0.3333

95 percent confidence interval:
 0.4358163 1.0000000
sample estimates:
probability of success
                  0.47
```

It is no surprise with the numbers obtained in combination with the large overall sample size that there is a statistically significant difference between the two products ($p<0.0001$). The estimated success rate in the tetrad test is 47%, with a one-sided lower confidence limit at 95% level of 43.6%. Hence, we can clearly prove differences based on the naïve but valid binomial test (Kunert and Meyners 1999).

The following code determines the intermediate measures for the $\chi^2$ test. In that, the vector p contains the probabilities for the different possible numbers of correct answers ($0-k=20$ here) based on the binomial distribution under the null hypothesis of no product differences, i.e. $p_0 = 1/3$. The vector x tabulates the observed data in the same way, but warranting that all possible numbers of correct answers (even if not observed, like 3 and 4 for this data) are included with a count of 0. The following lines determine the expected

8 Note that we use the common notation of marking R code with a ">" at the start of a code line; if the code breaks into multiple lines, it is marked with a "+"; this is how it appears in the R console. Note that these signs need to be omitted when writing or pasting the code into R. Output lines are not preceded by any sign; the output is truncated at times for brevity.

**Table 8.2** Observed and expected numbers of panelists out of $n = 30$ with the various numbers of correct answers out of $k = 20$ tetrad tests, along with the auxiliary statistics to determine the $\chi^2$ test; the $\chi^2$ statistic is the sum of the values in the last column.

| # Correct | Observed | Expected ($E$) | Difference ($D$) | Ratio $E^2/D$ |
|---|---|---|---|---|
| 0 | 0 | 0.01 | −0.01 | 0.01 |
| 1 | 0 | 0.09 | −0.09 | 0.09 |
| 2 | 1 | 0.43 | 0.57 | 0.76 |
| 3 | 0 | 1.29 | −1.29 | 1.29 |
| 4 | 0 | 2.73 | −2.73 | 2.73 |
| 5 | 1 | 4.37 | −3.37 | 2.60 |
| 6 | 2 | 5.46 | −3.46 | 2.20 |
| 7 | 6 | 5.46 | 0.54 | 0.05 |
| 8 | 2 | 4.44 | −2.44 | 1.34 |
| 9 | 4 | 2.96 | 1.04 | 0.37 |
| 10 | 4 | 1.63 | 2.37 | 3.46 |
| 11 | 2 | 0.74 | 1.26 | 2.15 |
| 12 | 4 | 0.28 | 3.72 | 49.94 |
| 13 | 1 | 0.09 | 0.91 | 9.80 |
| 14 | 2 | 0.02 | 1.98 | 183.43 |
| 15 | 0 | 0.00 | 0.00 | 0.00 |
| 16 | 0 | 0.00 | 0.00 | 0.00 |
| 17 | 0 | 0.00 | 0.00 | 0.00 |
| 18 | 1 | 0.00 | 1.00 | 152,927.14 |
| 19 | 0 | 0.00 | 0.00 | 0.00 |
| 20 | 0 | 0.00 | 0.00 | 0.00 |

numbers of assessors for each number of correct answers, the difference of those with the observed numbers, and the ratio of this squared difference divided by the expected numbers. We encourage the reader to run the code line by line to see the results by printing the output.[9] The last row will print the results; for improved readability, these results are transposed and after re-formatting provided in Table 8.2 instead of as raw R output.

```
> k <- 20
> p0 <- 1/3
> p <- dbinom(0:k, k, p0)
> x <- table(factor(exmpl1$correct, levels=0:k))
> x2 <- round(p*sum(x), 2)          # expected under H0
> x3 <- round(x-p*sum(x), 2)        # difference observed -
                                      expected
> x4 <- round((x-p*sum(x))**2 / (p*sum(x)), 2)
> out <- rbind(x,x2,x3,x4)
```

9 To print any object in the R console, just type its name, e.g. p or x4, and hit "Enter."

```
> rownames(out) <- c("observed", "expected",
+                     "difference", "ratio difference^2/
                      expected")
> print(out)
```

Two important conclusions can be drawn from these results. The first one is that for smaller numbers of correct answers (up to about 8), the expected number is usually larger than the observed, as indicated by the sign of the difference *D*. For larger numbers, we find the opposite. This implies that overall, we see too few assessors with low numbers of correct answers and too many with large numbers, relative to what is expected under the null hypothesis of no product differences. The second finding is that the by far most important contributor to the $\chi^2$ statistic is the one assessor with 18 out of 20 correct answers. This high number of correct answers is unexpected to occur even with 30 assessors; the likelihood for at least one out of 30 assessors getting 18 or more correct answers by chance alone is about 1 in 150,000. In fact, a binomial test based on the data only for this assessor alone would suffice for a significant result.

As a rule of thumb, the $\chi^2$ test should be used with the common approximation to the $\chi^2$ distribution only if all expected values *E* are at least 5. From Table 8.2, it becomes clear that this only holds true for two of the 21 cells, namely for six and seven correct answers. The approximation might be unreliable, which is why the use of Monte Carlo simulations is strongly recommended. The results from the $\chi^2$ test using such simulations can be easily obtained in R as

```
> set.seed(20210624)
> chisq.test(x, p=p, simulate.p.value = TRUE, B=9999)

        Chi-squared test for given probabilities with
simulated p-value (based on 9999 replicates)

X-squared = 153187, df = NA, p-value = 1e-04
```

Note that the first line of code sets a seed for the random number generator. This is not strictly necessary but warrants the reproducibility of the analysis; without that, repeated runs of the same line of code might lead to different results, so it is highly recommended to set a seed prior to such an analysis (though trying different seeds until the results are nice would result in a huge violation of the type I error rate and hence must not be done). The argument `simulate.p.value = TRUE` warrants that simulations are used rather than the approximation, while argument `B` specifies the number of random draws. For the validity of the approach, it is important that the observed data is considered as one of those simulations (see, e.g., Edgington and Onghena 2007), which is automatically done by the function, but not counted into *B*. Hence, if `B =10,000` were specified, in fact 10,001 draws would be considered, which could lead to p values with more than four decimals and falsely suggest higher accuracy than actually provided. Specifying `B =9999` results into the use of 10,000 observations instead, which assures that the resulting p value will never have more than four (non-zero) decimals. Using 10,000 simulations as done here should be more than enough for all practical purposes, and it takes only a fraction of a second to run. The smallest possible p value is 1/10,000 = 0.0001, which is what we in fact observe, so we would reject the null hypothesis of no product differences based on the $\chi^2$ test.

It is worth mentioning that the p value from the $\chi^2$ test would equally have been statistically significant without the assessor with 18 correct answers (with $p \approx 0.001$). Hence, it is not solely due to this assessor that a significant result was found.

Next, we apply the approach of Meyners (2002, 2005) to this data to determine the minimum number of discriminators among the 30 assessors. We have seen above that the binomial test with all assessors gives a significant result, and it does as well if we drop the data from the assessor with 18 correct answers (leaving 264 out of 580 correct answers), and if in addition we drop the data from the assessor giving 14 correct answers (leaving 250 out of 560 correct). This continues until we drop the data from the 11 best assessors; the remaining ones still have 143 out of 380 correct, which gives a one-sided p value of 0.043. Only if we drop the best 12 assessors, the p value becomes non-significant at level 5%. We therefore conclude that at least 11 out of 30 assessors discriminated the products at least at times, which gives a lower bound for the proportion of discriminators of about 36.7%.

In order to fit the cBB model, we use the package `sensR` which needs to be loaded first. Note that the function only provides two-sided confidence intervals; therefore, the confidence level is set to 0.9 here, such that the lower bound corresponds to the lower bound from a one-sided confidence interval at level 95%.

```
> require(sensR)
> summary(betabin(exmpl1, method="tetrad", corrected=TRUE),
+                 level=0.9)

Chance-corrected beta-binomial model for the tetrad protocol
with 90 percent confidence intervals
          Estimate Std. Error    Lower    Upper
mu         0.20560    0.04285 0.13512 0.27608
gamma      0.17544    0.08086 0.04244 0.30845
pc         0.47040    0.02856 0.42341 0.51738
pd         0.20560    0.04285 0.13512 0.27608
d-prime    0.91415    0.10720 0.72588 1.08324

log-likelihood:   -77.1468
LR-test of over-dispersion, G^2: 10.5809 df: 1 p-value:
0.001143
LR-test of association, G^2: 58.4587 df: 2 p-value: 2.022e-13
```

The key information extracted from this analysis is that the location parameter $\mu$ of the chance-corrected BB model is about 0.206, which corresponds to a proportion of true discriminations $p_d$ of 20.6% (cf. Section 8.4.7). The overall success probability $p_c$ is estimated to be about 47.0%, which is pretty close to the simple rate of correct answers from the previous analysis. The lower confidence bound for this rate is with 42.3% slightly lower than what is obtained from the naïve binomial test, so we do not gain on precision by using this model. The test for overdispersion gives a significant p value ($p = 0.0011$), and the overdispersion parameter $\gamma$ is estimated to be 0.175. We conclude that there are statistically significant and noteworthy differences in individual success rates between assessors. The last row in the output table provides the respective estimates for $d'$ in a Thurstonian context, taking the protocol used in the study into account.

Note that for the sake of analyzing a paired preference test, where the BB model instead of the cBB model would be applicable, we would need to specify `method="twoAFC"` and `corrected=FALSE` in the above code.

In order to dig deeper into the distribution of the individual success probabilities, we first apply the approach by Meyners and Duineveld (2008) based on earlier work by Carbonell et al. (2007). Based on Bayesian concepts using a non-informative prior, this approach determines the posterior probability for a given number of true discriminations in the population. We do not provide code details here as it is slightly more complex and of little educational value, but it is reasonably straightforward to implement with some slight programming experience. Instead, Figure 8.1 illustrates the posterior probabilities along with the fitted Beta distribution from the corresponding cBB model, properly scaled to match the discrete scale of the bar chart. We find that the cBB model overestimates the probabilities for very few discriminations (0–4, say), and underestimates those for medium (6 to about 12) as well as for many true discriminations (16 and more).

While the previous analysis only allows reasonable estimation of overall posterior probabilities, the approach by Duineveld and Meyners (2008) allows further investigation of the distribution of true discrimination rates by individual. This analysis is substantially more complex than all those discussed so far and requires good programming skills to implement along with good quantitative skills to evaluate the robustness of

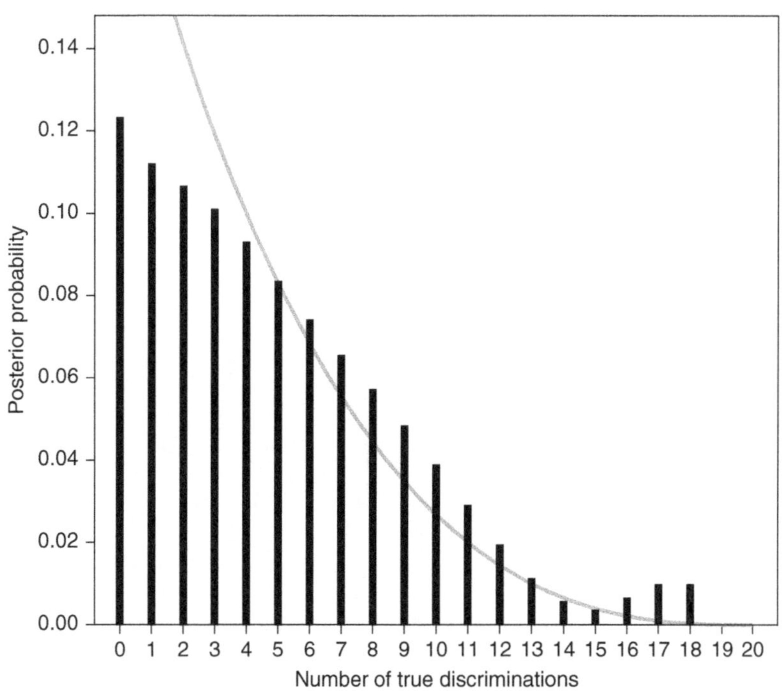


**Figure 8.1** Estimated distribution of posterior probabilities for any number of true discriminations rates for the replicated tetrad test data with $n = 30$ assessors and $k = 20$ replicates each. The solid grey line gives the density of the corresponding Beta distribution from the chance-corrected BB model (divided by 20 to match the scale).

the model. We will not provide details on the execution and evaluation of the sampling but refer to Duineveld and Meyners (2008) for details; we are following the same approach and criteria here, based on a non-informative prior. We confine our discussion to the results and interpretation thereof. This approach can provide additional insights regarding individual and panel performance, along with the associated uncertainties. Figure 8.2 shows the overall distribution of the posterior discrimination rates. It is seen that the individual discrimination rates are typically low (below 0.1, say) but with a rather heavy upper tail, suggesting that some assessors have a substantially larger discrimination rate (Figure 8.2a). Overall, 95% of the discrimination rates are expected between about 4% and 70%, which primarily indicates that there is substantial variation, such that a single mean number might not be very representative. The average discrimination rate is close to about 0.2, with a 95% highest posterior density (HPD) interval of [14.6%, 25.3%], as illustrated by the peaked distribution in Figure 8.2b.

Figure 8.3 shows the posterior distribution of individual discrimination rates for any of the observed numbers of correct answers out of 20 replications. On the extremes,

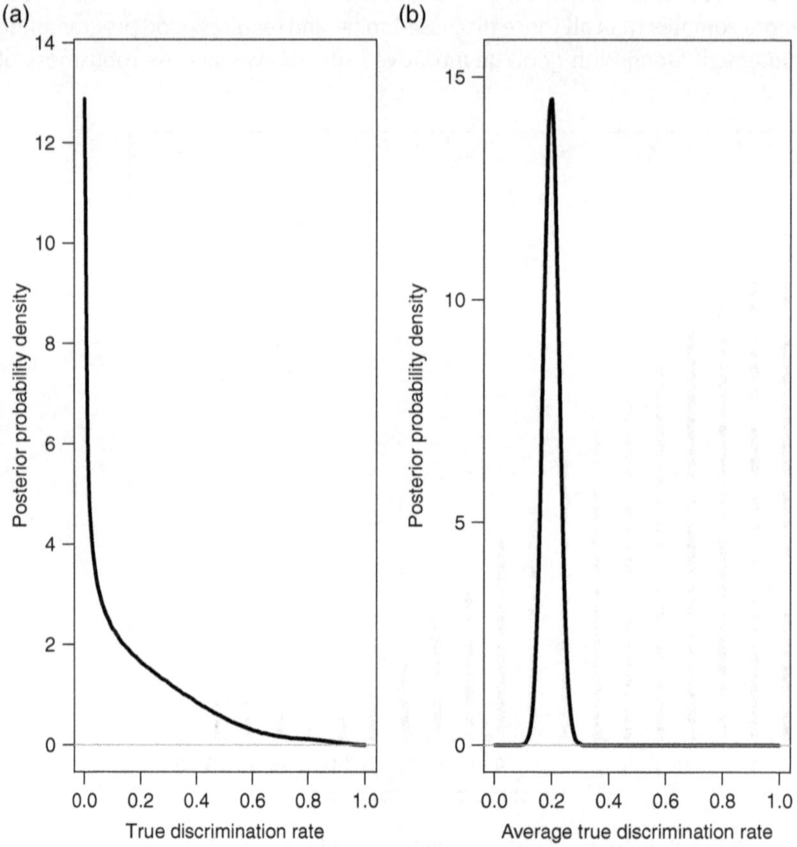


**Figure 8.2** Density of the posterior distribution of true discrimination rates (a) and of the posterior distribution of the mean discrimination rate (b), respectively.

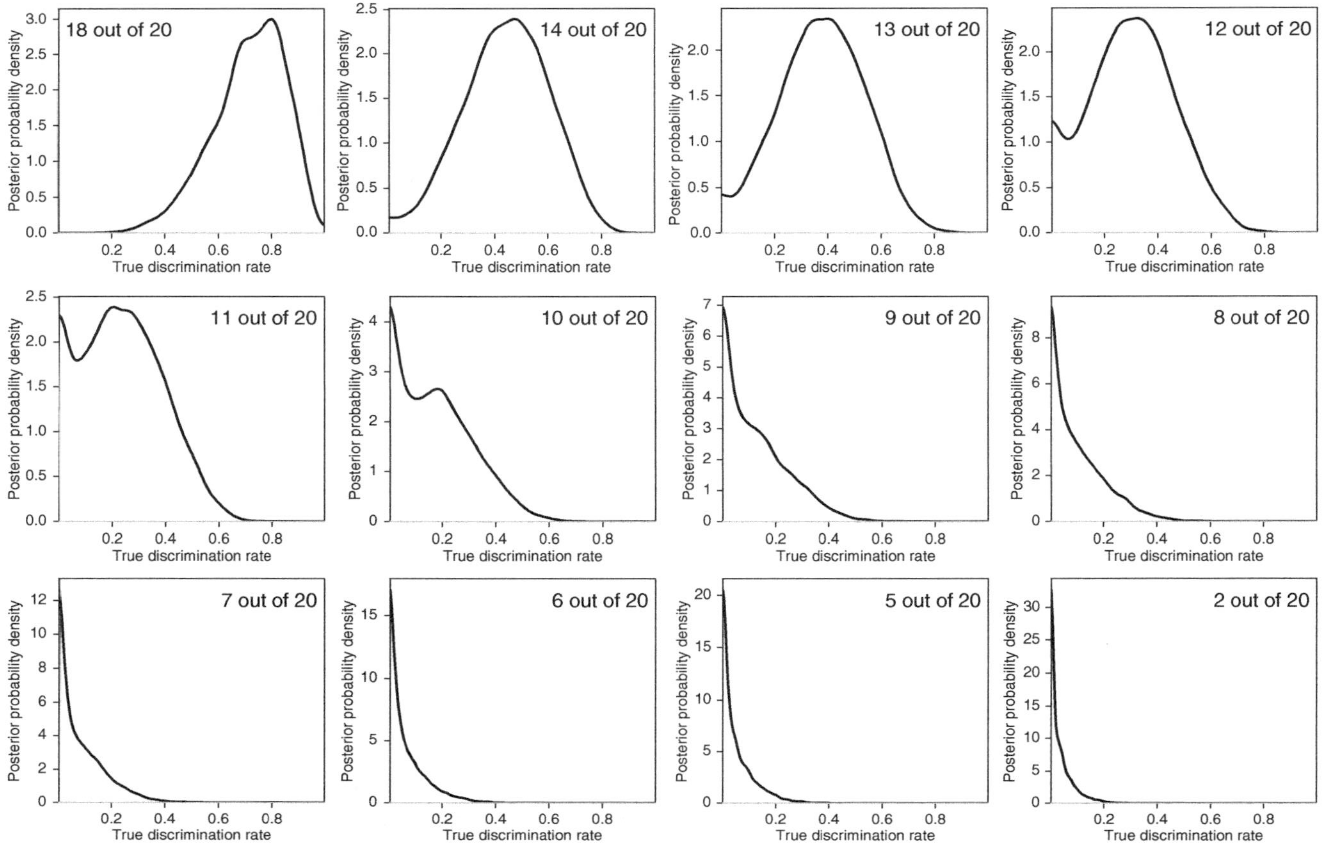


**Figure 8.3** Density of the posterior distribution of individual discrimination rates for the 12 unique numbers of observed correct answers out of 20. Note the different scales on the vertical axis, such that that absolute densities are not directly visually comparable between subplots.

it is found that the assessor with 18 correct answers is likely to truly discriminate in 50% or more of the cases, with a mode at a discrimination rate around 80%, while on the other end the assessor with only two correct answers has a very low probability to truly detect with a rate of 20% or more. Naturally, the other assessors are somewhere in between, while it is interesting to observe that with 12 out of 20 correct answers, a second peak of the distribution starts to be seen at a discrimination rate of 0. This peak is driven by the overall (across assessors) distribution of discrimination rates and starts dominating as of about 10 correct answers. Below that, the distribution resembles the one in Figure 8.2a except for a some stretching or shrinking to higher and lower discrimination rates, respectively.

As mentioned above, the data was not on true replications of the tetrad test but pooled across several tests, though we interpreted previous results as if they were exact replications. Formally, the analyses shown allow the conclusion that at least some of the product pairs tested were different (which is of little practical value), and that assessor performance varies substantially, with indications as to which assessors perform well overall and which might require some re-training. The latter is of potential value in panel selection and/or monitoring and shows the potential of the methods for this purpose (see also Sections 8.2.3 and 8.5.6).

**Example 8.2 Varying Numbers of Replications**

To exemplify the situation of different numbers of replications across panelists, we use the data published in Harries and Smith (1982) in their Table 2. However, to make the example more illustrative, we modified the number of correct answers for a few individuals without changing the total number of correct answers. Note that the data were collected over an extensive period (16 months, 28 sessions in total) and with expected variation in product characteristics over time, so subsequent analyses would again rather fall into panel performance monitoring than studying the differences between two specific products, but for sake of illustration, this differentiation is again not required. The data contain results from triangle tests to compare plaices (a type of flatfish) from fish farms with those trawled at sea by $n = 30$ assessors, whereby the number of replications varies from 1 to 22. The data are provided in Table 8.3.

The data can be loaded into R for further processing as

```
> exmpl2 <- data.frame(k = c(22,21,20,17,17,16, 14,12,
+                           11,11,11, 8, 8, 7, 6, 6,
+                            6, 5, 5, 5, 3, 3, 3, 3,
                             3, 3, 3, 1, 1, 1),
+                  correct = c(15,19, 8, 7, 5,11, 5, 5,
+                            8, 5, 2, 6, 1, 4, 5, 2,
+                            2, 5, 4, 3, 2, 2, 2, 1,
                             0, 0, 0, 1, 1, 0))
```

**Table 8.3** Modified triangle test results for the comparison of plaice from fish farms vs. plaice trawled at sea for 30 assessors and up to 22 replications. Data were originally published in Harries and Smith (1982) but partly modified to enhance illustration.

| Assessor | Trials | Correct | Assessor | Trials | Correct | Assessor | Trials | Correct |
|---|---|---|---|---|---|---|---|---|
| 1 | 22 | 15 | 11 | 11 | 2 | 21 | 3 | 2 |
| 2 | 21 | 19 | 12 | 8 | 6 | 22 | 3 | 2 |
| 3 | 20 | 8 | 13 | 8 | 1 | 23 | 3 | 2 |
| 4 | 17 | 7 | 14 | 7 | 4 | 24 | 3 | 1 |
| 5 | 17 | 5 | 15 | 6 | 5 | 25 | 3 | 0 |
| 6 | 16 | 11 | 16 | 6 | 2 | 26 | 3 | 0 |
| 7 | 14 | 5 | 17 | 6 | 2 | 27 | 3 | 0 |
| 8 | 12 | 5 | 18 | 5 | 5 | 28 | 1 | 1 |
| 9 | 11 | 8 | 19 | 5 | 4 | 29 | 1 | 1 |
| 10 | 11 | 5 | 20 | 5 | 3 | 30 | 1 | 0 |

In total, 252 triangle tests were conducted, in 131 of which the correct sample was chosen (note that we put output into italics here for readability; it is also truncated at times for brevity):

```
> colSums(exmpl2)
        k  correct
      252      131
```

The one-sided naïve binomial test based on all observations is obtained via

```
> binom.test(x=131, n=252, p=1/3, alternative= "greater")
number of successes = 131, number of trials = 252,
p-value = 8.438e-10
alternative hypothesis:
true probability of success is greater than 0.3333333
95 percent confidence interval:
 0.466121 1.000000
sample estimates:
probability of success
            0.5198413
```

We conclude that there is a statistically significant difference between the two products (types of plaice) with a p value <0.0001. The estimated success rate in the triangle test is about 52.0%, with a lower confidence limit (one-sided) at 95% level of 46.6%. So, we can clearly prove differences based on this valid test according to Kunert and Meyners (1999).

The $\chi^2$ test is unfortunately not straightforward to run for imbalanced scenarios with varying numbers of replications and would require dedicated programming, and a table like Table 8.2 cannot be reasonably produced. However, it is relatively easy to visually compare the empirical cumulative distribution function of success rates with what would be expected under the null hypothesis of no product differences. Figure 8.4 shows the respective display with the observed empirical distribution in grey, and the expected one in black. The slopes of the curves are close to each other for low success rates, with the rate of change of the observed function always being below the expected rate, indicating less than expected low success rates. This holds for success rates up to approximately 60%, where the rate of change of the distribution of observed rates catches up with and then exceeds the one for expected rates, indicating that there are more than expected success rates above 60% in the data, implying that at least some assessors must have been able to truly discriminate at times.

The cBB model is again fitted using package `sensR` (see Example 8.1). Note that in the second line, we need to reverse the order of columns in the dataset, as that is the required order by the function used. Remember that the function only provides two-sided confidence intervals, so the confidence level is set again to 0.9 and the lower bound then corresponds to the lower bound from a one-sided confidence interval at level 95%.

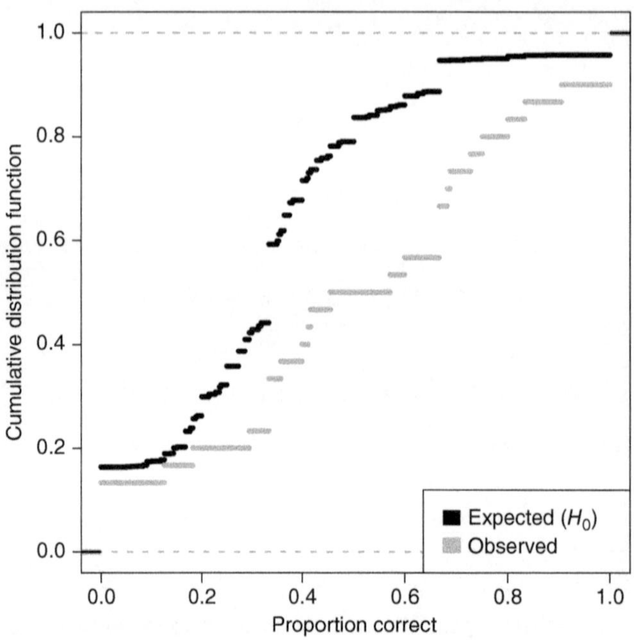


**Figure 8.4** Empirical distribution of success rates for the triangle test results for the comparison of plaice from fish farms vs. plaice trawled at sea for 30 assessors and up to 22 replications (grey) and the corresponding expected distribution under the null hypothesis of no product differences (black).

```
> summary(betabin(exmpl2[,2:1], method="triangle",
+          corrected=TRUE), level=0.9)

Chance-corrected beta-binomial model for the triangle
protocol
with 95 percent confidence intervals

         Estimate Std. Error   Lower   Upper
mu        0.26511    0.06822 0.15289 0.37732
gamma     0.33966    0.13236 0.12195 0.55737
pc        0.51007    0.04548 0.43526 0.58488
pd        0.26511    0.06822 0.15289 0.37732
d-prime   1.51879    0.23596 1.10665 1.89906

log-likelihood:  -56.7015
LR-test of over-dispersion, G^2: 15.2819 df: 1 p-value:
9.26e-05
LR-test of association, G^2: 52.2916 df: 2 p-value: 4.416e-12
```

We find that the location parameter $\mu$ of the chance-corrected BB model is about 0.265, which corresponds to a proportion of true discriminations $p_d$ of 26.5% (see Section 8.4.7). The overall success probability is estimated at 51.0%, which is slightly different though reasonably similar to the simple rate of correct answers from the previous analysis. The lower confidence bound for this rate is with 43.5% slightly lower than what is obtained from the naïve binomial test, so again we do not gain on precision by using this model. The test for overdispersion gives a significant p value ($p = 0.0001$), and the overdispersion parameter $\gamma$ is estimated to be 0.34. We conclude that there are statistically significant and noteworthy differences in individual success rates.

Most of the analyses from Example 8.1 can equally be performed for this data, though sometimes some more complex modifications are needed to account for the different numbers of replications. We do not show results for brevity; interpretation would be the same as in the previous section.

## 8.9 Conclusions

Replications in discrimination and preference tests gain more attention due to restrictions in the number of available assessors. While the analysis and interpretation of unreplicated tests with a larger number of independent assessors $n$ is typically easier, using replications offers potential further insights, which are unavailable from unreplicated tests. This includes information about consumer segmentation, assessor performance, or more detailed understanding of consumers' preferences (Wilke et al. 2006, including a motivating example). In contrast, to obtain comparable power for statistical inference, more assessments are needed overall when replications are used, and the maximum attainable power is bounded as long as the number of assessors $n$ is fixed.

When designing studies using replications, certain aspects need to be taken into consideration. Adequate representativeness of the pool of assessors matching the project objectives needs to be warranted in particular in consumer studies, which is not attainable via replications but only via independent assessors. Also, the models available generally rely on the assumption of fixed success probabilities per assessor. This requires careful consideration to rule out effects from either learning of fatigue, hence might impose an upper limit to the viable number of replications. Effective countermeasures (breaks, palate cleansing, . . .) should be carefully planned and adhered to. If needed, replications should be spread out over multiple sessions or even days. Potential fatigue in particular has to be carefully managed in studies to show equivalence for, e.g., claim substantiation. Decreasing discrimination ability of assessors over time could unduly reduce individual success probabilities and thereby lead to violation of the type I error rate, i.e., falsely concluding equivalence too often. In discrimination tests, fatigue would rather affect power, while the type I error rate would be respected.

It is worth emphasizing that the number of replicates $k_i$ of an assessor $i$ must not depend on their performance. In other words, we should not limit the number of replications of any assessor as we notice that they usually succeed (or fail) as they advance through their trials. This would bias the results as we would have more data from assessors with lower (or higher) success rates. Hence, the number of assessments should ideally be pre-determined, and deviations thereof and/or missing values should occur completely at random with no relation to an individual's performance.

We emphasize that the availability of replications should be leveraged beyond the pure boost in statistical power. Replications allow an assessment of the distribution of success rates across assessors or even individually, if the number of replications is large enough. Noting that differences in power among the various approaches to statistical testing were found to be small and negligible for most applications, the true benefit of more sophisticated models comes from the ability to describe the data in more nuances and estimate related parameters. This is, to date, largely under-leveraged.

## 8.10 Recommendations

We conclude with some general recommendations to take into account when considering replications in discrimination or preference testing:

- **Carefully determine the objective of the research**: For better representativeness of results, independent assessors (i.e., unreplicated testing) should be considered. If information about potential differences between individual assessors' success rates (assessor heterogeneity) are desired, replications are required; for robust results, at minimum four to six replications should be used in that case.
- **Sample sizes and power**: It is insufficient to replace independent observations by replications based on sample sizes used for unreplicated studies. If the number of assessors is too low to warrant decent power, replications might be used, but the total number of evaluations (i.e., assessors times replications) needs to be higher than the number of evaluations in a comparable experiment without replications.

- **Design**: We recommend aiming at balanced designs in which each assessor performs the same number of replications. The approaches discussed in this chapter, in most cases, can be applied also to imbalanced designs, but it is always helpful to obtain (about) the same amount of information from each subject. General good practices should be taken into account, e.g., reasonable number and length of breaks between evaluations, cleansing, masking, etc.
- **Randomization**: Ideally, for each assessment, the presentation order should be randomly selected from the potential options (e.g., from the six options, ABB, BAB, BBA, AAB, ABA, BAA, for a triangle test). No restrictions should be applied, neither within nor between assessors. This may lead to, e.g., the same presentation order in subsequent trials for an individual at times and will almost always lead to imbalances in usage of the various presentation orders both within and between assessors. However, it is the only way to warrant the validity of the statistical analysis methods. If complete randomization is not possible, every effort should be made to randomize as much as possible, and to avoid using the same presentation order throughout, or systematically using the same sequence of presentation orders for all assessors.
- **Proof of product differences**: Statistical testing to show differences is straightforward as the common binomial test can be used; a drop in power might be incurred for this test, but with reasonable numbers of replications and in common scenarios, this drop is limited and can be balanced by some increase in the total number of assessments (see previous recommendation). Alternative approaches might be used, e.g. a $\chi^2$ test, which has been shown to be reasonably powerful and offer further interpretation. More complex models like (chance-corrected) BB models do not seem to offer better power but have the downside that more than a single parameter needs to be considered in the tests. We rarely use these models for statistical testing of general product differences.
- **Proof of product similarity**: Absence of significant product differences must not be considered as evidence for product similarity. Unfortunately, the naïve binomial test is not appropriate to test for similarity using replications. However, dedicated approaches to allow for proofs of similarity have been developed recently and are straightforward to implement. Note that a prerequisite is always the careful choice of the equivalence margin.
- **Description and parameter estimation**: Next to statistical testing, understanding the data is highly recommended. Models including but not limited to the (chance-corrected) BB model can be used to gain further understanding of average success rates as well as the variability of individual success rates among assessors. This is useful information that is only available when replications are used. It is relevant to discrimination tests, but even more so for preference tests where different consumer groups may have systematically different true preferences. Simpler approaches (like cross-tabulating the number of correct answers) can also provide further insights into how the true success rates vary between assessors and deviate from what would be expected under assessor homogeneity. More complex models (e.g. Bayesian approaches) can be used for a more granular understanding of individual assessors' success rates, but require a more pronounced number of replications to give robust results.
- **Interpretation**: In case assessor heterogeneity is observed, a key question to address is what discriminates assessors with high success rates from those with low rates. Demographic, habits and practices data as well as consumption frequencies and the like

from individuals might be relevant to consider, thereby avoiding potentially missing a key consumer group for the product under consideration. Again, this is of particular importance for preference tests. Generally, make sure to bear in mind the potentially reduced representativeness of the assessors when using less assessors than usual when considering replications.

- **Panel performance**: Though we have not seen applications of the idea, the concepts in this chapter could also be applied to data from a sensory panel in which the same panelists perform various tasks (different products but same protocol). Though these do not constitute formal replications, in particular the estimation procedures discussed here might be helpful to distinguish between well- and less well-performing panelists, thereby offering an additional means to monitor and, through dedicated training, improve panelist and thereby panel performance.

## 8.11 Glossary and Notation

This section provides a quick access to the notation and some of the definitions used in this chapter. Table 8.4 holds the technical terms used with a corresponding brief description, while Table 8.5 covers the mathematical abbreviations and symbols.

**Table 8.4** Glossary of some key technical terms from this chapter.

| Term | Explanation |
|---|---|
| Replication | Repeated performance of the same task (same products in a discrimination or preference test) by the same assessor, though typically with different order unless by chance the same order was assigned again to this assessor. For ease of notation, we use the term for each assessment including the first; an unreplicated test would still have one replication, and if five assessments of the same task were taken by an assessor, we consider this as five replications rather than one initial assessment and four (formal) replications (Section 8.3.1). |
| Correct answer | In a forced-choice discrimination test, this is the correct answer, e.g., identifying the odd sample in a triangle test. In a two-way scenario like a paired preference test, there is no formally "correct" answer as a consumer may truly prefer A or B. For ease of notation, we refer to the choice of A (without loss of generalization) also as the correct answer, such that methods described for (one-sided) discrimination tests can be described the same way for two-way comparisons (Sections 8.3.2 and 8.3.6). |
| Success probability<br>Success rate | A success implies a correct answer has been given (see previous entry in this list). The success probability or success rate is the probability that a correct answer is given; as before, for two-way problems like paired preference tests, the success rate is the choice probability (preference rate) for one of the products, which we consider to be product A (Sections 8.3.3 and 8.3.6). |
| One-way test<br>One-sided test | Discrimination tests in which there is a correct answer like, e.g., the triangle or the 2-AFC tests. In these tests, the true success probability cannot fall below the guessing rate $p_0$ (see Table 8.5), and interest is only in whether the true success rate is larger than $p_0$. |

**Table 8.4** (Continued)

| Term | Explanation |
|---|---|
| Two-way test<br>Two-sided test | Discrimination or preference tests in which there is *no* correct answer, most notably the paired preference test where assessors may have a preference for either product. Though we use the term "correct answer" also in this context (see above) to denote choices for one of the products, the true choice probability can be anywhere between 0 and 1 and in particular also fall below the guessing rate $p_0$ (see Table 8.5). Interest is typically in both directions, i.e. whether the products are differently preferred and if so, which of the two is more preferred; in other words, we are interested in true choice rates larger but also smaller than $p_0$. |
| True discrimination | An assessor selects the correct sample based on sensory discrimination, not only by chance due to guessing correctly (Section 8.3.3). |
| Bernoulli distribution | Special case of a binomial distribution with just one observation, i.e. a binomial with parameters $n = 1$ and some arbitrary probability parameter (Section 8.3.5). |
| Homogeneity of assessors | All assessors truly have the exact same success rate; rarely used to contrast "assessor heterogeneity" (see next entry) (Section 8.3.7). |
| Heterogeneity of assessors | Assessors differ in their true success rates. Formally speaking, at least two assessors have different success rates, i.e. $p_i \neq p_j$ for at least one pair of assessors $(i, j)$. In practice, that situation is rare and of little interest; in contrast it typically matters if several assessors in a triangle test have a high success rate (e.g., $p_i = 0.8$), others a medium success rate (e.g., $p_i = 0.5$), while the rest does not sensorially discriminate and hence only guesses (i.e., $p_i = p_0 = 1/3$) (Section 8.3.7).<br>Assessor heterogeneity can only be investigated if true replications ($k \geq 2$) are available. |
| Randomization | Here: Random assignment of presentation order to assessors. Important statistical concept to warrant the validity of the analysis methods. Without proper randomization, no formally valid test is possible without strong additional and non-verifiable assumptions.<br>Recommendation is to randomly assign a presentation order to each assessor in each replication without any restrictions, and in particular independent from what the assessor (or other assessors) may have been assigned to before. This might lead to the same assessor (by chance) obtaining twice the same presentation order in sequence at times; the fact that this may (and will) happen is statistically acceptable and even desirable for valid tests. Many practitioners will attempt to balance presentation orders within and/or across assessors such that each presentation order is used (about) equally often. From a statistical validity point of view, this is not ideal, but it is certainly much better than always using the same presentation order, or systematically rotating among the various potential presentation orders (Sections 8.2.2, 8.4.1, and 8.10). |
| Validity of a statistical test | A statistical test is valid if the type I error rate (false positive rate) does not exceed the nominal level $\alpha$. So, if the test is conducted at 5% level, a test with a true false positive rate of 8%, say, would be invalid (Section 8.4.2). |
| Location<br>Location parameter | Alternative common term for the mean of a distribution (indicating where the distribution lies on average, thereby explaining the terminology); in this chapter, the mean of the distribution is typically equal to the average success rate (possibly after reparameterization in, e.g., the BB models). In a normal distribution with mean $\mu$ and variance $\sigma^2$, $\mu$ is the location parameter (Section 8.4.6). |

(*Continued*)

**Table 8.4** (Continued)

| Term | Explanation |
|---|---|
| Variation<br>Variance parameter | Alternative common term for the variability of a distribution (indicating how much the data varies around the average). Particularly relevant for the BB models in this chapter that have one parameter (after potential reparameterization) determining the variance. In a normal distribution with mean $\mu$ and variance $\sigma^2$, $\sigma^2$ is the variance parameter (Section 8.4.6). |
| Beta-Binomial (BB) model | Classical model to analyze replicated discrimination test data. At a higher level, assessors' success rates are assumed to follow a Beta distribution, whereas at the lower level, the number of correct answers from an assessor $i$ is assumed to follow a binomial distribution with chance parameters $p_i$, i.e., this assessor's individual success rate. A limitation of this model is that it generally allows for success rates to fall below the guessing probability $p_0$, which is not reasonable for one-way tests like triangle, tetrad or 2-AFC tests, for which the chance-corrected BB model is preferable. It is a more reasonable model for two-way tests like the paired preference test (Section 8.4.6). |
| Chance-corrected Beta-Binomial (cBB) model | Refined model to analyze replicated discrimination test data. Similar to the BB model but using a transformation of the distribution to warrant that success rates will not be modeled to fall below the guessing probability $p_0$. Only applicable to (one-way) discrimination tests, not for two-way tests like the paired preference test (Section 8.4.7). |
| Statistical testing | Formal statistical test, typically giving p values. This is used to statistically prove in a formal way that products are perceptibly different (or equivalent), without indicating the magnitude of potential differences. |
| Estimation<br>Parameter estimation | Estimation of key characteristics of distributions of relevance to the application. Usually, these characteristics can be expressed by parameters of the underlying probability distribution, e.g., the mean value or the variance, and therefore it is often referred to as "parameter estimation." It quantifies, e.g., the success rate (and hence the magnitude of potential product differences) and the variability, but without formally testing whether the difference is real or might as well have been observed due to chance alone. |

**Table 8.5** Explanation of mathematical notation and symbols.

| Notation | Explanation |
|---|---|
| $n$ | Number of assessors in a study (Section 8.3.1). |
| $i, j$ | Suffix indicating an individual assessor $i$ and $j$, respectively (Section 8.3.1). |
| $k_i$ | Number of replications of assessor $i$ (Section 8.3.1). |
| $k$ | Number of replications for each assessor if all assessors perform the same number of repeated tasks, i.e., $k = k_i$ for every assessors $i$ (Section 8.3.1). |
| $p_0$ | Success probability from pure guessing when the assessor cannot sensorially discriminate the products; examples: $p_0 = 1/3$ for the triangle test and $p_0 = 1/2$ for a paired preference or a 2-AFC test (Section 8.3.2). |
| $p_c$ | True (mean) success probability (see Table 8.4) for a population of interest (Sections 8.3.2 and 8.3.3). |
| $p_d$ | Probability of true discrimination (see Table 8.4) (Section 8.3.3). |

**Table 8.5** (Continued)

| Notation | Explanation |
|---|---|
| $p_i$ | Success probability of assessor $i$ (Section 8.3.5). |
| $P$ | Probability distribution of success rates across a population of interest (Sections 8.3.3 and 8.3.5). |
| $X_i$ | Random variable denoting the number of successes of assessor $i$ across their $k_i$ assessments (Section 8.3.3). |
| $X = \sum_{i=1}^{n} X_i$ | Random variable denoting the total number of successes across all assessors and assessments (Section 8.3.3). |
| $a, b$ | General parameters of a Beta distribution that are equally used in Beta-Binomial (BB) models. These are rarely used directly but are rather reparameterized into parameters $\mu$ and $\gamma$ (see subsequent entries) which have a much more straightforward interpretation (Section 8.4.6). |
| $\mu$ | Location parameter in a BB model determined from parameters $a$ and $b$ of a general Beta distribution as $\mu = a/(a+b)$. $\mu$ ranges between 0 and 1 and gives the average (mean) success rate of the respective BB distribution and thereby has an immediate and obvious interpretation (Section 8.4.6). |
| $\gamma$ | Variance parameter in a BB model determined from parameters $a$ and $b$ of a general Beta distribution as $\gamma = 1/(a+b+1)$. $\gamma$ ranges between 0 and 1 and represents the heterogeneity of the assessors (see Table 8.4); the larger $\gamma$, the more variable are the individual success rates. At times, an alternative parameter $\theta = 1/(a+b)$ is used instead of $\gamma$, which has the same information, but can take any positive value and is hence more difficult to interpret at times. In this chapter, we only use $\gamma$ (Section 8.4.6). |

## Acknowledgements

We are grateful to Susana Mugica for making the tetrad test data used for Example 8.1 available to us.

## References

Agresti, A. (2002). *Categorical Data Analysis*, 2e. Hoboken, NJ: Wiley.

Altham, P.M.E. (1971). The analysis of matched proportions. *Biometrika 58*: 561–576. https://doi.org/10.1093/biomet/58.3.561.

Altman, D.G. and Bland, J.M. (1995). Statistics notes: absence of evidence is not evidence of absence. *Br. Med. J. 311*: 485. https://doi.org/10.1136/bmj.311.7003.485.

ASTM International (2022). E1958-22 Standard Guide for Sensory Claim Substantiation, West Conshohocken, PA. https://doi.org/10.1520/E1958-22.

Bayarri, S., Carbonell, I., Izquierdo, L., and Tárrega, A. (2008). Replicated triangle and duo–trio tests: discrimination capacity of assessors evaluated by Bayes' rule. *Food Qual. Prefer. 19*: 519–523. https://doi.org/10.1016/j.foodqual.2008.02.008.

Berger, R.L. (1982). Multiparameter hypothesis testing and acceptance sampling. *Technometrics 24*: 295–300. https://www.tandfonline.com/doi/abs/10.1080/00401706.1982.10487790.

Bi, J. (2003). Difficulties and a way out: a Bayesian approach for sensory difference and preference tests. *J. Sens. Stud. 18*: 1–18. https://doi.org/10.1111/j.1745-459X.2003.tb00369.x.

Bi, J. (2007). Replicated similarity testing using paired comparison method. *J. Sens. Stud. 22*: 176–186. https://doi.org/10.1111/j.1745-459X.2006.00099.x.

Bi, J. (2011). Bayesian approach to sensory preference, difference and equivalence tests. *J. Sens. Stud. 26*: 383–399. https://doi.org/10.1111/j.1745-459X.2011.00354.x.

Bi, J. (2015). *Sensory Discrimination Tests and Measurements: Sensometrics in Sensory Evaluation*, 2e. Oxford: Wiley.

Bi, J. and Ennis, D.M. (1998). A Thurstonian variant of the beta-binomial model for replicated difference tests. *J. Sens. Stud. 13*: 461–466. https://doi.org/10.1111/j.1745-459X.1998.tb00100.x.

Bi, J. and Ennis, D.M. (1999a). The power of sensory discrimination methods used in replicated difference and preference tests. *J. Sens. Stud. 14*: 289–302. https://doi.org/10.1111/j.1745-459X.1999.tb00118.x.

Bi, J. and Ennis, D.M. (1999b). Beta-binomial tables for replicated difference and preference tests. *J. Sens. Stud. 14*: 347–368. https://doi.org/10.1111/j.1745-459X.1999.tb00121.x.

Bi, J. and Ennis, D.M. (2001). Exact beta-binomial tables for small experiments. *J. Sens. Stud. 16*: 319–325. https://doi.org/10.1111/j.1745-459X.2001.tb00304.x.

Bi, J., Templeton-Janik, L., Ennis, J.M., and Ennis, D.M. (2000). Replicated difference and preference tests: how to account for inter-trial variation. *Food Qual. Prefer. 11*: 269–273. https://doi.org/10.1016/S0950-3293(99)00032-4.

Bilder, C.R. and Loughin, T.M. (2015). *Analysis of Categorical Data with R*. Boca Raton, FL: CRC Press.

Bretz, F., Hothorn, T., and Westfall, P. (2011). *Multiple Comparisons Using R*. Boca Raton, FL: CRC Press.

Brockhoff, P.B. (2003). The statistical power of replications in difference tests. *Food Qual. Prefer. 14*: 405–417. https://doi.org/10.1016/S0950-3293(03)00003-X.

Brockhoff, P.B. and Christensen, R.H.B. (2010). Thurstonian models for sensory discrimination tests as generalized linear models. *Food Qual. Prefer. 21*: 330–338. https://doi.org/10.1016/j.foodqual.2009.04.003.

Brockhoff, P.B. and Linander, C.B. (2017). Analysis of the data using the R package sensR. In: *Discrimination Testing in Sensory Science* (ed. L. Rogers), 303–344. Duxford: Woodhead Publishing.

Brockhoff, P.B. and Schlich, P. (1998). Handling replications in discrimination tests. *Food Qual. Prefer. 9*: 303–312. https://doi.org/10.1016/S0950-3293(98)00014-7.

Carbonell, L., Carbonell, I., and Izquierdo, L. (2007). Triangle tests. Number of discriminators estimated by Bayes' rule. *Food Qual. Prefer. 18*: 117–120. https://doi.org/10.1016/j.foodqual.2005.09.002.

Christensen, R.H.B. and Brockhoff, P.B. (2023). *sensR – an R-package for sensory discrimination. R Packag. Vers.* 1: 5–3. http://www.cran.r-project.org/package=sensR/.

Cochran, W.G. (1952). The chi-square test of goodness of fit. *Ann. Math. Stat. 23*: 315–345. https://doi.org/10.1214/aoms/1177729380.

Cochrane, C.Y.C., Dubnicka, S., and Loughin, T. (2005). Comparison of methods for analyzing replicated preference tests. *J. Sens. Stud. 20*: 484–502. https://doi.org/10.1111/j.1745-459X.2005.00042.x.

Cubero-Castillo, E., Ramirez-Gutierrez, M., Araya-Quesada, Y., and O'Mahony, M. (2019). The beta-binomial: a preliminary comparison of smaller samples having many replications versus larger samples having fewer replications. *J. Sens. Stud. 34*: e12477. https://doi.org/10.1111/joss.12477.

Dubnicka, S.R. (2013). A Bayesian approach to analyzing replicated preference tests. *J. Sens. Stud.* 28: 171–187. https://doi.org/10.1111/joss.12033.

Duineveld, K. and Meyners, M. (2008). Hierarchical Bayesian analysis of true discrimination rates in replicated triangle tests. *Food Qual. Prefer. 19*: 292–305. https://doi.org/10.1016/j.foodqual.2007.10.005.

Edgington, E. and Onghena, P. (2007). *Randomization Tests*, 4e. Boca Raton, FL: Chapman and Hall/CRC.

Ennis, D.M. (1993). The power of sensory discrimination methods. *J. Sens. Stud. 8*: 353–370. https://doi.org/10.1111/j.1745-459X.1993.tb00225.x.

Ennis, D.M. and Bi, J. (1998). The beta-binomial model: accounting for inter-trial variation in replicated difference and preference tests. *J. Sens. Stud. 13*: 389–412. https://doi.org/10.1111/j.1745-459X.1998.tb00097.x.

Ennis, D.M. and Bi, J. (1999). The Dirichlet-multinomial model: accounting for inter-trial variation in replicated ratings. *J. Sens. Stud. 14*: 321–345. https://doi.org/10.1111/j.1745-459X.1999.tb00120.x.

Gelman, A., Carlin, J.B., Stern, H.S., and Rubin, D.B. (2004). *Bayesian Data Analysis*, 2e. Boca Raton, FL: Chapman & Hall/CRC.

Harries, J.M. and Smith, G.L. (1982). The two-factor triangle test. *Int. J. Food Sci. Technol. 17*: 153–162. https://doi.org/10.1111/j.1365-2621.1982.tb00172.x.

Hartung, J., Cottrell, J.E., and Giffin, J.P. (1983). Absence of evidence is not evidence of absence. *Anesthesiology 58*: 298–300. https://doi.org/10.1097/00000542-198303000-00033.

Hunter, E.A., Piggott, J.R., and Lee, K. Y.M. (2000). Analysis of discrimination tests. In: *Actes des 6èmes Journées Européennes Agro-Industrie et Méthodes Statistiques* (ed. Société Française de Statistique) (19–21 January 2000). France: Pau

International Organization for Standardization (2021). Sensory analysis – methodology – triangle test (ISO Standard No. 4120:2021). https://www.iso.org/standard/76666.html (accessed 13 September 2023).

Jesionka, V., Rousseau, B., and Ennis, J.M. (2014). Transitioning from proportion of discriminators to a more meaningful measure of sensory difference. *Food Qual. Prefer. 32*: 77–82. https://doi.org/10.1016/j.foodqual.2013.04.007.

Kim, M.-A., Chae, J.-E., van Hout, D., and Lee, H.-S. (2012). Discriminations of the A–Not A difference test improved when "A" was familiarized using a brand image. *Food Qual. Prefer. 23*: 3–12. https://doi.org/10.1016/j.foodqual.2011.06.007.

Kunert, J. (2001). On repeated difference testing. *Food Qual. Prefer. 12*: 385–391. https://doi.org/10.1016/S0950-3293(01)00029-5.

Kunert, J. and Meyners, M. (1999). On the triangle test with replications. *Food Qual. Prefer. 10*: 477–482. https://doi.org/10.1016/S0950-3293(99)00047-6.

Lee, H.-S. and O'Mahony, M. (2006). Sensory difference testing: the problem of overdispersion and the use of the beta-binomial statistical analysis. *Food Sci. Biotechnol. 15*: 494–498.

Lehmann, E.L. and Romano, J.P. (2005). *Testing Statistical Hypotheses*, 3e. New York: Springer.

Liggett, R.E. and Delwiche, J.F. (2005). The beta-binomial model: variability in overdispersion across methods and over time. *J. Sens. Stud. 20*: 48–61. https://doi.org/10.1111/j.1745-459X.2005.00003.x.

Linander, C.B., Christensen, R.H.B., Cleaver, G., and Brockhoff, P.B. (2019). Individual differences in replicated multi-product experiments with Thurstonian mixed models for binary paired comparison data. *Food Qual. Prefer. 75*: 220–229. https://doi.org/10.1016/j.foodqual.2019.01.010.

Marchisano, C., Lim, J., Cho, H.S. et al. (2003). Consumers report preference when they should not: a cross-cultural study. *J. Sens. Stud.* 18: 487–516. https://doi.org/10.1111/j.1745-459X.2003.tb00402.x.

Marin, J.-M. and Robert, C.P. (2014). *Bayesian Essentials with R*, 2e. New York: Springer.

McNemar, Q. (1947). Note on the sampling error of the difference between correlated proportions or percentages. *Psychometrika 12*: 153–157. https://doi.org/10.1007/BF02295996.

Meyners, M. (2002). On the number of perceivers in a triangle test with replications. In: *Actes des 7èmes Journées Européennes Agro-Industrie et Méthodes Statistiques* (eds. Duby, C. and Cassar, J.P.), 85–89, Lille, France (16–18 January 2002).

Meyners, M. (2005). Deriving a lower bound for the proportion of perceivers in replicated difference tests by means of multiple test theory. *Stat. Pap. 46*: 225–246. https://doi.org/10.1007/BF02762969.

Meyners, M. (2007a). Proper and improper use and interpretation of beta-binomial models in the analysis of replicated difference and preference tests. *Food Qual. Prefer. 18*: 741–750. https://doi.org/10.1016/j.foodqual.2007.01.001.

Meyners, M. (2007b). Easy and powerful analysis of replicated paired preference tests using the $\chi^2$ test. *Food Qual. Prefer. 18*: 938–948. https://doi.org/10.1016/j.foodqual.2007.03.002.

Meyners, M. (2008). On the definition of equivalence in beta-binomial models. In: *Actes des 10èmes Journées Européennes Agro-Industrie et Méthodes Statistiques* (ed. Société Française de Statistique), 245–251. Belgium: Louvain-la-Neuve (23–25 January 2008).

Meyners, M. (2012). Equivalence tests – a review. *Food Qual. Prefer. 26*: 231–245. https://doi.org/10.1016/j.foodqual.2012.05.003.

Meyners, M. and Brockhoff, P.B. (2003). The design of replicated difference tests. *J. Sens. Stud. 18*: 291–324. https://doi.org/10.1111/j.1745-459X.2003.tb00391.x.

Meyners, M. and Duineveld, K. (2008). Approximating the distribution of discrimination rates in replicated difference tests using Bayes' rule. *Food Qual. Prefer. 19*: 135–138. https://doi.org/10.1016/j.foodqual.2007.07.003.

Meyners, M. and Hasted, A. (2021). On the applicability of ANOVA models for CATA data. *Food Qual. Prefer.* 92: 104219. https://doi.org/10.1016/j.foodqual.2021.104219.

Meyners, M. and Hasted, A. (2022). Reply to Bi and Kuesten: ANOVA outperforms logistic regression for the analysis of CATA data. *Food Qual. Prefer.* 95: 104339. https://doi.org/10.1016/j.foodqual.2021.104339.

Meyners, M. and Hasted, A. (2023). On the choice of appropriate models for CATA data – A further reply to Bi and Kuesten. *Food Qual. Prefer.* 106: 104818. https://doi.org/10.1016/j.foodqual.2023.104818.

Meyners, M., Carr, B.T., and Kunert, J. (2023). Equivalence and non-inferiority tests using replicated discrimination and preference data. *J. Sens. Stud.* e12882. https://doi.org/10.1111/joss.12882.

R Core Team (2022). *R: A Language and Environment for Statistical Computing*. Vienna: R Foundation for Statistical Computing https://www.R-project.org/.

Robert, C.P. and Casella, R. (2010). *Introducing Monte Carlo Methods with R*. New York: Springer.

Schlich, P. (1993). Risk tables for discrimination tests. *Food Qual. Prefer. 4*: 141–151. https://doi.org/10.1016/0950-3293(93)90157-2.

Schuirmann, D.J. (1981). On hypothesis testing to determine if the mean of a normal distribution is contained in a known interval. *Biometrics 37*: 617. https://www.jstor.org/stable/2530576.

Skellam, J.G. (1948). A probability distribution derived from the binomial distribution by regarding the probability of success as variable between the sets of trials. *J. R. Stat. Soc. Ser. B 10*: 257–261. https://doi.org/10.1111/j.2517-6161.1948.tb00014.x.

Thurstone, L. (1927). A law of comparative judgment. *Psychol. Rev. 34* (4): 273–286. https://doi.org/10.1037/h0070288.

Wilke, K.D., Cochrane, C.-Y.C., and Chambers, E. IV (2006). Multiple preference tests can provide more information on consumer preferences. *J. Sens. Stud.* 21: 612–625. https://doi.org/10.1111/j.1745-459X.2006.00086.x.

# Section 2

# Applications

# 9

# Sensory Quality Measurement Based on SDT Discrimination

*Hye-Seong Lee*[1], *Min-A Kim*[2] *and Danielle van Hout*[3]

[1] *Department of Food Science and Biotechnology, College of Engineering, Ewha Womans University, Seoul, South Korea*
[2] *Division of Food and Nutrition, Chonnam National University, Gwangju, South Korea*
[3] *Aigora, Richmond, VA, USA*

## 9.1 Introduction

The food and beverage industry faces significant challenges to meet consumer demands for healthier and more sustainable products. While companies are working to reformulate products and reduce excess ingredients like sugar, salt, and fat, maintaining sensory quality remains a crucial factor for maintaining consumer preference. To effectively manage product reformulation and ingredient and process changes, a sensitive sensory evaluation method employing appropriate discrimination test formats and data analysis is essential for accurately measuring sensory differences between existing and improved products.

To ensure successful product development, a company must first define its sensory quality standards. This involves establishing a business action standard (AS) that outlines the allowable size of sensory difference – in other words, the threshold at which a change in sensory experience will impact consumer behavior. This standard should be based on a successful reference product that has been on the market for an extended period. Consistently evaluating the product's sensory quality based on these standards is critical, as product and market changes can impact consumer preferences and expectations over time.

Signal detection theory (SDT) sensory discrimination analysis using variants of A-Not A and 2-Alternative Forced Choice (2-AFC) methods (O'Mahony 1979; O'Mahony et al. 1980; Lee et al. 2007; Hautuset al. 2009; Lawless and Heymann 2010; van Hout et al. 2011; Bi et al. 2013; Hautus et al. 2020; Kim et al. 2020, 2012) has recently been actively studied and used to define such quality standards and measure the overall sensory differences of two or more test products from a reference product. This method can be used with both in-house trained sensory panels and groups of consumers to accomplish a sensory quality management program.

Various formats of discrimination test methods exist, and selecting the appropriate test format and discrimination method depends on the experimental situation, such as the

*Discrimination Testing in Sensory Evaluation*, First Edition. Edited by Lauren Rogers, Joanne Hort, Sarah E. Kemp and Tracey Hollowood.

purpose of the evaluation and the sample characteristics. The quality of sensory analysis results is influenced not only by appropriate statistical analysis but also by the proper selection and application of the test method for collecting raw data. A successful in-house sensory quality management program requires an understanding of the differences between discrimination test formats and their corresponding models in the SDT and Thurstonian modeling (TM) context.

Thus, in this chapter, we will first explain the various formats of discrimination test methods and how their instruction should differ to collect the data more accurately according to the SDT models. To illustrate the differences and similarities of these various discrimination methods more clearly, a new way of classifying various sensory discrimination test methods will be introduced: Reference-based discrimination test methods (specified methods) vs. Classification test methods (unspecified methods). Sensory specification for comparative evaluation in the specified sensory discrimination tests does not necessarily need to be described as a single characteristic or attribute, and the unique taste of a specific product (especially the golden reference product) can also be used as the specification for the comparison/evaluation. Reference-based discrimination test methods can be considered as specified methods and included the standard signal detection methods such as A-Not A and 2-AFC and those with the reminder (Hautus et al. 2009). Thus, in this chapter, the reference-based discrimination tests methodology will be focused on as a new/leading-edge sensory quality measurement methodology for sensory quality management in the SDT and TM context.

## 9.2 A New Classification of Sensory Discrimination Tests and Their Relative Performance

### 9.2.1 A New Way of Classification of Sensory Discrimination Tests

Various formats and test procedures are used for sensory discrimination tests to collect sensory discrimination data. These discrimination test formats can be divided into two types – specified tests (or reference-based methods) vs. unspecified tests (or classified methods), depending on the information available about which sensory characteristics are compared (or the reference for the comparisons) at the time of evaluation.

Table 9.1 shows these two broad classifications of sensory discrimination test formats and representative discrimination test methods belonging to two different classes (Lee 2016). When specified sensory discrimination test methods are used, usually taste characteristics such as sweetness and saltiness, or overall taste strength such as stronger taste or weaker taste are traditionally specified, so it is also called attribute-specified discrimination tests.

When a company has a successful product that has been on the market for an extended period, the unique sensory characteristics of this "golden reference product" can serve as a specification for comparison. In this case, test subjects must familiarize themselves with the sensory characteristics of the reference product. On the other hand, when sensory characteristics or attributes are not specified, the test format is called an unspecified sensory discrimination test or an overall sensory discrimination test. However, various test

**Table 9.1** Classification of various sensory discrimination test formats.

| | | Reference-based discrimination test methods (specified methods) | Classification test methods (unspecified methods) |
|---|---|---|---|
| Forced-choice method | | 2-AFC<br>2-AFC with reminder (2-AFCR)<br>3-AFC<br>4-AFC<br>Specified tetrad<br>Specified two-out-of-five | Triangle<br>duo-trio<br>Dual-pair<br>Unspecified tetrad<br>Two-out-of-five |
| Unforced-choice methods | Three-way response | 2-Alternative choice (2-AC) | |
| | Binary response | A-Not A<br>A-Not A with reminder (A-Not AR) | Same-different |
| | Rating response | Ratings of A-Not A<br>Ratings of A-Not AR | Ratings of same-different |

methods employ different cognitive decision processes to respond, so it is not sufficient to merely label them as overall discrimination tests. It is crucial to understand the psychological aspects of the different test formats and methods when measuring differences and discriminations. Therefore, in this chapter, we will use the name of the classification tests for these unspecified test methods, instead of the name of the overall discrimination tests, to clarify the evaluation task of these test methods.

Specifically, for example, different from the traditional classification for the 2-AFC method as an attribute-specified test, according to the new classification, the 2-AFC test can be used as an overall discrimination test method, with specifying the overall sensory properties of one of the two products to be compared. The 2-AFC with Reminder (2-AFCR) method uses a reminder before the paired comparisons to trigger the sensory memory of the specified product (Hautus et al. 2009). The 2-AFCR and constant-reference duo-trio tests (Kim and Lee 2012; Kim et al. 2014), belonging to the discrimination test designs that use a constant (fixed) reference, are in fact procedurally equivalent. Therefore, these two methods have been considered jointly and referred to as the constant-ref. DTF/2-AFCR test (Jeong et al. 2016; Shin et al. 2016). The traditional version of the duo-trio has been denoted by DTF (the traditional version, duo-trio with the reference presented first), indicating the position of the reference in the sample presentation (see Rousseau et al. 2002; Kim et al. 2010; Kim and Lee 2012 for various versions of duo-trio with the reference in different sample positions). In Table 9.1, also it should be noted that although the duo-trio test method uses a reference in a test format, the balanced-reference version of the duo-trio test method is considered an unspecified method. This is because, in the traditional balanced-reference version of the duo-trio test method, the subjects are meant to compare the size of the difference between the reference and the test samples in the test pair, not identifying the sensory characteristics of the reference. Only in the fixed-reference or the constant-reference version of duo-trio, in which subjects are allowed and required to remember what is the reference, the duo-trio test method can be considered a specified reference-based discrimination test method (Table 9.1).

The sensory discrimination measurement using these various test methods can be performed by using both a trained panel and a group of consumer subjects (Rousseau et al. 2002; van Hout 2014; Kim, M.A., Sim, H.M., and Lee, H.S. 2015,Kim et. 2016). When selecting the test methods among the various test formats shown in Table 9.1, the following questions should first be considered.

1) What is the purpose of the sensory evaluation?
2) Who will perform the evaluation, the sensory panel or consumer subjects?
3) Are the subjects more familiar with the overall sensory characteristics of a specific product among the two products included in the evaluation?

If the purpose of the test is to evaluate the potential issues of sensory quality changes of a specific product, a test method must be selected from among the reference-based test methods or specified methods. Conversely, when the sensory panel or consumer subjects have the same familiarity with the two products and they will be able to compare the size of the overall sensory difference between them without identifying a product with a specific sensory characteristic, the test method can be selected from among classification test methods or unspecified methods.

Discrimination test methods can also be divided into forced-choice methods and unforced-choice methods according to the response format used in the evaluation (Table 9.1). For using unforced-choice methods, binary responses, three-way responses, or rating responses can be used. A binary response is to choose one of two possible responses – "A (Yes)" or "Not A (No)". A three-way response is a response format that is used for 2-Alternative Choice (2-AC), which is a case where the "no difference" response option is included in the same evaluation as 2-AFC.

Another variant of an unforced choice method is the use of a rating option in addition to the binary response, for example, a response to the "A (Yes)" or "Not A (No)" question. When ratings are added to the binary response, the response options are extended, providing more opportunities for sensory panelists to use different response options for different products when evaluating multiple test products (Choi et al. 2023b). This is analogous to the case of using a category scale with a larger number of categories for sensory intensity measurements but it is dealing with a much smaller range of sensory differences. When applying SDT sensory discrimination analysis, the use of ratings has greater meaning because it makes the test capable of producing sufficient data points that receiver operating characteristic (ROC) curves could be fitted to compute $d'$. In experiments in psychology using visual or auditory stimuli other than taste or flavor stimuli, many repetitions are possible. Thus, the data obtained from different experimental sessions can be treated as independent data points for producing ROC curves. On the other hand, for sensory experiments carrying out taste and flavor discrimination tests, a large number of repetitions are not feasible in an experimental session. Therefore, a common practice in sensory experiments is to conduct an experiment consisting of several sessions and several days of repetitions using a sensory panel and perform that data analyses using pooled data over repetitions and panelists. In this approach, it was assumed that the pooled data were produced from one experiment. In this case, a rating option is vital to producing multiple data points to fit the ROC curves for SDT analysis and statistics based on TM. For A-Not A sensory discrimination tests, the three-point sureness ratings using categories of "sure," "not sure," and "I do not know, but I guess" (Lee et al. 2007a, b; Christensen et al. 2011; Kim et al. 2012) are commonly used for SDT analysis.

### 9.2.2 Conducting the Discrimination Tests and Collecting the Data

Figure 9.1 shows the test instructions and sample presentation forms for each discrimination test method given in Table 9.1. During sensory discrimination tests, all samples are usually presented as blind samples using a three-digit random number, except for the reminder sample of the reference "A" used in A-Not A tests and 2-AFC tests with reminder methods. Test subjects generally taste the samples in the order they were presented, from left to right. After tasting all samples, subjects are asked to choose one response option from the provided response options according to the test instructions.

Although the number of samples presented for each test method and the test instructions for comparing samples varies, in general, only two types of products are presented in one evaluation (Lawless 2013; Ennis et al. 2014). The test methods included in the forced-choice method can be generalized and expressed as an "$M + N$" test, where $N$ is the number of samples to be selected (specified) and $M$ is the number of samples different from it. Namely, the "$M + N$" contains two groups of methods: specified "$M + N$" and unspecified "$M + N$." Considering two classes of samples, A and B, the "$M + N$" test is formulated with $M$ samples in class A and $N$ samples in class B. In the case of the 2-AFC, $M = N = 1$, in the case of the 3-AFC $M = 2$ and $N = 1$, and in the case of the unspecified Tetrad, $M = N = 2$ (Bi et al. 2014).

Typically, subjects are instructed to rinse their mouths before each test and each sample tasting. However, in some cases, not rinsing between samples can improve test sensitivity, as subjects can compare samples better without the interference of interstimulus rinsing (Lee et al. 2009). Therefore, mouth-rinsing may be performed only before each test, depending on the nature of the sample, memory loss, and change in the state of the mouth. When interstimulus rinsing is not instructed between samples, a test sample of weak concentration is often presented as a primer before the first test sample to ensure that the sense of the first test sample is made in a similar situation to that of the other samples (Rousseau et al. 2002). To prepare for discrimination tests, particularly unforced-test methods like A-Not A and same-different tests, it is recommended to conduct a familiarization procedure (FP) to help subjects detect the differences between test samples and learn about the evaluative sensory dimension of characterizing and distinguishing the samples.

During the FP, subjects can taste one sample (only the reference sample) or compare both samples to establish evaluation criteria across the range of possible samples (Lee et al. 2007). This helps the sensory panel perform consistent evaluations even if repeated tests are conducted.

When using specified sensory discrimination tests in the forced choice method, basic taste characteristics or overall taste intensity can be specified as the taste of "A." A specific food's unique taste or golden reference can also be used as the standard for comparison. If it is difficult to specify the sensory characteristics, a sensory warm-up procedure can be performed. This involves tasting two different samples alternately with almost no time

| Test method | Sample presentation form | Test instruction |
|---|---|---|
| 2-AFC | 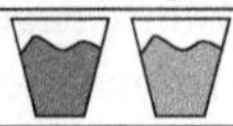 | Choose "A" of the two samples. |
| 2-AFC with reminder (2-AFCR) |  | Taste "A" first, recall the characteristics of "A," taste the two test samples, and choose the same sample as "A" tasted earlier. |
| 2-AC |  | Choose which of the two samples is more like "A" or is the same. |
| 3-AFC |  | Choose "A" out of three samples. |
| 4-AFC | 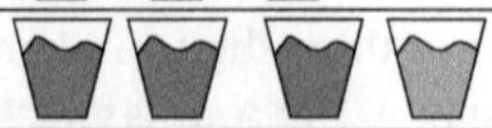 | Choose "A" out of four samples. |
| Triangle |  | Choose the most different sample. |
| Duo-trio |  | Choose a sample more similar to the reference sample (R). |
| Dual-pair | 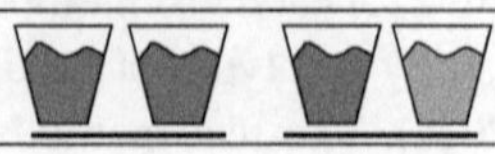 | Choose a different pair of samples from the two pairs of samples. |
| Specified tetrad | 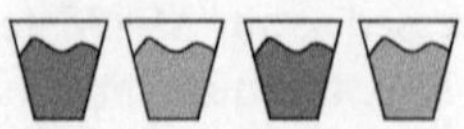 | Among the four samples, separate two samples of "A" characteristics into one group and the other two samples into another group. |
| Unspecified tetrad | 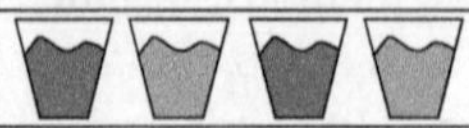 | Divide the four samples into two groups of two samples with the same characteristics. |
| Two-out-of-five | 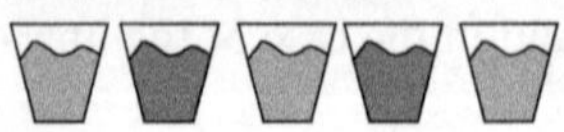 | Divide the five samples into one group of two samples and another group of three samples by grouping the samples with the same characteristics. |
| A-Not A |  | Determine whether it is an "A" or "Not A." |
| A-Not A with reminder (A-Not AR) | 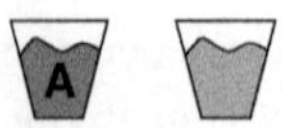 | Taste "A" first, and then taste the test sample and determine if the test sample is "A" or "Not A" referring to the sensory characteristics of "A" you tasted earlier. |
| Ratings of A-Not A |  | Determine if it is "A" or "Not A," and indicate the degree of certainty(sureness) by grade**. |
| Same-different | 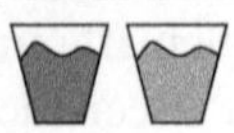 | Classify whether the two samples are the same sample (belonging to the same class) or different samples. |
| Ratings of same-different | 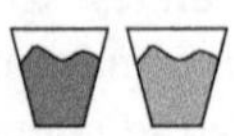 | Classify whether the two samples are the same sample (belonging to the same class) or different samples, and then indicate the degree of certainty by grade. |

**Figure 9.1** Sample presentation format and test instructions for various sensory discrimination test methods. *"A" can also be a specific characteristic, such as a stronger taste, stronger sweetness, or the unique taste of a specific food. **Various rating methods can also be used for sureness judgments (confidence gradings).

interval to increase sensitivity to sensory differences. The warm-up can help detect sensory differences even when subjects are initially unable to recognize them during the tests (Pfaffmann et al. 1954; O'Mahony et al. 1988; Thieme and O'Mahony 1990).

In the context of laboratory sensory analysis, discrimination tests are typically conducted using a trained sensory panel, where a limited number of panelists perform multiple tests, and the results are pooled for analysis. Conversely, in the case of consumer sensory analysis, a larger number of target consumers are typically recruited to participate in the test, with only one or a few repetitions for each, and the results are pooled for analysis. Thus, to ensure the accuracy of sensory discrimination tests, it is important to balance the order of sample presentation across subjects due to significant sample order effects. Assuming that the two products used in each test method are A and B, Table 9.2 shows the possible sample presentation order and the probability of getting the correct answer by chance for each test method given in Table 9.1.

Depending on the experimental design, the presentation order may need to be balanced not only across all subjects but also within individual subjects when multiple evaluations are possible per person and investigation for individual differences is important. Therefore, for analysis of inter-consumer differences, it is vital to select the test methods less affected by the order effects because it is necessary to balance out all the possible order by individual subjects through repeated tests (Kim et al. 2016, 2022).

### 9.2.3 Performance Differences Between the Various Designs of Discrimination Tests

Different test methods have varying capabilities in detecting small sensory differences, resulting in different levels of correct response rates even when the same subjects and exactly identical sample pairs are used. Reference-based and specified sensory discrimination test methods generally outperform classification or unspecified discrimination test methods, which compare the size of the absolute sensory difference to select the most different sample. For instance, the probability of obtaining a correct answer by chance is the same for 3-AFC and triangle tests (1/3), but the cognitive decision strategy in the 3-AFC test yields a higher probability of a correct answer (Frijters 1979; O'Mahony et al. 1994). In a similar matter, the 2-AFC test and the duo-trio test both have a probability of 1/2 for obtaining the correct answer by chance. However, the 2-AFC test has a higher probability of obtaining correct answers in evaluations compared to the duo-trio test (O'Mahony and Rousseau 2003). These differences in performance between various sensory discrimination test methods have been explained by psychometric models studied in SDT and TM contexts. Early psychometric models developed over 50 years ago, include 2-AFC (David et al. 1962; Green and Swets 1966), 3-AFC (Green and Swets 1966), triangle and duo-trio methods (David et al. 1962) for the forced-choice methods, and A-Not A and same-different methods (Macmillan et al. 1977) for the unforced-choice methods. The psychometric models that have been relatively recently introduced in the field of the sensory discrimination approach include 2-AFC with reminder (2-AFCR) (Hautus et al. 2009), 4-AFC (Bi et al. 2014), dual-pair (Rousseau and Ennis 2001), specified tetrad and unspecified tetrad methods (Ennis et al. 1998) for the forced-choice methods, and A-Not A with reminder (A-Not AR) (Hautus et al. 2009) and ratings of A-Not A (Bi et al. 2010) for the unforced-choice methods.

**Table 9.2** Sample presentation of various sensory discrimination test methods and probability of correct answer by chance.

| Test methods | Possible sample presentation orders | Number of samples required in a test | Chance probability ($P_{chance}$) |
|---|---|---|---|
| 2-AFC | <AB>, <BA> | 2 | 1/2 |
| 2-AFC with reminder (2-AFCR) | <A: AB>, <A: BA> | 3 | 1/2 |
| 3-AFC | <ABB>, <BAA>, <AAB>, <BBA>, <ABA>, <BAB> | 3 | 1/3 |
| 4-AFC | <ABBB>, <BAAA>, <AABA>, <BBAB>, <ABAA>, <BABB>, <AAAB>, <BBBA> | 4 | 1/4 |
| Triangle | <ABB>, <BAA>, <AAB>, <BBA>, <ABA>, <BAB> | 3 | 1/3 |
| Duo-trio | <R: AB>, <R: BA> | 3 | 1/2 |
| Dual-pair | <AA, AB>, <AA, BA>, <BB, AB>, <BB, BA>, <AB, AA>, <BA, AA>, <AB, BB>, <BA, BB> | 4 | 1/2 |
| Specified tetrad | <AABB>, <ABAB>, <ABBA>, <BAAB>, <BABA>, <BBAA> | 4 | 1/6 |
| Unspecified tetrad | <AABB>, <ABAB>, <ABBA>, <BAAB>, <BABA>, <BBAA> | 4 | 1/3 |
| Two-out-of-five | <AABBA>, <ABABA>, <ABBAA>, <BAABA>, <BABAA>, <BBAAA>, <AAABB>, <AABAB>, <AABBA>, <ABAAB>, <BBABA>, <ABBAB>, <AABBB>, <ABABB>, <ABBBA>, <BAABB>, <BABBA>, <BBABA>, <BABAB>, <BBBAA> | 5 | 1/10 |
| A-Not A | <A>, <B> | 1 | — |
| A-Not A with reminder (A-Not AR) | <A: A>, <A: B> | 2 | — |
| Same-different | <AA>, <BB>, <AB>, <BA> | 2 | — |

While the psychometric models are effective in predicting performance differences between different discrimination tests, sensory experiments involving food and beverage products face additional challenges that these models yet do not account for: factors such as fatigue and the complexity of tasting and preparing numerous samples can significantly impact test performance levels.

Furthermore, in various unspecified discrimination test methods like the triangle, same-different, unspecified tetrad, and duo-trio tests, which are classified as classification test methods, the test designs involve a pair of products presented randomly in the corresponding order of the possible sample presentation sequences shown in Figure 9.1. In contrast to this classification method, the reference-based discrimination test designs including the A-Not A, A-Not A with reminder (A-Not AR), 2-AFC with reminder (2-AFCR), and constant-reference duo-trio tests use a fixed reference and thus have a lower number of possible sample presentation sequences (Kim and Lee 2012; Jeong et al. 2016). A-Not AR and 2-AFCR, in particular, use the reminder paradigm and present a fixed reference as a reminder before each test (see Hautus et al. 2009 for more explanation of the reminder paradigm).

The two reminder methods that use a fixed reference scheme have been found to be the optimal test designs for discrimination tests involving untrained or consumer subjects when discriminating multiple types of samples against a fixed reference. When compared to the commonly studied balanced reference classification methods, such as the triangle and unspecified tetrad tests, it has been observed that the reminder methods are less susceptible to operational interference. As a result, they have resulted in $d'$ estimates that are closer to the theoretically predicted values (Jeong et al. 2016). Additionally, it has been discovered that the constant-ref. DTF/2-AFCR test is a superior design compared to the triangle test and the same-different test (Choi et al. 2014; Kim et al. 2014) due to its optimal test sequences.

To elaborate further, the two reminder methods require a lower number of samples in a test compared to the unspecified tetrad test. This means that when equalizing the number of samples required, the total number of tests would be much larger for the reminder methods than for the tetrad test. As a result, the relative theoretical power deficiency of the reminder methods may be compromised when the number of tests increases. In a sensory discrimination test, the number of test samples required often determines the operational test power and applied efficiency (Jeong et al. 2016). Although the test power has typically been demonstrated in terms of the number of tests required to detect the given sensory difference, the number of test samples is equally important to consider (Ennis 1993; Schlich 1993; Bi 2011; Ennis and Jesionka 2011; Jesionka et al. 2014; Kim et al. 2014).

Figure 9.2 illustrates the differences in power curves computed according to the number of tests (a) and the number of samples used (b). The number of tests is equalized as 30 tests across different methods in Figure 9.2a. In this situation, it is natural to think that 30 consumer subjects performed each test once. Yet, due to the sequence effects of the many classification discrimination tests including the triangle and duo-trio tests (cf. O'Mahony and Odbert 1985; O'Mahony 1995; Kim et al. 2014; Lau et al. 2004), when studying consumer discrimination, it is not desirable to use only one test sequence, rather replicated tests using all possible test sequences per subject in a balanced design is more appropriate. In situations with replicated tests per consumer subject, subjects' responses are generally pooled over many subjects and the number of tests can be quite large. In Figure 9.2b, the power curves are shown for such a situation in which the number of test stimuli used for the discrimination experiment is equalized at 1440, assuming 60 subjects who perform a different number of replications for each method: 12, 8, 8, and 6 replications for A-Not AR, 2-AFCR, triangle, and unspecified tetrad tests, respectively. Although the shapes of the power curves are based on the assumptions of unbiased responses and no replication effects, these reflect a reasonable portrayal of the relationships between the test methods evaluated in this study, by equalizing the number of stimuli required for the tests. In this illustration, for small sensory differences (for example, $d' < 0.9$), with a large number of tests, A-Not AR

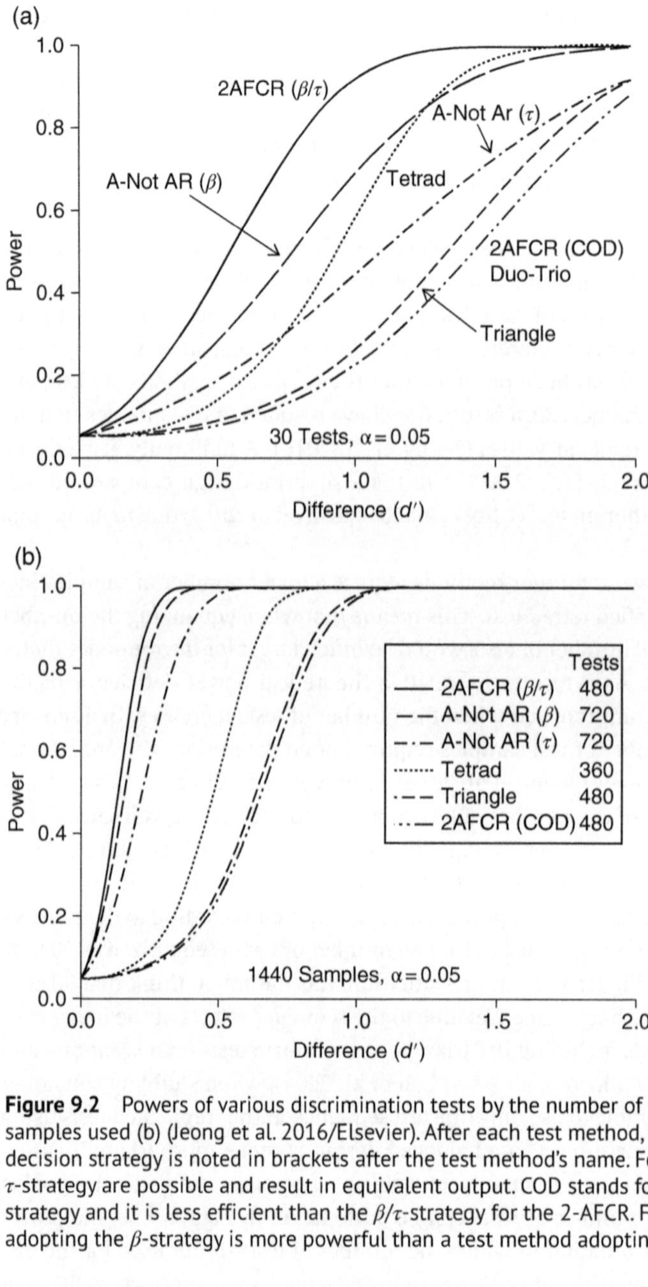


**Figure 9.2** Powers of various discrimination tests by the number of tests (a) and by the number of samples used (b) (Jeong et al. 2016/Elsevier). After each test method, the possible cognitive decision strategy is noted in brackets after the test method's name. For 2-AFCR both $\beta$-strategy and $\tau$-strategy are possible and result in equivalent output. COD stands for comparison of distances strategy and it is less efficient than the $\beta/\tau$-strategy for the 2-AFCR. For A-Not AR, a test method adopting the $\beta$-strategy is more powerful than a test method adopting the $\tau$-strategy.

would be much more powerful and thus more precise than the other three methods – the constant-ref. DTF/2-AFCR, triangle, and unspecified tetrad tests (Figure 9.2b). For $d' > 0.9$, it is apparent that the theoretical test power does not matter under the condition of equalizing the number of stimuli and subjects but does when varying the number of tests. In this situation, operational issues including physiological effects and cognitive factors (Lee et al. 2009; Kim and Lee 2012; Kim et al. 2014) that can be largely affected by the sample configuration in the replicated tests, may play a major role in determining the applied discrimination performance and the operational test power. In other words, although the theoretical test powers are all equally sufficient, some methods could be more susceptible to operational hindrances such as sensory adaptation and memory degradation and thus result in lower discrimination performance than the other methods in application/operation. The discrimination performance of methods with lower operational power is reflected by the lower confidence limit of the $d'$ estimates. This means that when using the same set of stimuli and subjects with equal sensitivity, methods with poor operational power will result in lower $d'$ estimates and associated confidence intervals. Therefore, comparisons of the practical efficiency and operational test power of sensory discrimination test methods can also be made by comparing the $d'$ estimates obtained from different test methods when equalizing the number of samples required, assuming the theoretical power is sufficient.

## 9.3 Reference-based Discrimination Test Methodology: SDT Sensory Quality Measurements

### 9.3.1 Introduction

When people consume food and beverages, the taste can be perceived slightly differently each time depending on factors such as the eating process, portion size, and psychological attention. Despite this, frequent consumers learn the sensory quality and develop an evaluative criterion for the product. This means that when they make a repeat purchase, they have an expectation of the sensory quality. To ensure successful in-house quality management, companies must be able to predict consumers' evaluative criteria and set quality standards that are stricter than those of the consumers. However, if the quality standards are too strict, companies may miss the opportunity to lower costs through reformulation or ingredient replacements. Therefore, companies must consider the risks and benefits and implement different quality standards for different categories of products and business objectives.

The quality standard (business AS) can be decided based on the meaningful size of the sensory difference that consumers would perceive. The concept of the meaningful size of sensory difference has been referred to in many different terms including "allowed difference" (Bi 2005), "similarity limit" (Bi 2011), "consumer relevance" (Ishii et al. 2007), and "meaningful difference" (Ennis et al. 2014; van Hout 2014; Kim, M.A., Sim, H.M., and Lee, H.S.2015) in applications of sensory discrimination test methodology. In literature, the Thurstonian discriminal distance $d' = 1$ was suggested as a practical difference by Kuesten (2001), while $d' = 0.2, 0.4$, and 1.0 was suggested as a strict, a common and a liberal criterion for similarity testing, respectively, by Bi (2011). The meaningful size of difference can be determined by studying target consumers' discrimination ability and/or preference behavior relevant to the sensory difference perceived.

Therefore, to use a sensory panel to predict consumer perception, the sensory panel should be trained to be familiar with the reference product as much as consumers and be able to produce good quality data that can be linked to consumer perception. In addition, the test methods have to be efficient and effective to save money and time to perform the methodology; namely, the implementation of sensory quality management and consistently occurring business-decision making have to be most efficient and effective based on the applied discrimination methodology. In this section the reference-based discrimination test methodology will be introduced as the advanced SDT sensory quality measurement methodology for sensory quality management that can achieve these goals.

### 9.3.2 Test Design

This test methodology can be used by both a trained sensory panel and a group of consumer subjects. For in-house sensory quality management, a small number of trained panelists is generally used. They need to acquire familiarity with the reference product before they can participate in the sensory tests. Thus, the test designs compose of several test sessions of a FP and main tests using reminder test designs, A-Not AR or 2-AFCR. The first session is generally performed as a practice session.

Figure 9.3 illustrates an example of each A-Not AR (a) and 2-AFCR (b) sample presentation and response format using a two-step sureness rating. Assuming three test products ("$T_1$," "$T_2$," and "$T_3$") being discriminated from a reference product (R), the reminder (R) is always presented before the presentation of a blind sample (in the case of A-Not AR) or a pair of blind samples (in the case of 2-AFCR). In the case of A-Not AR, as shown in

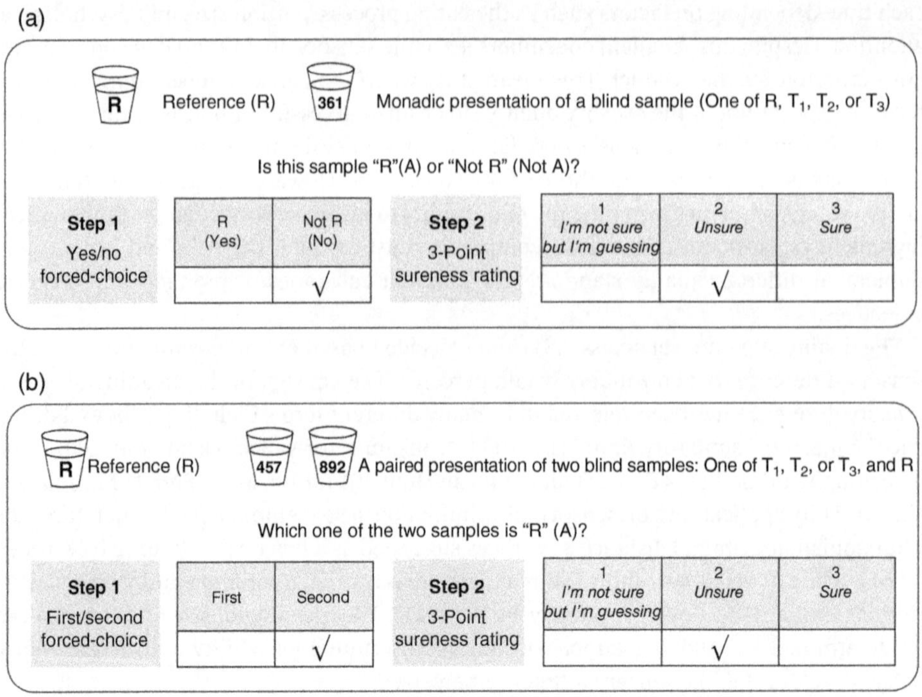


**Figure 9.3** An example of A-Not AR (a) and 2-AFCR (b) response format using a two-step rating with a 3-point sureness rating. (a) Response format of A-Not AR with a two-step rating. (b) Response format of 2-AFCR with a two-step rating.

Figure 9.3a, a monadic presentation of a blind sample is evaluated and this blind test sample is one of the test products or the reference product, presented in random order. In the case of 2-AFCR, as shown in Figure 9.3b, a pair of blind samples is evaluated consisting of one test sample and one reference sample. The product order of the blind sample pair is randomized. In the example shown in Figure 9.3, the two-step rating procedure using a 3-point sureness rating following the forced choice has been used.

Figure 9.4. illustrates various experimental designs of A-Not AR test protocols to be performed in one day (two sessions), as examples of the reference-based discrimination experiment. Each protocol has adopted a different type of FP. Depending on the experiment objectives, replications of such experimental design with the same panelists over several days can be conducted to increase the sample size.

For each main test session, for Protocol 1 (Figure 9.4a), eight test samples are paired with the reminder, providing two responses per each product, while for Protocols 2 and 3 (Figure 9.4b, c), six blind samples are paired with the reminder, providing a single response per each of test products and three responses for the reference product. For the former case, the reference presentation probability (RPP) is 0.25, while for the latter case, the RPP is 0.5. If 18 tastings of samples in one experimental session are feasible, such designs can be applied to obtain the maximum number of data points. Recent studies found that higher RPP (0.5 RPP) tends to result in faster learning and improvement of the discrimination performance over repetitions (Choi et al. 2023a; Lee et al. 2023).

### 9.3.3 Familiarization Procedure

Using reference-based discrimination tests such as the A-Not AR test reliably with a group of subjects requires the subjects to be familiar with the reference product. When subjects are not sufficiently familiarized, they have no stable memory of the reference and will have difficulty deciding consistently whether a test sample is the same as the reference product or not. This reduces the test performance (Choi et al. 2023a, b; Kim et al. 2012). Through a FP, recognition of both the reference and test samples being different from the reference is enhanced and the decision criteria are stabilized (Choi et al. 2023a, b; Rousseau et al. 1999).

#### 9.3.3.1 Repeated Reference Tastings

The conventional approach to familiarize subjects with the reference in A-Not A is to provide the reference before starting the A-Not A test (Lee et al. 2007a, b; Mun et al. 2019; van Hout et al. 2011; Jeong et al. 2016; Choi et al. 2023a, b) as indicated in Protocol 1 in Figure 9.4a.

#### 9.3.3.2 Specified Categorization Task

A discrimination task can also be used as a FP to learn about the nature and degree of difference between the products (Kim and Lee 2012; Kim et al. 2012, 2015; Shin et al. 2016; Jeong et al. 2017). In Protocol 2 in Figure 9.4b, a specified tetrad test was used as the FP. Shin et al. (2016) and Lee et al. (2023) reported that specified tetrad tests can be used as an effective FP for conducting 2-AFC or a fixed reference duo-trio method and A-Not AR method. It is especially efficient if there was only a limited number of replication possible. Further, Choi et al. (2023a) showed that a categorization task similar to a specified hexagon test could be an effective FP for conducting A-Not A with a two-step rating method.

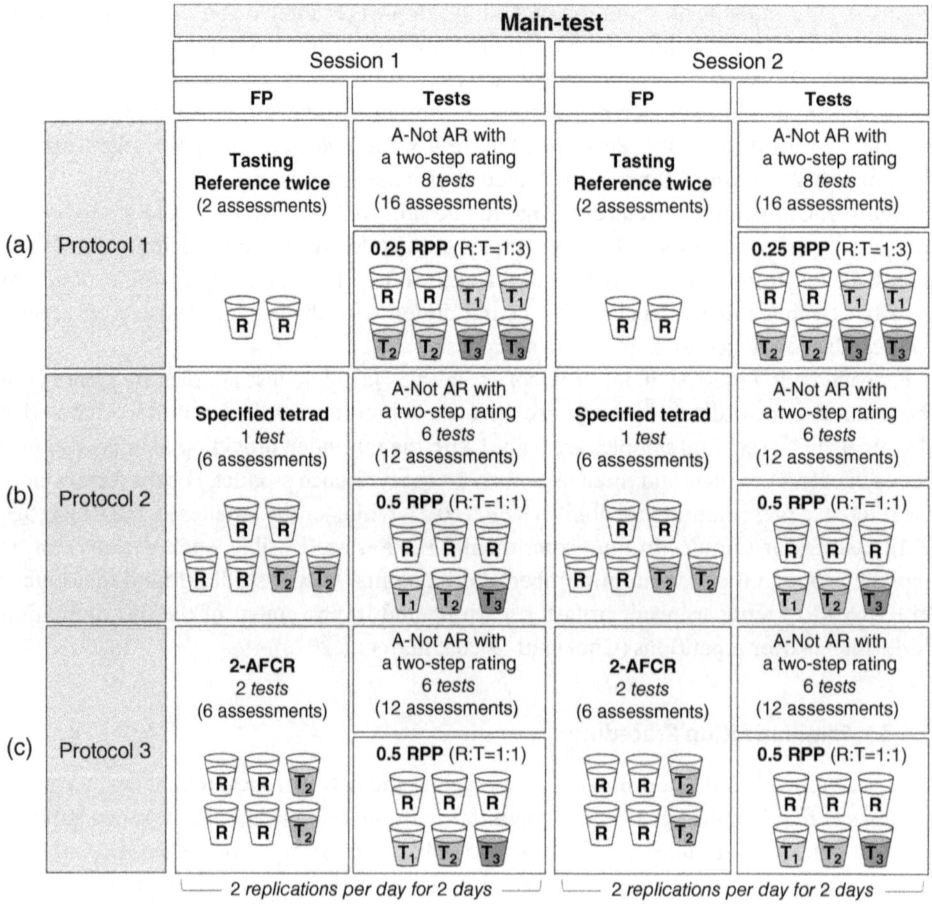


**Figure 9.4** Examples of A-Not AR test protocols with three different types of familiarization procedures (adapted from Lee et al. 2023). "FP" stands for familiarization procedure, and "RPP" stands for reference presentation probability. "R" stands for a reference sample, and "$T_1$," "$T_2$," and "$T_3$" stands for the three test samples. One of R, $T_1$, $T_2$, and $T_3$ is presented randomly as a blind sample

#### 9.3.3.3 2-AFC with Reminder

When more replications are possible, the 2-AFCR test method shown in Protocol 3 in Figure 9.3c, might be the optimal option for the FP performed with the reference-based discrimination methods as it facilitates the learning the difference of the test samples from a reference better than other procedures (Lee et al. 2023). For more information on the test format and instructions, see Figures 9.1 and 9.3b, respectively.

### 9.3.4 Utilizing Signal Detection Theory Indices: Data Processing Approach and Applications

For accurate and reliable business decision-making in quality management, the sensory test methods and data analysis used should be robust and provide informative results.

The unique design of the reference-based discrimination test methods such as the A-Not AR test, in which separate responses are given for each sample, including the (blind) reference, makes it possible to calculate an index indicating the degree of familiarity/recognition with samples called recognition $d'$ ($d'_{Rec}$) in addition to the middle criterion location ($c_0$) indicating subjects' evaluative criterion and discrimination performance $d'$ indicating the size of product difference in the SDT context. The subject's familiarity with the sensory characteristics of the reference is very important for SDT analysis using the reference-based discrimination test methodology because it influences the stability of the decision criteria and directly affects the quality of the SDT results. The subject's familiarity with products influences the results of every discrimination test, yet, for most sensory discrimination tests, it is not possible to investigate the subject's familiarity because it cannot be measured separately from the discrimination performance using conventional tests.

Thus, the SDT analyses using the reference-based discrimination test methods provide quantitative indices that reflect not only the product differences but also panel performance and data quality. The computation of three different SDT indices, product discrimination $d'$, judgment criterion $c_0$, and recognition $d'$ ($d'_{Rec}$), and their meanings and applications will be described and discussed in this section. These three different SDT indices $d'$, $c_0$, and $d'_{Rec}$ are closely related but have different meanings and are complementary in terms of the information they provide for assessing sensory panel performance and data quality. All SDT indices can be computed by using SDT Assistant software (Hautus 2021).

#### 9.3.4.1 Discrimination $d'$: A Measure of the Size of Product Difference

***9.3.4.1.1 Data Process*** Figure 9.5 illustrates how the data are collected and pre-treated for the SDT analysis using the A-Not AR with a two-step rating. The test samples, including samples of the reference and test products, are presented randomly in a blinded manner,

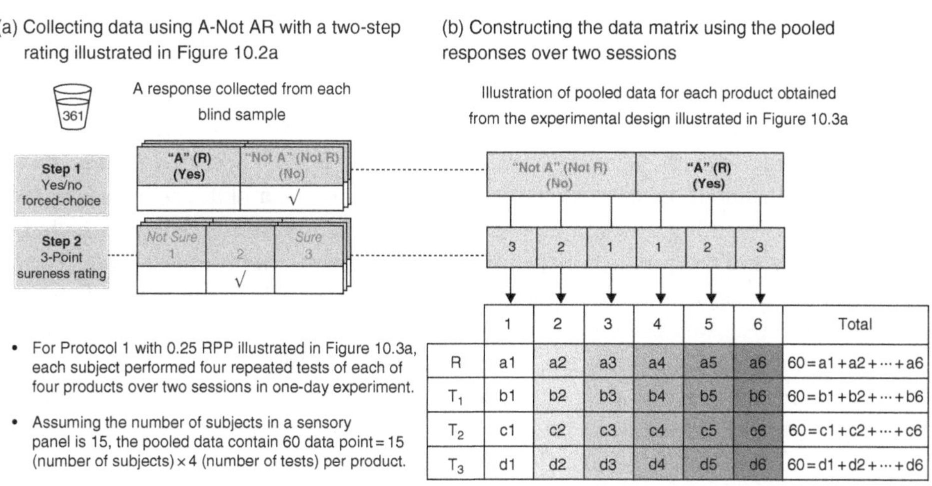


**Figure 9.5** Illustration of the data process on the pooled data obtained from A-Not AR with a two-step rating; (a) how to collect the data using a two-step rating format, (b) how to construct data matrix using the obtained pooled responses for signal detection analysis (adapted from Choi et al. 2023a). a1 refers to the value in the cell identified by a1.

one by one, and one response is collected from each of these blind test samples. In the A-Not AR with a two-step rating test, the subjects are instructed to answer "A (Yes)" or "Not A (No)" in step 1 and then to answer a three-point sureness rating in step 2 (Figure 9.5a). As this method is commonly applied when there is a golden standard or reference (R) product, in step 1, the subject is often instructed to answer if the presented blind test sample is "R" or "Not R," as shown in Figure 9.5a. Repeated tests are performed per subject and in general, the data points are pooled over all the subjects. Let's assume an experiment consisting of two sessions each session having eight tests for discriminating three test products (coded as "$T_1$," "$T_2$," and "$T_3$") from a reference (coded as "R"), as illustrated in Protocol 1 in Figure 9.4a. In this case, the RPP is $R:T_1:T_2:T_3 = 1:1:1:1$, presenting each sample at equal frequency. From such a session, assuming the number of subjects (sensory panelists) is 15, a total of 60 data points is pooled for a six-point data matrix composed for each product, including the reference as shown in Figure 9.5b.

Figure 9.6 illustrates the data matrices used for pair-wise comparisons between the reference and a test product and how the signal detection decision space is drawn from the data matrices to compute the discrimination $d'$. The data obtained from tests of each blind samples of test product is compared with the data obtained from tests of the blinded reference samples, and the size of the sensory difference $d'$ is calculated. In SDT, the data matrix of each product is treated as a perceptual distribution based on a Gaussian probability, reflecting the one-dimensional representation of sensory evidence in the brain as illustrated at the bottom of Figure 9.6 (Hautus et al. 2021).

As the A-Not AR rating test uses a single sample monadic test, evaluating the difference/similarity of the overall sensory properties, there is only one dimension of variability for the sensory perception (Figure 9.6c). Along this one dimension, subjects' five decision criteria locations (c) used to decide on the two-step response options are illustrated in Figure 9.6b. Although the illustration used a two-step rating response option, only a single step "A" or "Not A" response option can also be used. In this case the decision criterion (response bias) located in the middle of the five criteria is the criterion (response boundary) between step "A" and "Not A" and it was named "the middle criterion location ($c_0$)" (Choi et al. 2023a, b). This middle criterion location ($c_0$) indicates a subject's response tendency between "A" and "Not A" (or between "R" and "Not R"). When the subject is familiar enough with the sensory characteristics of the reference product and understands the discrimination dimension, regardless of where the middle criterion is located, the same decision criteria would be used for each of all (blind) test samples including both reference and test products, separating response distributions of test products from the distributions of the reference. The $d'$ measure is based on this separation between the distributions of sensory evidence and reflects the degree of sensory difference between each test product and the reference, not contaminated by the subject's decision criteria. In other words, the important advantage of SDT is that both the sensory discrimination performance ($d'$) and the criteria location (c) of the subject can be observed (Lawless and Heymann 2010; Hautus et al. 2021). Looking at Figure 9.6, the size of the sensory difference is lager between R and $T_3$ than between R and $T_1$ because the two distributions are more separated between R and $T_3$, producing higher hit rates compared with the false alarms rates for the $R-T_3$ product pair than the $R - T_1$ product pair (Figure 9.6).

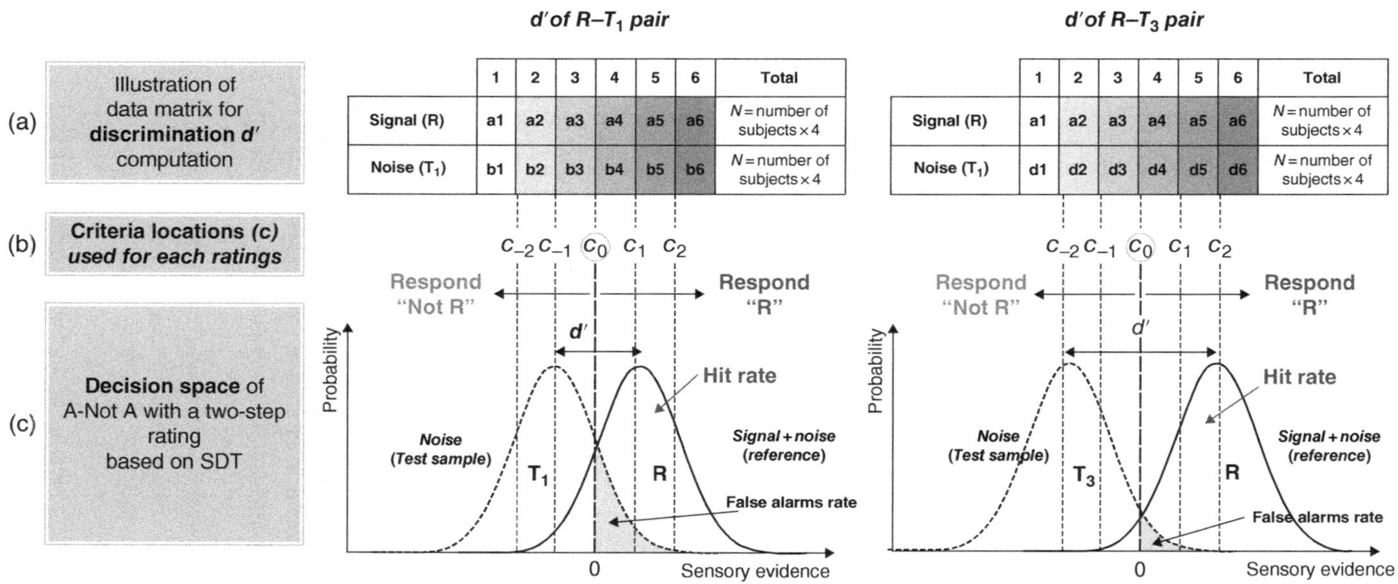


**Figure 9.6** Illustration of the data matrix for pair-wise comparisons of the responses obtained from the A-Not AR with a two-step rating for computing the discrimination $d'$ and criteria location ($c$), and decision space based on signal detection theory; (a) how to construct a data matrix for computing $d'$ for each pair, (b) illustration of subjects' judgement criteria used to provide responses to the A-Not AR rating test, (c) signal detection decision space, indicating the size of discrimination $d'$ (adapted from Choi et al. 2023a).

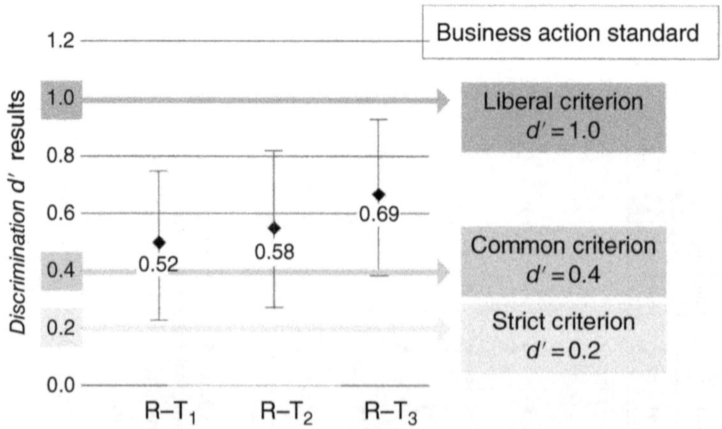


**Figure 9.7** Illustration of the results of discrimination *d′* and their comparison to the business action standard (examples of the quality criterion).

The *d′* estimates were calculated to compare the subjects' discriminability between a reference (R) and test samples ($T_1$, $T_2$, and $T_3$). The responses to a reference sample are treated as a signal distribution, and the responses to each test sample are treated as a noise distribution. The *d′* estimates of <R – $T_1$>, <R – $T_2$>, and <R – $T_3$> sample pairs are calculated to check the sensory differences between the reference and each test sample. In addition, <R – $T_{1+2+3}$> can also be analyzed to observe the subjects' general discrimination performance by pooling the responses of three test samples.

***9.3.4.1.2 Applications*** The discrimination *d′* is one of the most widely used SDT measures. It indicates the degree of sensory difference between the samples and can be used as an index for panel discrimination performance. By comparing the size of *d′* considering 95% confidence interval (the discrimination performance) with a pre-determined AS of *d′* – an internal sensory quality standard for quality management in a company, it can be determined whether the size of the difference is acceptable or not. If the observed *d′* result of a sample pair is larger than the pre-determined AS *d′*, the difference between the test product and the reference product is interpreted as significant. In turn, the test product would be rejected as an alternative to the reference. However, if the observed *d′* result is smaller than the pre-determined *d′*, such differences would be considered insignificant, and the test product would be accepted. Bi (2011) suggested that $d' = 0.2$, 0.4, and 1.0 as strict, common, and liberal criteria for similarity testing. Figure 9.7 illustrates the *d′* results of three test products where all the test products would be rejected based on the strict criterion ($d' = 0.2$) or common criterion ($d' = 0.4$), while they will be accepted as alternatives to the reference product if the company has a liberal criterion ($d' = 1.0$).

#### 9.3.4.2 Criterion Location, c: A Measure of the Subject Judgements

***9.3.4.2.1 Data Process*** When performing an A-Not AR test, subjects are required to make a decision between two response options, "A" or "Not A," for a given blind sample. To examine whether a subject's responses are skewed towards "A" or "Not A" among these two response options, for each subject, the middle criterion location $c_0$ which is the response boundary located in the middle among all possible response boundaries can be computed (Figure 9.6b). For this computation, the SDT Assistant software is also used (Hautus 2021); an appropriate model for the test has to be selected first; for example, for A-Not A test, the equal-variance normal model for Yes–No can be adopted. Then, the responses to each sample are summed, and the responses obtained from the reference and the test samples are treated as a "signal + noise" probability and a "noise" probability, respectively.

Figure 9.8 illustrates two different subject groups having discrepant criterion locations – decision criteria. Subjects with loose $c$ ($c_0 < 0$) tend to respond more "R" (a) and subjects with strict $c$ ($c_0 > 0$) tend to respond more "Not R." Thus, regardless of discrimination performance, discrepant $c_0$ values can be computed for the same pair of products. These values could be shifted by appropriate test instructions suggesting to use the whole range of the response options and 0.5 RPP of sample presentation (Choi et al. 2023a).

#### *9.3.4.2.2 Application*

When using a sensory panel as an instrument for quality management, the data across all panelists are pooled for the discrimination analysis. Therefore, it is important to make all the panelists use the equivalent judgments. To check for this, it is necessary to check for inter-subject variability of this criterion measure. In other words, this measure can be used to classify panelists having the similar response criteria and find out panelists having an extreme criterion. Such panelists who have an extreme decision criterion or being biased towards certain response options need to be trained for using response options more appropriately.

### 9.3.4.3 Recognition $d'$ ($d'_{Rec}$): A Measure to Check the Panel Performance

***9.3.4.3.1 Data Process*** Figure 9.9 illustrates how to compute recognition $d'$ ($d'_{Rec}$). Previously, for measuring consumer satisfaction using a response format of the A-Not A with a two-step rating, the signal distribution was constructed by each sample response, and the noise distribution was generated hypothetically by mirror effects (Kim et al. 2018b). Similar to this approach, the $d'_{Rec}$ values can be computed to monitor the subjects' recognition of each sample. Namely, the pair-wise data matrix is composed for each $d'_{Rec}$ value, by matching the reversed data set across the rating categories (Figure 9.9a). The better the subjects recognize the samples, the bigger the $d'_{Rec}$ values are. On the other hand, the $d'_{Rec}$ values are small, even negative, when subjects have unstable recognition of the samples, meaning they do not correctly respond to each sample.

***9.3.4.3.2 Application*** The $d'_{Rec}$ values of the reference (R) reflect the degrees of recognizing R (familiarity with R). And this can be used as a quality measure for assessing the nature of panel performance; that is, how well panelists learned about sensory characteristics of the reference and trained to consistently perform the sensory discrimination tests.

The $d'_{Rec}$ results of the test products ($T$) reflect the degrees of recognizing $T$ differentiating from R. This can also be used as a quality measure for assessing the nature of panel performance; that is, how well panelists learned to discriminate sensory characteristics of the test products from the reference product. While trained, this should be proportionally increased with $d'_{Rec}$ results of R for un-biased subjects.

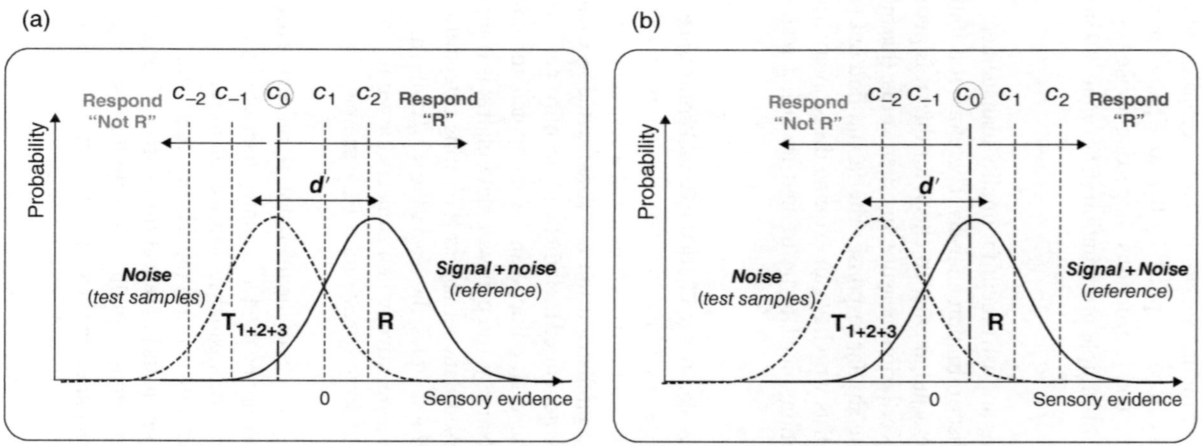


**Figure 9.8** Illustrations of two different subject groups having discrepant decision criteria based on the middle criterion location ($c_0$). "$c_0$" is a middle criterion location which is the response boundary between response "R" and "Not R" in the A-Not A with a two-step rating (adapted from Choi et al. 2023a). (a) Subjects with **loose c ($c_0 < 0$)**. (b) Subjects with **strict c ($c_0 > 0$)**.

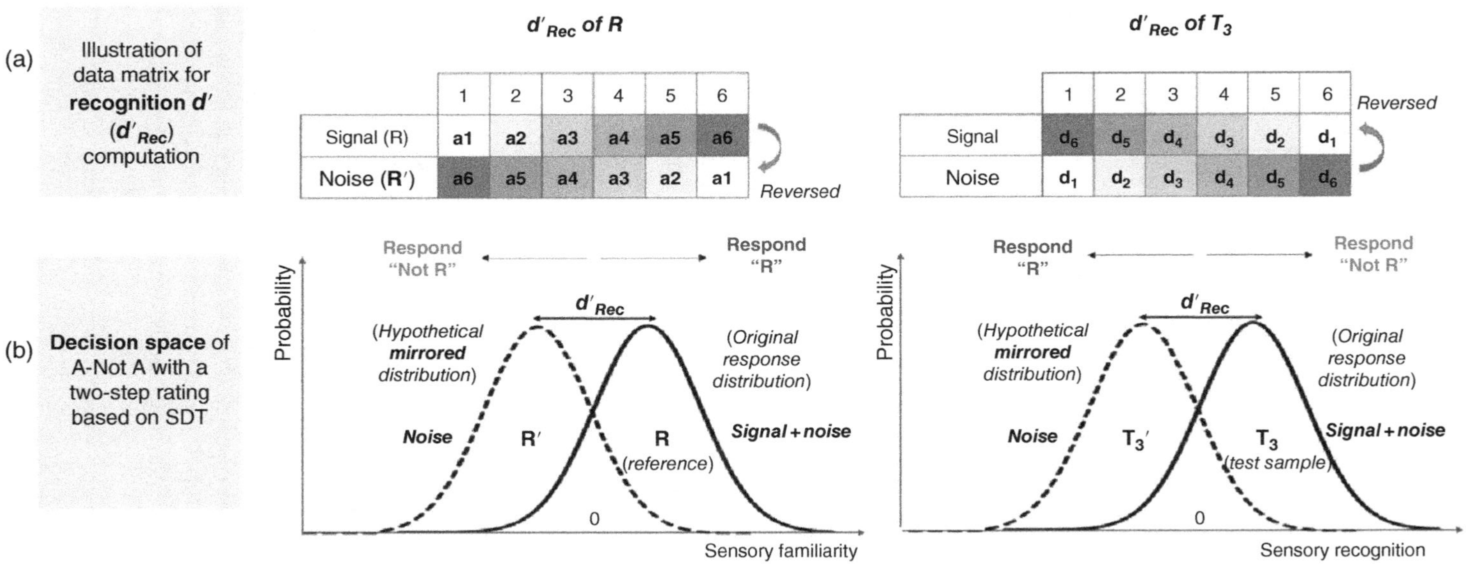


**Figure 9.9** Illustration of recognition $d'$ ($d'_{Rec}$) computation; (a) data matrix composed from the pooled responses obtained from the A-Not AR with a two-step rating for computing recognition $d'$ ($d'_{Rec}$), (b) decision space based on signal detection theory (adapted from Choi et al. 2023a).

To be informative for making better business decisions, a sensory analysis should be based on quantitative indices, and SDT analyses can be used for such purposes. The three different SDT indices, including the $c_0$, $d'$, and $d'_{Rec}$, which are closely related to each other, have different meanings and provide different information on monitoring the panel performance and checking the data quality of the discrimination test.

When conducting sensory discrimination using a sensory panel, optimal data quality management can be achieved when by considering the situation below. In these situations, discrimination $d'$ will be increased and the data quality will be improved based on pooled data analysis.

1) *The decision criterion (criterion location) measure, $c_0$ is close to zero,* having non-significant inter-subject variability.
2) Both the $d'_{Rec}$ values of the reference (R) and the $d'_{Rec}$ results of the test products ($T$) are positive.

### 9.3.5 A Strategy for Defining and Maintaining Action Standards

To effectively manage the sensory product quality, it is essential to establish consumer-relevant AS as part of a comprehensive sensory evaluation program. These standards must strike a delicate balance, ensuring that product differences that consumers would reject are identified and addressed, while also allowing for acceptable variation to avoid unnecessarily limiting product reformulation opportunities.

The following steps outline a strategic approach for establishing such consumer-relevant AS for product quality management using discrimination testing with sensory panels and calibrating the sensory panelists so that it can predict consumers' sensitivity. Detailed explanations of data analysis and interpretation of these steps can be found in the preceding section of this chapter and are illustrated in the case study presented in Section 9.3.6.

1) **Implement the sensory panel**: Screen and select panelists, train the panelists in the basic skills of sensory evaluation of products, practise discrimination tests, and familiarize them with the product type.
2) **Design the discrimination test**: Experiment to determine the ideal discrimination test design (A-Not A, A-Not AR, or 2-AFCR) for this specific product type, which will serve as a quality control measure. Establish the context under which the panel will evaluate the products, such as with or without host food, and under red light conditions. Introduce the reference product and test method to the panel to ensure familiarity.
3) **Monitor the panelists' quality**: Perform a discrimination test monitoring study. Include four to five products that differ from the reference product in different ways, similar to what would be expected when products are being reformulated. Monitor the panelists' quality using the SDT quality parameters $d'$, $d'_{Rec}$, *and* $c$. In case the panelists' quality is not up to the required standards, familiarize the panelists further and perform a new monitoring test.
4) **Select the product set and conduct the study**: Determine which products should be included in the main experiment, representing the diversity in product differences that can be expected through reformulation. Perform sensory discrimination tests with the sensory panel and with representative consumers on the same set of products. Ensure that

the consumers evaluate the products in a context that represents their normal usage. E.g. margarine on bread, bouillon as basis in a soup.

5) **Data analysis and modeling**: Analyze the results to explore the relationship between the panel and consumers, through correlation analysis and the development of a linear regression model to predict consumer responses from sensory panel results. Decide on the critical value for consumers, below which the size of the difference should remain without it impacting consumers' perception. As a rule of thumb, $d' < 1$ can be used as it is universally regarded as a just noticeable difference (JND).
6) **Define action standards (AS)**: After processing all relevant information, establish AS for the sensory panel. These standards determine the level of difference below which the reference can vary without impacting consumer perception of the product. Once the AS is established, the sensory panel can conduct discrimination tests and utilize the AS as a critical value to assess consumer acceptance. This process is essential for maintaining product quality and meeting consumer expectations in a competitive market environment.

### 9.3.6 Case Study: Driving Consumer Relevance in Sensory Quality Control Through Action Standard Optimization: Bridging the Gap Between Sensory Panel and Consumer Sensitivity

#### 9.3.6.1 Introduction

A CPG company produces a specific margarine for many years. Many consumers use the product daily and rely on its consistently high quality. For quality control they have an internal panel familiarized with the reference product and performing A-Not A tests of reformulated products vs. the reference product. The current AS is $d' < 1$ in the sensory panel, as this is commonly regarded as a JND. To remain flexible in the competitive market environment, the company wants to understand the consumer relevance of this AS, by determining how much it can change the processing factors and ingredients without affecting the sensory quality as perceived by consumers. Determined to optimize their sensory panel-based AS into more consumer-relevant AS, they want to explore the relationship between the sensitivities of consumers and the internal sensory panel (Ishhi et al. 2007; Van Hout 2014). Utilizing these optimized AS, the sensory panel can serve as an evaluative instrument to assess if consumers will perceive the sensory differences caused by the alterations of the product formulation.

#### 9.3.6.2 Establishing the Current Sensory Panel-based AS

The current AS is measured by the sensory panel, with $d' < 1$ indicating acceptance. In practice, a $d'$ value greater than 0.65 would fail. Historical data shows that when $d'$ exceeds 0.65, the upper confidence interval is likely greater than $d' = 1$. Given a usual sample size of $N = 40$ and an average standard error (SE) of 0.21, we can estimate a typical confidence interval. First, we calculate the margin of error (ME) by multiplying the $z$-score for a 90% confidence interval (1.645) by the SE (0.21), resulting in an ME = 0.345. For a $d' = 0.65$, the 90% confidence interval is $d' \pm \text{ME}$ (0.305, 0.995). Since we are only concerned with the upper level of the CI being below $d' = 1$, we can be 95% confident that if the sensory panel $d' = 0.65$ the true $d'$ value is significantly below 1, meeting the acceptance criteria. Note that a two-sided CI is used for comparing different $d'$ values. If the focus was solely on whether the difference is significantly below 1, a 1-sided upper-CI at 95% would yield a similar conclusion.

#### 9.3.6.3 Methods and Materials

***9.3.6.3.1 Products*** The study consisted of two experiments that contained different product sets.

- In Experiment 1, seven commercial brands of margarine from the European market were used with small variations in texture and flavor. These products will be referred to as A, B, C, D, E, F, and G.
- In Experiment 2, two sets of products were used. The first product set consisted of two commercial products from Experiment 1 (A and D), and two mixtures of these two products. These products will be referred to as A, D, Ad (mixture of two-thirds of A and one-third of D), and Da (mixture of one-third of A and two-thirds of D). The second product set included one of the commercial products (F) selected from Experiment 1 and two products from the same base product but with different levels of salt added. Products will be referred to as F, F1, and F2.

***9.3.6.3.2 Test Method*** Multiple A-Not AR tests in which each test product (including blind references) is compared to the reference.

***9.3.6.3.3 Sensory Panel*** The internal panel consists of 12 panelists that are screened and trained to evaluate margarine products regularly and consistently, and are familiarized with the reference by tasting it a few times. After familiarization, they perform the replicated A-Not AR tests. The panel performs one replication of one product set on a test day. First, eight repetitions of the product set 1 were completed, after which the panel continued with product set 2. The study was completed in 16 test days, with a total of eight replications per panelist. Note that there were a few missing data points due to the absence of panelists on some days.

***9.3.6.3.4 Consumers*** Consumers that frequently use the reference product (at least 2 times a week) are selected. The consumers first evaluated product set 1, and one month later product set 2, a total of 105 consumers participated in the study. Note that there were a few missing data points due to the dropout of consumers in the second study. In the consumer study, the respondents were first given a reference product to refresh their memory, after which they performed the multiple A-Not AR study in a randomized balanced order in two sessions.

#### 9.3.6.4 Data Analysis

***9.3.6.4.1 Sensory Panel*** For the data of the internal sensory panel, two approaches are possible for calculating $d'$ for a group of subjects. The first method is to calculate $d'$ for each subject and then average over individuals. The second method is to pool the data from all subjects and calculate a pooled $d'$. It depends on the variation in the group of subjects and the number of repetitions which of the two methods is optimal. Hautus (1997) found that when the number of repetitions is low, a pooled $d'$ would be less biased. With many replications per subject, the average $d'$ is less biased. The general recommendation is to calculate $d'$ in both ways and use the highest, as it has the greatest likelihood of being the least biased. In this study, pooling resulted in the highest $d'$ values and would be the most accurate estimate of the true $d'$.

***9.3.6.4.2 Consumers*** The consumer responses in the A-Not AR test were pooled. After which $d'$ estimates and standard deviations were calculated by fitting ROC curves. This was accomplished using the software, SDT Assistant (Hautus 2021), to fit the appropriate SDT model to the data using maximum-likelihood estimation. The likelihood function employed was that developed by Dorfman and Alf (1969).

#### 9.3.6.5 Results

The following tables contain the results of the studies: the results of the sensory panel are in Table 9.3 and the consumer study results can be found in Table 9.4.

Table 9.5 presents a summary of the sensory panel and consumer results, including $d'$ values and standard errors (SE). To determine significant differences between two $d'$ values, 90% confidence intervals are employed. In each row, letters are assigned to indicate

**Table 9.3** Sensory panel results.

| Sensory panel study | Different sure | Different unsure | Guess different | Guess same | Same unsure | Same sure | Total | $d'$ |
|---|---|---|---|---|---|---|---|---|
| **Experiment 1** | | | | | | | | |
| Ref(A)-B | 88 | 2 | 1 | 0 | 1 | 0 | 92 | 2.77 |
| Ref(A)-C | 72 | 9 | 1 | 2 | 4 | 4 | 92 | 1.77 |
| Ref(A)-D | 77 | 3 | 1 | 2 | 4 | 5 | 92 | 1.81 |
| Ref(A)-E | 88 | 1 | 0 | 0 | 3 | 0 | 92 | 2.62 |
| Ref(A)-F | 81 | 4 | 4 | 2 | 0 | 1 | 92 | 2.25 |
| Ref(A)-G | 83 | 3 | 1 | 1 | 1 | 3 | 92 | 2.17 |
| Ref(A)-A (blind reference) | 16 | 3 | 4 | 4 | 21 | 44 | 92 | |
| **Experiment 2** | | | | | | | | |
| Ref(A)-D | 92 | 12 | 1 | 1 | 4 | 1 | 111 | 1.79 |
| Ref(A)-Ad | 46 | 12 | 2 | 3 | 23 | 26 | 112 | 0.7 |
| Ref(A)-Da | 95 | 7 | 0 | 1 | 8 | 1 | 112 | 1.83 |
| Ref(A)-A (blind reference) | 22 | 9 | 1 | 2 | 21 | 57 | 112 | |
| Ref(F)-F1 | 56 | 16 | 1 | 0 | 19 | 7 | 99 | 0.63 |
| Ref(F)-F2 | 79 | 9 | 2 | 2 | 4 | 5 | 101 | 1.28 |
| Ref(F)-F (blind reference) | 32 | 14 | 1 | 3 | 29 | 20 | 99 | |

**Table 9.4** Consumer study results.

| Consumer study | different sure | different unsure | guess different | guess same | same unsure | same sure | Total | $d'$ |
|---|---|---|---|---|---|---|---|---|
| **Experiment 1** | | | | | | | | |
| Ref(A)-B | 73 | 13 | 5 | 5 | 7 | 2 | 105 | 1.24 |
| Ref(A)-C | 45 | 17 | 3 | 6 | 13 | 13 | 97 | 0.53 |
| Ref(A)-D | 60 | 14 | 6 | 6 | 8 | 11 | 105 | 0.78 |
| Ref(A)-E | 56 | 15 | 4 | 9 | 16 | 5 | 105 | 0.77 |
| Ref(A)-F | 56 | 16 | 11 | 9 | 7 | 6 | 105 | 0.84 |
| Ref(A)-G | 73 | 10 | 4 | 3 | 8 | 7 | 105 | 1.08 |
| Ref(A)-A (blind reference) | 21 | 22 | 10 | 10 | 21 | 21 | 105 | |
| **Experiment 2** | | | | | | | | |
| Ref(A)-D | 47 | 22 | 11 | 7 | 8 | 5 | 100 | 0.72 |
| Ref(A)-Ad | 34 | 20 | 12 | 6 | 24 | 5 | 101 | 0.38 |
| Ref(A)-Da | 43 | 21 | 2 | 9 | 18 | 7 | 100 | 0.53 |
| Ref(A)-A (blind reference) | 25 | 11 | 13 | 12 | 21 | 19 | 101 | |
| Ref(F)-F1 | 47 | 15 | 7 | 10 | 17 | 5 | 101 | 0.49 |
| Ref(F)-F2 | 58 | 12 | 7 | 5 | 5 | 13 | 100 | 0.74 |
| Ref(F)-F (blind reference) | 28 | 13 | 9 | 15 | 26 | 10 | 101 | |

**Table 9.5** Sensory differences, are measured by consumers and the internal sensory panel.

| | | Experiment 1 | | | | | |
|---|---|---|---|---|---|---|---|
| | | **Ref(A)-B** | **Ref(A)-C** | **Ref(A)-D** | **Ref(A)-E** | **Ref(A)-F** | **Ref(A)-G** |
| Sensory panel | $d'$ | 2.77a[a] | 1.77b | 1.81b | 2.62ab | 2.25ab | 2.17ab |
| | SE | 0.26 | 0.17 | 0.2 | 0.26 | 0.22 | 0.22 |
| Consumers | $d'$ | 1.24a | 0.53b | 0.78ab | 0.77ab | 0.84ab | 1.08ab |
| | SE | 0.16 | 0.15 | 0.15 | 0.15 | 0.15 | 0.16 |
| | | **Experiment 2** | | | | | |
| | | **Ref(A)-D** | **Ref(A)-Ad** | **Ref(A)-Da** | **Ref(F)-F1** | **Ref(F)-F2** | |
| Sensory panel | $d'$ | 1.79a | 0.70b | 1.83a | 0.63a | 1.28a | |
| | SE | 0.21 | 0.15 | 0.21 | 0.17 | 0.21 | |
| Consumers | $d'$ | 0.72a | 0.38a | 0.53a | 0.49a | 0.74a | |
| | SE | 0.16 | 0.15 | 0.15 | 0.16 | 0.17 | |

[a] Different letters are used to indicate significant differences ($p = 0.05$) between $d'$ values in a row.

significant differences ($p = 0.05$) between $d'$ values. If two $d'$ values in a row share the same letter, it signifies that their confidence intervals overlap and the difference between the two values is not statistically significant.

The $d'$ analysis of results revealed that for all products, differences detected by the sensory panel are larger than those detected by consumers. The panel was significantly more sensitive than consumers and able to detect more subtle differences in the product. The correlation between the two measures is 0.70, indicating that the $d'$ value of consumers can be reasonably predicted from the $d'$ of a trained panel. The replicated products A and D, incorporated in both experiments, exhibited highly similar outcomes (resp. 1.81 and 1.79 in the sensory panel, and 0.78 and 0.72 for consumers). This consistency indicates the robustness and repeatability of the data, providing greater confidence in merging the datasets for enhanced learning. Moreover, it suggests that integrating new datasets could further refine the relationship between the internal sensory panel and consumers.

Linear regression was used to investigate the relationship between consumer and sensory panel sensitivity and to ascertain the degree of consistency in the observed relationship. A linear regression analysis was conducted to establish the relationship between the $d'$ value of consumers from the $d'$ value of a trained panel, Figure 9.10. shows the $d'$ values of the product differences and regression line with a 90% two-sided CI, illustrating the relationship between the sensory panel and consumers.

The linear regression model describing the relationship between the sensory panel ($x$) and consumers ($y$) is $y = 0.28x + 0.23$. The regression coefficient of 0.28 (95% confidence interval = 0.11–0.46) indicates that the panel is more sensitive than consumers. The model quality is expressed by $R^2$, which indicates the proportion of the variance in the consumer data that can be explained by the sensory data(s) in the model. $R^2$ values range from 0 to 1, with higher values signifying a better fit and a stronger relationship between the variables. This model's $R^2$ value is 0.49, which can be considered as moderate and means that the linear regression model explains 49% of the variance in the dependent variable. In other

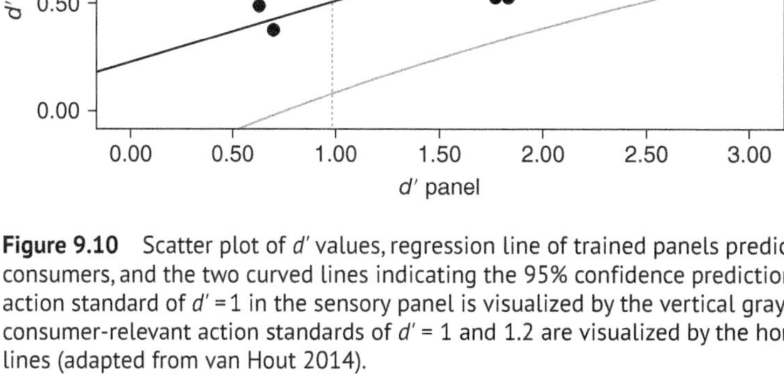


**Figure 9.10** Scatter plot of $d'$ values, regression line of trained panels predicting the sensitivity of consumers, and the two curved lines indicating the 95% confidence prediction interval. The original action standard of $d' = 1$ in the sensory panel is visualized by the vertical gray dotted line. The consumer-relevant action standards of $d' = 1$ and 1.2 are visualized by the horizontal gray dotted lines (adapted from van Hout 2014).

words, 49% of the variation in consumer sensitivity can be accounted for by sensory panel sensitivity. The remaining 51% of the variance is unexplained, which could be due to other factors not included in the model, random error, or non-linear relationships between the variables. Taking into account that the product set is small, two different reference products were included, together with a broad range of product differences in the test, this is a reasonably good basic model that can be improved by adding more data from future tests.

#### 9.3.6.6 Optimizing the Sensory Panel-based AS to Become More Consumer-Relevant

With this new information, the AS can be optimized and become more consumer-relevant. For the new AS the assumption is that for the products to be acceptably similar, the difference as perceived by consumers should be $d' < 1$. Translated to differences in the sensory panel, we can use the information from Figure 9.10. When the upper confidence interval reaches $d' = 1$ for consumers, this difference in the sensory panel is $d' = 1.3$. This means that differences of $d' < 1.3$ in the sensory panel, are likely to be $d' < 1$ for consumers. The new sensory panel AS for products to be regarded as similar can therefore be changed to $d' < 1.3$. Given the usual sample size of $N = 40$ and the average standard error (SE) of 0.21, the ME = 0.345. For a $d' = 0.90$, the 90% confidence interval is $d' \pm$ ME (0.56, 1.245). Since we are only concerned with the upper level of the CI being below $d' = 1.3$, we can be 90% confident that when the difference in the panel is below $d' = 0.90$, the true $d'$ value is significantly below 1.3, meeting the acceptance criteria.

Using the updated AS, the sensory panel now serves as a predictive tool for evaluating the consumer relevance of perceived differences. The new sensory panel's established AS is set at $d' < 1.3$, indicating that any $d'$ values less than 0.90 are considered acceptable. In contrast, the previous AS would reject any differences with a $d'$ value equal to or greater than 0.65. This increased tolerance for product variations from the reference will provide product development teams with more opportunities for reformulation.

## 9.4 Further Development

### 9.4.1 Further Development of the Sensory Panel as a Predictive Tool for Consumer Perception

To enhance the sensory panel's ability to predict consumer perceptions more accurately, the model can be refined using additional data from both the sensory panel and consumers. Incorporating more data will help fine-tune the predictive model. Furthermore, the model can be adapted for other consumer groups, as variations in sensitivity may exist between consumers from different countries.

For maintaining sensory panel performance and sensory quality data check, it would also be needed to utilize regular training on the sensory stimuli representing the sensory space of the sensory map of products under interests to have them to learn and maintain the familiarity to the sensory perception of the product. Thus, more investigations of effective and efficient panel training procedure for developing sensory panels as heavy consumers are still required.

### 9.4.2 Further Development of the Reference-Based Methodology for Consumer Research for Affective Tests

In this chapter, the reference-based SDT discrimination test methodology using A-Not AR with a two-step rating was demonstrated focusing on the efficient and effective use of sensory panels. This reference-based discrimination approach could also be applied to consumer research to quantify consumer perception and affects including hedonic, preference, and satisfaction (acceptance) evaluation. Kim et al. (2018a and 2018b) developed a new measurement method for a satisfaction measure as an alternative to the hedonic scaling, referred to as the "degree of satisfaction-difference (DOSD)" test. In this test, the discrimination measure $d'$, reflecting the satisfaction difference between a test sample and the reference, and the satisfaction measure $d'_{SAT}$, reflecting the degree of satisfaction for each product, can be computed and linked with the results of the degree of sensory differences between samples contained from sensory panels. In particular, the satisfaction measure $d'_{SAT}$ is calculated by comparing each sample to a neutral sample, neither satisfying nor not satisfying. Here, criterion ($c$) measures could be used to determine if a sample was satisfying or not. The DOSD test was the original consumer sensory research methodology that utilizes the affective framing of the reference using cognitive warm-up and a series of paired comparisons of multiple samples with a fixed reference (Kim et al. 2018a, b, 2019). Utilizing the DOSD format, yet, without adopting the affective familiarization or cognitive warm-up, a different version of signal detection reminder-preference test has also recently been proposed and successfully applied to test the increase in the perceived saltiness of food without adding more salt (Esfanjnai and Mohebbi 2023). More research is needed for comparing and refining the different approaches of the reminder-preference tests and the affective procedures to further improve the signal detection framework to be able to study the sensory perception/discriminability and preference simultaneously.

In this chapter, it has been explained that by implementing the reminder-based discrimination test methodology using an internal sensory panel, it is possible to predict consumer discrimination. But there could be cases in which even if consumers perceive sensory differences between the reference and a test sample (for example, a prototype – one of the reformulated samples), the perceived difference could be negligible and may not be large enough to change their preference. In this case, the reformulated sample can still be considered a candidate to replace the original product, although there are perceivable sensory differences between them. Moreover, when the reformulated sample is perceived differently from the reference, it is possible that it could be preferred over the reference. Therefore, to make efficient business decisions to introduce reformulated food products, it would be useful to study not only the sensory differences between the reference and the reformulated samples but also consumer preferences of them. For this reason, an affective difference-preference test using an A-Not A with a reminder (A-Not AR) design with a difference question followed by a preference question was also introduced and named the reminder-preference test (Kim et al. 2022). Based on SDT and TM, this test can quantify the degree of sensory difference and preference of each sample from the reference product with and without marketing/brand effects, by utilizing the affective cognitive warm-up for the reference (Kim et al. 2023). The reminder-preference test is developed as an economical approach using consumer sensory research methodology that can be easily conducted in a laboratory. More research on this reference-preference test methodology is also required to advance sensory quality management.

The SDT discrimination approach could also be applied to consumer perception and concept measurement; holistic softness (Kim et al. 2013), and Goso flavor (Kim et al. 2020) have been studied in SDT context, in a similar manner as for computing recognition $d'$ ($d'_{Rec}$). This approach was found to be a useful quantitative measure because it doubles the data points, increases the test power, and allows multiple product comparisons. More research on consumer affect and memory measures using such SDT indices would enhance not only the understanding of consumers' quality perception of products but also open up innovative approaches to sensorial and emotional segmentation of the food market.

## References

Bi, J. (2005). Similarity testing in sensory and consumer research. *Food. Qual. Prefer.*, 16(2), 139–149.

Bi, J. (2011). Similarity tests using forced-choice methods in terms of Thurstonian discriminal distance, *d'*. *J. Sens. Stud. 26* (2): 151–157.

Bi, J. (2015). *Sensory Discrimination Tests and Measurements: Sensometrics in Sensory Evaluation*, 2e. Wiley: Wiley.

Bi, J., Lee, H. S., and O'Mahony, M. (2010). d' and variance of d' for four-alternative forcedchoice (4-AFC). *J. Sens. Stud.*, 25(5), 740–750.

Bi, J., Lee, H. S., and O'Mahony, M. (2013). Statistical analysis of receiver operating characteristic (roc) curves for the ratings of the A-Not A and the same–different methods. *J. Sens. Stud.*, 28(1), 34–46.

Bi, J., Lee, H.S., and O'Mahony, M. (2014). Estimation of Thurstonian models for various forced-choice sensory discrimination methods as a form of the "$M+N$" test. *J. Sens. Stud. 29* (5): 325–338.

Choi, E.S., Lee, H.L., Kwon, B.K. et al. (2023a). Sensory test A-Not A rating signal detection: panel's sensory learning and stability of decision criteria. *Food Res. Int.* 167: 112592.

Choi, E.S., Van Hout, D., and Lee, H.S. (2023b). Improving the performance of A-Not A sensory discrimination ratings by modifying sample presentation probability. *Food Qual. Prefer.* 104: 104748. https://doi.org/10.1016/j.foodqual.2022.104748.

Choi, Y.J., Kim, J.Y., Christensen, R.H.B. et al. (2014). Superior performance of constant-saltier-reference DTF and DTFM to same-different tests by consumers for discriminating products varying sodium contents. *Food Qual. Prefer. 37*: 100–108.

Christensen, R.H.B., Cleaver, G., and Brockhoff, P.B. (2011). Statistical and Thurstonian models for the A-Not A protocol with and without sureness. *Food Qual. Prefer.* 22 (6): 542–549.

David, H. A., Trivedi, M. C., and Harshbarger, B. (1962). Pair, triangle and duo-trio tests. Technical Report 55. Department of Statistics, Virginia Polytechnic Institute, Blacksburg, Virginia.

Dorfman, D.D. and Alf, E. Jr. (1969). Maximum-likelihood estimation of parameters of signal-detection theory and determination of confidence intervals – rating-method data. *J. Math. Psychol.* 6 (3): 487–496.

Ennis, D.M. (1993). The power of sensory discrimination methods. *J. Sens. Stud.* 8 (4): 353–370.

Ennis, J. M., Ennis, D. M., Yip, D., & O'Mahony, M. (1998). Thurstonian models for variants of the method of tetrads. *Br. J. Math. Stat. Psychol.*, 51, 205–215.

Ennis, J.M. and Jesionka, V. (2011). The power of sensory discrimination methods revisited. *J. Sens. Stud. 26* (5): 371–382.

Ennis, J.M., Rousseau, B., and Ennis, D.M. (2014). Sensory difference tests as measurement instruments: a review of recent advances. *J. Sens. Stud.* 29: 89–102.

Esfanjnai, A.F. and Mohebbi, M. (2023). A signal detection reminder methodology to evaluate effects of odor–taste-taste interactions to enhance the perceived saltiness. *Food Res. Int.* 165: 112464.

Frijters, J.E.R. (1979). Variations of the triangular method and the relationship of its unidimensional probabilistic models to three-alternative forced-choice signal detection theory models. *Br. J. Math. Stat. Psychol.* 32 (2): 229–241.

Green, D. M. and Swets, J. A. (1966). Signal detection theory and psychophysics. New York: Wiley.

Hautus (2021). Signal Detection Assistant software V2.0. https://hautus.org/sdt-assistant.php.

Hautus, M.J. (1997). Calculating estimates of sensitivity from group data: pooled versus averaged estimators. *Behav. Res. Methods Instrum. Comput.* 29 (4): 556–562.

Hautus, M.J., Macmillan, N.A., and Creelman, C.D. (2021). *Detection Theory: A User's Guide.* Routledge.

Hautus, M.J., van Hout, D., and Lee, H.S. (2009). Variants of A Not-A and 2AFC tests: signal detection theory models. *Food Qual. Prefer.* 20 (3): 222–229.

Hautus, M.J., van Hout, D., Lee, H.S. et al. (2020). Observed discriminability is more variable than predicted by signal detection theory. *Food Qual. Prefer.* 79: 103774.

Ishii, R., Kawaguchi, H., O'Mahony, M., and Rousseau, B. (2007). Relating consumer and trained panels' discriminative sensitivities using vanilla flavored ice cream as a medium. *Food Qual. Prefer.* 18 (1): 89–96.

Jeong, Y.N., Kang, B.A., Jeong, M.J. et al. (2016). Sensory discrimination by consumers of multiple stimuli from a reference: Stimulus configuration in A-Not AR and constant-ref. duo-trio superior to triangle and unspecified tetrad? *Food Qual. Prefer. 47*: 10–22.

Jeong, Y.N., van Hout, D., Groeneschild, C., and Lee, H.S. (2017). Comparative categorization method: Using 2-AFC strategy in constant-reference duo-trio for discrimination of multiple stimuli from a reference. *Food Qual. Prefer.* 62: 284–295.

Jesionka, V., Rousseau, B., and Ennis, J.M. (2014). Transitioning from proportion of discriminators to a more meaningful measure of sensory difference. *Food Qual. Prefer. 32*: 77–82.

Kim, I.A., Cho, H.Y., Hautus, M.J., and Lee, H.S. (2020). Exploration of a new consumer test method based on metacognitive certainty. *Food Qual. Prefer.* 81: 103857.

Kim, I.-A., Kim, H.-L., Cho, H.-Y., and Lee, H.-S. (2015). Optimal difference test sequence and power for discriminating soups of varying sodium content: DTFM version of dual-reference duo-trio with unspecified tetrad tests. *Food Res. Int.* https://doi.org/10.1016/j.foodres.2015.06.019.

Kim, I.A., Seo, J.E., Kim, H. J, and Lee, H.S. (2013). A signal detection discrimination using only one stimulus class for measurement of a conceptual attribute. In: *2013 Annual Meeting of Korean Society of Food Science and Technology*, Choenan, Korea (28–30 August 2013).

Kim, M.A. and Lee, H.S. (2012). Investigation of operationally more powerful duo-trio test protocols: effects of different reference schemes. *Food Qual. Prefer. 25* (2): 183–191.

Kim, M.A., Chae, J.E., van Hout, D., and Lee, H.S. (2012). Discriminations of the A-Not A difference test improved when "A" was familiarized using a brand image. *Food Qual. Prefer.* 23 (1): 3–12.

Kim, M.A., Chae, J.E., van Hout, D., and Lee, H.S. (2014). Higher performance of constant-reference duo-trio test incorporating affective reference framing in comparison with triangle test. *Food Qual. Prefer. 32*: 113–125.

Kim, M.A., Lee, Y.M., and Lee, H.S. (2010). Comparison of $d'$ estimates produced by three versions of a duo-trio test for discriminating tomato juices with varying salt concentrations: the effects of the number and position of the reference stimulus. *Food Qual. Prefer. 21* (5): 504–511.

Kim, M.-A., Lee, Y.-J., Kim, M.-S., and Lee, H.-S. (2022). Reminder–preference test, affective difference-preference test using reference framing with a brand: 1. Sensitivity comparisons with the same-different difference–preference test. *Food Res. Int. 155*: 111065.

Kim, M.-A., Kim, M.-S., Lee, Y.-J., and Lee, H.-S. (2023). *Reminder–Preference Test, Affective Difference-Preference Test Using Reference Framing with a Brand: 2. Performance Comparisons with Difference Tests and Preference Tests, Respectively*. Food Research International.

Kim, M.A., Sim, H.M., and Lee, H.S. (2015). Affective discrimination methodology: determination and use of a consumer-relevant sensory difference for food quality maintenance. *Food Res. Int. 70*: 47–54.

Kim, M.A., Sim, H.M., and Lee, H.S. (2016). Idiographic duo-trio tests using a constant-reference based on preference of each consumer: sample presentation sequence in difference test can be customized for individual consumers to reduce error. *Food Res. Int. 89*: 454–462.

Kim, M.-A., van Hout, D., Dessirier, J.M., and Lee, H.-S. (2018a). Degree of satisfaction-difference (DOSD) method for measuring consumer acceptance: a signal detection measurement with higher reliability than hedonic scaling. *Food Qual. Prefer. 63*: 28–37.

Kim, M.-A., van Hout, D., and Lee, H.-S. (2018b). Degree of satisfaction-difference (DOSD) method for measuring consumer acceptance: comparative and absolute measures of satisfaction based on signal detection theory. *Food Qual. Prefer. 68*: 167–172.

Kim, M.-A., van Hout, D., Zandstra, E.H., and Lee, H.-S. (2019). Consumer acceptance measurement focusing on a specified sensory attribute of products: can the attribute-specified degree of satisfaction-difference (DOSD) method replace hedonic scaling? *Food Qual. Prefer. 75*: 198–208.

Kuesten, C. L. (2001). Sequential use of the triangle, 2-AC, 2-AFC, and same-different methods applied to a cost-reduction effort: consumer learning acquired throughout testing and influence on preference judgements. *Food. Qual. Prefer.*, 12(5–7), 447–455.

Lau, S. and O'mahony, M., & Rousseau, B. (2004). Are three-sample tasks less sensitive than two-sample tasks? Memory effects in the testing of taste discrimination. *Percept. Psychophys. 66*: 464–474.

Lawless, H.T. (2013). *Quantitative Sensory Analysis: Psychophysics, Models and Intelligent Design*. Wiley.

Lawless, H.T. and Heymann, H. (2010). *Sensory Evaluation of Food: Principles and Practices*, vol. 2. New York: Springer.

Lee, H.L., van Hout, D., and Lee, H.S. (2023). Improving the performance of A-Not AR discrimination test using a sensory panel: effects of the test protocols on sensory data quality. *Food Qual. Prefer.* 104: 104740.

Lee, H. S. (2016). Affective discrimination methodology for assessing consumers: Strategic research for sensory quality management. In: *The 2nd Asian Sensory and Consumer Research Symposium*, Shanghai, China (15–17 May 2016).

Lee, H.S., van Hout, D., and Hautus, M.J. (2007a). Comparison of performance in the A-Not A, 2-AFC, and same-different tests for the flavor discrimination of margarines: the effect of cognitive decision strategies. *Food Qual. Prefer.* 18 (6): 920–928.

Lee, H.S., van Hout, D., and O'Mahony, M. (2007b). Sensory difference tests for margarine: a comparison of R-Indices derived from ranking and A-Not A methods considering response bias and cognitive strategies. *Food Qual. Prefer.* 18 (4): 675–680.

Lee, Y.M., Chae, J.E., and Lee, H.S. (2009). Effects of order of tasting in sensory difference tests using apple juice stimuli: development of a new model. *J. Food Sci. 74* (6): S268–S275.

Macmillan, N. A., Kaplan, H. L., and Creelman, C. D. (1977). The psychophysics of categorical perception. *Psychol. Rev.*, 84(5), 452–471.

Mun, J.W., Kim, M.A., Sim, H.M., and Lee, H.S. (2019). Investigation of test performance of the dual reminder A-Not A (DR A-Not A) in comparison to 3-AFC for discriminating samples of drinking water. *Food Qual. Prefer.* 77: 43–50.

O'Mahony, M., Garske, S., and Klapman, K. (1980). Rating and ranking procedures for short-cut signal detection multiple difference tests. *J. Food Sci.* 45 (2): 392–393.

O'Mahony, M. (1995). Who told you the triangle test was simple? *Food Qual. Prefer. 6* (4): 227–238.

O'Mahony, M.A.P.D. (1979). Short-cut signal detection measures for sensory analysis. *J. Food Sci.* 44 (1): 302–303.

O'Mahony, M. and Odbert, N. (1985). A comparison of sensory difference testing procedures: sequential sensitivity analysis and aspects of taste adaptation. *J. Food Sci. 50* (4): 1055–1058.

O'Mahony, M., Masuoka, S., and Ishii, R. (1994). A theoretical note on difference tests: models, paradoxes and cognitive strategies. *J. Sens. Stud. 9* (3): 247–272.

O'Mahony, M. and Rousseau, B. (2003). Discrimination testing: a few ideas, old and new. *Food Qual. Prefer. 14* (2): 157–164.

O'Mahony, M., Thieme, U., and Goldstein, L.R. (1988). The warm-up effect as a means of increasing the discriminability of sensory difference tests. *J. Food Sci.* 53 (6): 1848–1850.

Pfaffmann, C., Schlosberg, H., and Cornsweet, J. (1954). Variables affecting difference tests. *Food Accept. Test. Methodol.* 4–20.

Rousseau, B. and Ennis, D. M. (2001). A Thurstonian model for the dual pair (4IAX) discrimination method. *Percept. Psychophys.*, 63, 1083–1090.

Rousseau, B., Rogeaux, M., and O'Mahony, M. (1999). Mustard discrimination by same-different and triangle tests: aspects of irritation, memory and s criteria. *Food Qual. Prefer.* 10 (3): 173–184.

Rousseau, B., Stroh, S., and O'Mahony, M. (2002). Investigating more powerful discrimination tests with consumers: effects of memory and response bias. *Food Qual. Prefer.* 13 (1): 39–45.

Schlich, P. (1993). Risk tables for discrimination tests. *Food Qual. Prefer. 4* (3): 141–151.

Shin, H.K., Hautus, M.J., and Lee, H.S. (2016). Unspecified duo-trio tests can be as powerful as the specified 2-AFC: effects of instructions and familiarization procedures on cognitive decision strategies. *Food Res. Int.* 79: 114–125.

Thieme, U. and O'Mahony, M.I.C.H.A.E.L. (1990). Modifications to sensory difference test protocols: the warmed up paired comparison, the single standard duo-trio and the A-Not A test modified for response bias. *J. Sens. Stud.* 5 (3): 159–176.

van Hout, D., Hautus, M.J., and Lee, H.S. (2011). Investigation of test performance over repeated session using signal detection theory: comparison of three nonattribute-specified difference test 2-AFCR, A-Not A and 2-AFC. *J. Sens. Stud.* 26 (5): 311–321.

van Hout, D.H.A. (2014). Measuring meaningful differences – sensory testing-based decision making in an industrial context: applications of signal detection theory and Thurstonian modeling. Ph.D. thesis. Erasmus University Rotterdam, Research Institute of Management.

# 10

# Discrimination Testing in Flavors and Fragrances: A Practical View

*Karine Miot[1], Carlos Gómez-Corona[2] and Isabelle Cayeux[3]*

[1] *Sensory Department, Perfumery & Beauty BU, dsm-firmenich, London, United Kingdom*
[2] *Human Insights Department, Taste Texture & Health BU, dsm-firmenich, Neuilly-sur-Seine, France*
[3] *Perception & Cognitive Neurosciences Department, Science & Research, dsm-firmenich, Geneva, Switzerland*

## 10.1 Introduction to Discrimination Testing for Flavors and Fragrances

This chapter deals with the application of discriminative tests in the flavor and fragrance (F&F) industry, which is often seen as a very specialized industry. Several examples are used to illustrate common problems, such as flavor integration (merging two different flavors into a single formula) and sugar and salt reduction. For these examples, different discriminative tests are used, and general aspects of the tests are discussed. Throughout the text, several specificities of the flavor industry, and specifically of the perfume industry, are explained. Toward the end of the chapter, the reader will find the addition of two non-common aspects (in terms of number of publications) of the discriminative tests: social responsibility and machine learning (ML) to substitute the use of discriminative tests in certain conditions.

### 10.1.1 General Considerations

The purpose of discriminative tests is to determine whether two physically different products are perceptually different or whether they can be considered perceptually similar (Dacremont and Sauvageot 2019). In the F&F industry, the nature of that difference can be a chemical variation (variation in the nature or in the proportion of different raw materials), a product-base variation (changes in the nature of the product when the flavor or fragrance is added), or a production variation (changes in the production of either the flavor or the fragrance or the product base). In any of these cases, the perceptual difference generated by the variations is small enough to question the possibility of perceiving it.

Different types of tests can be used. Table 10.1 presents seven different tests that can be used for the F&F industry. Discrimination tests are usually conducted by sensory panelists

*Discrimination Testing in Sensory Evaluation*, First Edition. Edited by Lauren Rogers, Joanne Hort, Sarah E. Kemp and Tracey Hollowood.

**Table 10.1** Discriminative tests that are typically used in the F&F industry.

| Test | Test description<br>Number of samples presented<br>Description<br>Question or task asked of panelist | Flavor applications | Fragrance applications |
|---|---|---|---|
| Triangular | Three samples presented.<br>Two samples are similar, and one is different.<br>*Question: Which of the three samples is different?* | ✓ | ✓ |
| Tetrad | Four samples presented.<br>Two samples come from one product and the other two from a different product.<br>*Task: Sort the four samples into two groups of two, based on their similarities.* | ✓ ✗ Category specific | |
| 2-AFC | Two samples presented.<br>Each sample comes from a different product.<br>*Question: Which sample is the most. ...* | ✓ | ✓ |
| 2-AFC R | 2AFC R procedure is identical to the constant reference duo-trio procedure. Two samples are presented $P_1$ and $P_2$.<br>*Task: Indicate whether $P_2$ was presented first or second after the reminder in the sequence.* | ✓ | ✓ |
| 3-AFC | Three samples presented.<br>Two samples come from the same product and one is different.<br>*Question: Which sample is the most. ...* | ✓ | ✓ |
| Duo-trio | Three samples presented.<br>One sample is coded as reference. Of the other two samples, one is the reference and one is different.<br>*Question: Which sample is the same as the reference?* | ✓ | ✓ |
| A-not A | A set of samples is presented, usually including between three and five samples.<br>One sample is coded as reference A, and of the rest of the samples, one is A and the others are different.<br>*Question: Please identify whether the samples are similar to or different from A.* | ✓ | ✓ ✓ ✗<br>Rarely used |
| R-index | Similar to A-not A but with the following answer options:<br>• I am sure it is the reference.<br>• It is the reference, but I am not sure.<br>• It is not the reference, but I am not sure.<br>• I am sure it is not the reference. | ✓ | ✗ ✗ ✗<br>Rarely used |

AFC, alternative forced choice. Adapted from Hautus et al. (2009).

in the F&F industry although some applications are also found with consumers (Rousseau et al. 2002), especially for final validations and to measure whether two products are perceived differently by consumers. In selecting a test, many facets must be considered: the nature of the perceptual difference that is of interest, the skills of the participants, and the experience level of the sensory analyst.

There is always a common question for the sensory scientists: Which discriminative test should I use for this project? The answer is never simple, as it depends on the nature of the difference between the samples, the material available (sometimes there is not enough sample for the test we would like to run), the number of panelists available, and, in industry, sometimes habit (certain industries are simply used to running specific discrimination tests). Figure 10.1 is a simplified way at looking at the different options that a sensory scientist can have when working with flavors and fragrances. The first axis horizontally separates the tests into less versus more panelist fatigue and vertically separates the tests into less versus more guessing probability. This can give the sensory scientist a quick view of the most common tests used in the F&F industry.

In addition to test selection, a suitably set up sensory discrimination testing program will involve five components that are essential to its success: sensory protocol, sample size, size of the underlying difference of interest $\delta R$, Type I error level ($\alpha$, the likelihood of wrongly concluding a difference, usually set at 5%), and Type II error ($\beta$, the likelihood of missing a difference of size $\delta R$, usually set at 10% or 20%; test power $= 1-\beta$) (Rousseau 2015).

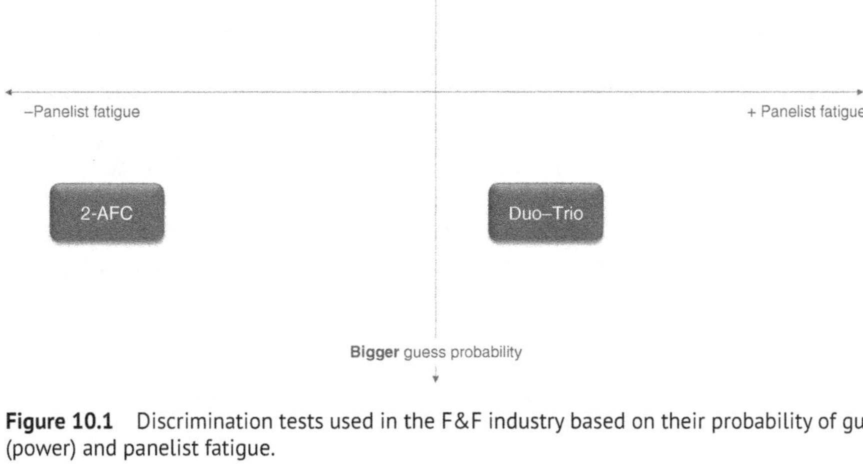


**Figure 10.1** Discrimination tests used in the F&F industry based on their probability of guess (power) and panelist fatigue.

### 10.1.2 Methodology

In the F&F industry, the use of discriminative tests is fundamental in the work of flavorists and perfumers, production, and of course sensory evaluation. The decision to use one test over another depends on the experience of the sensory leader, the number of panelists and their training, and the objective of the study or project. Traditionally, the triangular test (Table 10.1) is the iconic test in the F&F industry, probably due to the ease of test execution and communication of the results. The triangular test has been the target of multiple publications of standards (ISO 4120:2021) and case studies (Gomez-Corona et al. 2020). In the last five to seven years, the tetrad test has been more widely used and has gained interest for sensory evaluation, with publications suggesting its advantages, and one suggesting that both the triangular and the tetrad test give similar results and have similar power (Burns et al. 2018).

#### 10.1.2.1 Panelists

Performing a discriminative test in the F&F industry is a complex task, and novice participants need to perform multiple tests to master the "mechanics" of the tests. The sensory scientist must take into consideration that the more samples that are compared, the more difficult it will be to master the test. In the specific case of fragrance, unlike flavor, the evaluation is done in cabins (not a standard sensory booth, but a special place called a cabin; see Figure 10.3) and they are prepared with the fragrance for the evaluation. To complete a discriminative test, the panelist needs to move from one cabin to another to evaluate the fragrances (e.g., fragrance is applied to the floor of the cabin or sprayed in the cabin).

When the discriminative test of a fragrance is performed in a cabin, the panelist may need to involve more of their working, or short-term, memory (see Valentin et al. 2010). Imagine a triangular test in which the panelist walks into three different cabins and then rates the different fragrances. It is not at all the same principle as in flavor evaluation, in which the location of the test is the same (e.g., sensory booth). The protocols for discriminative tests in the fragrance industry must take these complexities into account when the type of tests to run are being selected.

When an overall recommendation is required – a test in which the participants have the habit of performing sensory evaluations and have already performed discriminative tests – it is possible to perform a triangular test, a tetrad test, or a three-alternative forced choice (3-AFC) test. For a test with novice participants who have never performed discriminatory tests, we may favor two-alternative forced choice (2-AFC) tests or a duo-trio test. Novice participants will be able to run tests that include more samples, provided that prior familiarization is arranged when the test is proposed with another pair of products. Taking two or three familiarization tests may be sufficient (Dacremont and Sauvageot 2019). The authors do not recommend the use of tetrad for fragrances when the test needs to be performed in cabins. The number of cabins needed is higher, and performing the task is much more complex. On the other hand, although habituation occurs in all sensory modalities, some of the most noticeable instances take place in the olfactory system. Habituation is a filter that optimizes the processing of information by our brain. It results in an unconscious reduced responsiveness to continuous stimulation, meaning a perceptual decrement in odor intensity. Therefore, the set-up of the sensory protocol must take into account these phenomena, especially for perfumery tests by, for example, increasing the time between two stimulations (Sinding et al. 2017). The increased time can cause other issues, such as panelists relying more on short sensory memory; however, that effect is yet to be studied.

Regarding the number of panelists required for a discrimination test, this is defined by the ISO Standard guidance (ISO 4120:2021 for triangular tests), and it takes into consideration the risks that the sensory scientist wants to define as a target for a discriminative test. It is common practice in the industry to have a minimum number of 20–30 panelists for a test. According to ISO 4120:2021, 20 panelists will have $\alpha = 0.2$, $\beta = 0.2$, for a $P_d$ (proportion of discriminative panelists) of 30%. For 30 panelists, there will be $\alpha = 0.1$, $\beta = 0.2$, for a $P_d = 30\%$.

#### 10.1.2.2 Test Location

Several locations can be used to perform a discrimination test. Figures 10.2 and 10.3 show the most common places for discriminative evaluation in the F&F industry. Some depend on the development process of the flavor or fragrance. For example, during and after the creation process, flavorists or perfumer/fragrance development managers like to

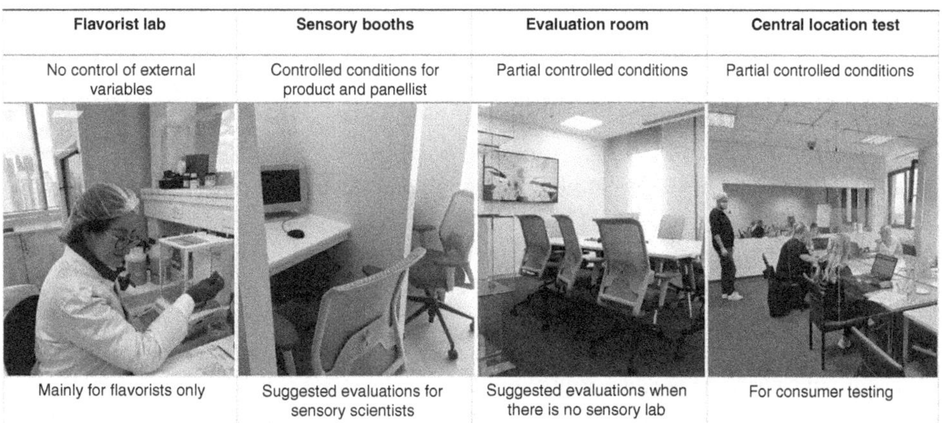


**Figure 10.2** Different locations for the execution of discrimination tests for flavor applications (image from dsm-firmenich, 2021).

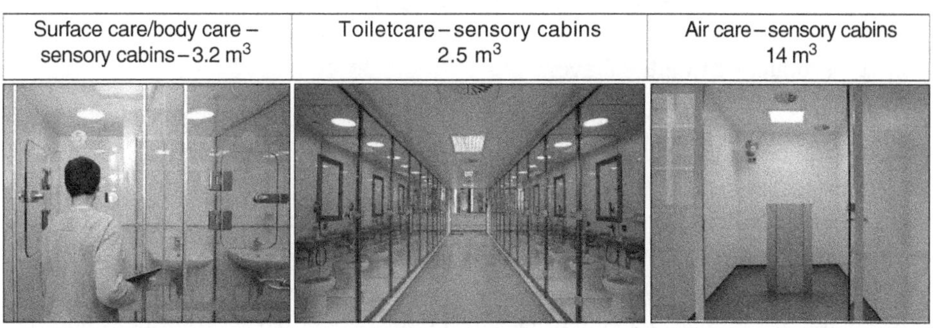


**Figure 10.3** Different locations for the execution of discrimination tests for fragrance applications (image from dsm-firmenich).

individually run, or run with their colleagues, a set of discriminative tests in the creation lab. These discrimination tests are usually triangular tests. The objective of these explorative tests is to understand the possible perceptual differences across two or more sets of samples. The evaluation can be done by using only smelling strips or even diluted flavor solutions (Figure 10.2). The execution of this type of test lacks external validity, as it is usually done with a low number of highly trained persons.

Besides the creation lab where the samples are concentrated flavors and fragrances, the most common place for a discriminative test is the sensory booth, in which standard conditions are required, such as maintaining samples at the same temperature, using coded samples, controlling the quantity of sample per test and panelists, and controlling external conditions (air extraction, temperature, noise, light, etc.). In a sensory lab, at least 5–10 sensory booths are expected in order to enable the correct and rapid execution of a discriminatory test.

Specific to perfumery, discrimination tests are also often run in sensory cabins, in which the blooming or the diffusion of fragrances can be evaluated under controlled conditions. In a sensory perfumery lab, different cabin sizes are usually expected in order to reflect the products' usage condition. For example, air fresheners are tested in large cabins (15–25 $m^3$), whereas toilet cleaning products are tested in smaller size cabins (~2.5 $m^3$).

Another common location is the evaluation rooms or even meeting rooms. They have less controlled conditions than a sensory lab does, but they can be as good as a lab for simple evaluations where temperature control of the samples or their preparation is not complex. Evaluation of samples that require specific temperature controls (carbonated drinks, soups, ready meals, sausages, etc.) is not recommended in this type of room or of sample fragrances that need to be sprayed in the room.

Finally, the central location test (CLT) can be used to run discriminative tests with consumers. These types of studies are less common and are usually done as a final validation process, as mentioned earlier. In a CLT, the conditions to control variables such as temperature and light are usually the same as in a sensory lab. Although it is not common to do discriminative tests at the consumer's house (home use test), it is done for special types of products (e.g., personal care products). However, it is not usually done in the F&F industry as there is no control of the test conditions.

## 10.2 Discrimination Testing for Flavors

### 10.2.1 Product – Stimuli – Flavor

In the flavor industry or academic projects that use flavors, discriminatory testing is performed with the same standards of quality and care as in any other industry. There are no limitations that are as specific as in the case of fragrances, but it is crucial to properly define the taste evaluation matrix. This matrix should be defined on the basis of the test objectives. Figure 10.4 shows seven different ways to evaluate a flavor.

The first form is the flavor per se, usually in the source solvent, for example, propylene glycol in the case of many liquid flavors. Direct evaluation of flavors is normally done in a dilution of the same flavor (evaluations are not performed on concentrated flavors). These evaluations are done solely by flavorists and using smelling strips.

| **Flavor in solvent** | **Tasting solution** sweet tonalities | **Tasting solution** savoury tonalities | **Fondant** for mint and non-soluble | **Fats and oils** | **Tasting base** | **Final product** |
|---|---|---|---|---|---|---|
| Dilution of flavour (e.g., propylene glycol) | Solution of water, sugar (3–5° Bx) and citric acid | Solution of water and salt | Sugar glass | Neutral oil | Simple product application (broth, bread, water...) | Final matrix where the flavour is used |
|  | | | | | | |
| Mainly for flavourists only | For initial screening | For initial screening | For initial screening | For initial screening or final evaluation | To test the relationship with a complex matrix | To validate in final product |

**Figure 10.4** Seven different forms of tasting flavors, depending on where the base is, from tasting solution to final product (First six images from dsm-firmenich. Last image modified from webandi/Pixabay, royalty free images, and webandi/Pixabay).

A common way to evaluate flavors is by using a tasting solution, in which a certain dose of flavor is added (depending on its concentration). For fruit flavors (berries, pip, tropical, stone fruits), a solution is made with sugar at 3–5° Bx (3–5% sugar), sometimes balanced with 0.05% citric acid. In the case of citrus flavors (lemon, lime, mandarin, orange, etc.), the same sweet solution of 3–5° Bx is made, but with the addition of 0.1% of citric acid. In the case of brown notes (e.g., vanilla, almond, chocolate), the tasting solution consists of only 5° Bx. The same principle is used for salty flavors (meat, cheese, fat) in which a standard solution of 1% salt is made. Tasting solutions can be tested exclusively with a trained sensory panel. On the other hand, for flavors that are insoluble in water, a fondant is made with icing sugar and water (100 g of sugar and 20 ml of water). A certain dose of flavor is added (e.g., 0.1% for a mint extract), and the test can also be performed with a trained sensory panel. In the case of fat-soluble flavors, the flavor can be applied directly into a neutral vegetable oil that does not provide additional flavor notes, especially to evaluate margarines or flavored oils. The oil sample can be served at room temperature, or it can be slightly warmed (3–35 °C). It is important to limit the amount of oil that the panelist ingests, as it can cause digestive problems. For the evaluation of margarines, it is recommended to have the samples at 5 °C (Lee et al. 2007). There is a specific standard for olive oil measurement (ISO 16657:2006).

The discriminative evaluation of flavors, applied to a simple or more complex tasting base, is done to simulate both types of interactions. On the physical–chemical level, the flavor may interact with the ingredients of a base and so affect flavor release and finally its perception. On a perceptual level, global flavor perception arises from the central integration of multiple sensory inputs: taste, smell, and trigeminal and tactile sensations. Congruence (generally defined as the extent to which two stimuli are appropriate for combination in a food product) between taste (e.g., a sweet and sour taste solution) and flavor (e.g., citrus flavor) is important for better integration of the neural flavor signal, thus allowing meaningful flavor perception (Shepard et al. 2015). That is why it is important to use a base that is most similar to the final product or to use the final product itself.

It is common to see discriminative tests performed on a standard base of flavored water, bakery products, and broths for savory flavors. These bases consist of a simple formulation and are not the final product that can be found in the market.

Evaluation of the final product is in many cases the last part of the validation of a project, when it is necessary to see the differences between two products. For example, in cases of a flavor match (when matching the profile of one flavor to another is required), projects may aim to reduce sugar, salt, or some other ingredient. The advantage of evaluating taste in the final product is the possibility of using discriminating tests with consumers or with panelists who have a lower level of training; however, this can be a time-consuming step or there may be additional costs linked to sample production.

### 10.2.2 Applications

There are numerous different applications of discriminative tests in the flavor industry. Table 10.2 summarizes 10 common uses for flavors only and for products that contain the flavor.

For the evaluation of flavors in a tasting solution, there can be different objectives, such as testing for a raw material or process change, day-to-day quality control, shelf life, batch comparisons, and the integration of two or more flavors. The process of flavor integration – the combination of two or more flavors into a single formula – is common in the flavor industry. Why? Sometimes during product development, two flavors are used together, for example, the development of a flavored water with two different lemon profiles (green and yellow). The final formula will consist of only one flavor in which both formulas are merged. The process to validate the integration of the two flavors into one is most of the time a discrimination test to ensure that the profile of two flavors used together is the same as that of the integrated flavor.

**Table 10.2** Different objectives in a project in which discrimination tests are used in flavor applications.

| Test objective | Type of tests | Product application |
|---|---|---|
| [a]Raw material change in flavor composition | Triangular | Tasting solution |
| [a]Process change in flavor manufacture | Triangular | Tasting solution |
| [a]Quality control process in day-to-day process | Triangular, duo-trio | Tasting solution or base |
| [a]Shelf-life evaluation of flavor | Triangular, R-index | Tasting solution |
| [a]Batch comparison | Triangular, tetrad | Tasting solution |
| [a]Two or more flavors integration process | Triangular, tetrad | Tasting solution |
| [b]Change in final product | Triangular, R-index | Final product |
| [b]Sugar reduction projects | Triangular, 2-AFC, tetrad | Final product |
| [b]Salt reduction projects | Triangular, 2-AFC, tetrad | Final product |
| [b]Dosage changes in final product | Triangular, 2-AFC | Final product |

[a] Flavor evaluation only.

[b] Final product that contains the flavor (e.g., soup, biscuit, beer).

Regarding the application of discriminative tests to evaluate a final product, the most common approach is to do so when there is a change in the final product that can change the profile of the flavor (e.g., change in the acid type of a flavored beverage), and this has become more common in recent years due to sugar and salt reduction projects. When reducing salt and sugar, the profile of the flavor is susceptible to large changes.

Some product changes might be the use of non-caloric sweeteners (e.g., stevia or sucralose) that do not have the exact sweet profile, and therefore, the profile of a product changes considerably. Imagine a full-sugar vanilla biscuit that is changed to a 30% reduced-sugar biscuit. The vanilla profile will change because the vanilla flavor needs the sweet base of the product to develop a typical sweet and creamy vanilla profile.

The same happens with salt reduction projects in which the flavor of chicken changes, depending on the percentage by which salt is reduced in the final product. In most cases, sugar and salt reduction projects have a reduction target above 15%, which then require changes to the flavor itself, and therefore, the changes in the product are validated with the help of discriminative tests, at least at the early stages of the process.

The final validation steps might include a consumer study with heavy users of the brand or the product category to verify that the consumer does not see any difference in terms of overall liking (hedonics). Some companies even perform discriminative tests with consumers; however, it is not good practice to do so without explaining the task to the naïve consumer and performing at least a warm-up task (performing a discriminative task with fake stimuli, just to explain the nature of the test). However, performing discriminative tests with consumers is always a debate in the sensory domain (Chollet et al. 2005; Ishii et al. 2007; Ares and Varela 2017).

### 10.2.3 Case Studies

#### 10.2.3.1 Integration of Two Flavors: Evaluation in Tasting Solution

Imagine we have developed a flavored alcohol beverage (FAB) with tequila, in which the flavor is a perfect balance of green lemon and cucumber. During the development phases, two flavors were used to make the prototypes that were then tested with consumers. For the final product, only one flavor should go into the commercialized product to facilitate production logistics. The flavorist developed a new integrated flavor in which he or she unifies the two flavor formulas into one. To validate that there are no significant differences between the original FAB with two flavors (green lemon and cucumber) and the FAB with the integrated flavors, a discriminative test is planned (Figure 10.5).

The triangular test that focusses on assessing differences (and not similarities) is designed with 30 panelists and performed in traditional sensory booths to evaluate food and beverages (Figure 10.2). Usually in the flavor industry, the use of the difference test is preferred versus the similarity test. Although a more "technical view" can suggest the use of a similarity test, in the case of a flavor's integration project, it is accepted that as long as there are no significant differences (in a difference test) across the samples, then the integration process of the flavors can continue. An additional point is that during industrial production, small differences can occur linked to raw material origins, process of mixing the raw materials, and production of flavor. Therefore, if the two samples studied are not significantly different in a discriminative difference test, they will fulfil the objective.

**Figure 10.5** A FAB was developed with two flavors (reference with test codes 458 and 981) and a FAB was developed with the integrated flavors (test codes 125 and 392). *Source:* stevepb / Pixabay; 2691517199qqcom / Pixabay.

The order of presentation is balanced among samples and panelists. Each panelist receives 25 ml of the FAB at 9 °C and all samples are coded (see Figure 10.5 for codes between samples). Table 10.3 shows the details regarding the order of samples presented to the panelists, their answers, and, at the far right, the correct answer. In this example, 14 correct responses were obtained by the panelists, which is equivalent to a 0.467 proportion of correct responses. In a triangular test, the guess probability is 0.333. The results with the 14 correct responses have a $p$ value of 0.090, and because the $\alpha$ was fixed at 5%, we can conclude that there are no significant differences across the FAB with the integrated flavor and the FAB with two flavors. At this moment of the project, the integrated flavor of lemon-cucumber is considered suitable for final production and commercialization, and no further consumer validation should be required.

#### 10.2.3.2 Sugar Reduction in Beverages in Laboratory Samples

Sugar reduction might be one of the trending projects in the food industry, from small reductions (−15%) to products containing no added sugar. In this example, a traditional full-sugar strawberry yogurt is evaluated versus a prototype that is reduced in sugar by 45%. For a strict evaluation, the samples are screened first by using a tetrad test. This relatively new discriminative test (Ishii et al. 2014) is an alternative to triangular tests. While it has been present in the literature for some time, its greater statistical power has been investigated only in the past 10 years (Ennis and Jesionka 2011), and therefore, the food industry is particularly interested in this discriminative test. What has been reported in previous research (Ennis 2012) is that the tetrad method seems to be more powerful than the triangular test by a factor of approximately 3. The guessing probability for the tetrad test is like that of the triangular test 1/3; direct comparisons between the two methods can readily be made in terms of performance, and the same binomial tables can be used to investigate whether the results of an experiment show a statistically significant difference.

**Table 10.3** Details of the triangular test: order of presentation by panelist, answer, and correct test code (in italics and in the last column of the table).

| Panelist | Sample 1 | Sample 2 | Sample 3 | Assessor's answer | Correct answer |
|---|---|---|---|---|---|
| J1 | 125 | 458 | 392 | 0 | 458 |
| J2 | 458 | 981 | 392 | 1 | *392* |
| J3 | 392 | 981 | 458 | 0 | 392 |
| J4 | 125 | 981 | 458 | 0 | 125 |
| J5 | 981 | 392 | 125 | 0 | 981 |
| J6 | 458 | 392 | 125 | 1 | *458* |
| J7 | 981 | 392 | 458 | 1 | *392* |
| J8 | 125 | 392 | 458 | 0 | 458 |
| J9 | 458 | 981 | 125 | 0 | 125 |
| J10 | 458 | 392 | 981 | 1 | *392* |
| J11 | 458 | 125 | 981 | 1 | *125* |
| J12 | 981 | 458 | 392 | 0 | 392 |
| J13 | 458 | 125 | 392 | 0 | 458 |
| J14 | 125 | 458 | 981 | 1 | *125* |
| J15 | 981 | 458 | 125 | 0 | 125 |
| J16 | 392 | 458 | 981 | 1 | *392* |
| J17 | 981 | 125 | 392 | 0 | 981 |
| J18 | 125 | 392 | 981 | 0 | 981 |
| J19 | 125 | 981 | 392 | 0 | 981 |
| J20 | 392 | 125 | 981 | 0 | 981 |
| J21 | 392 | 458 | 125 | 0 | 458 |
| J22 | 392 | 125 | 458 | 1 | *458* |
| J23 | 981 | 125 | 458 | 1 | *125* |
| J24 | 392 | 981 | 125 | 1 | *981* |
| J25 | 125 | 458 | 392 | 0 | 458 |
| J26 | 458 | 981 | 392 | 1 | *392* |
| J27 | 392 | 981 | 458 | 1 | *392* |
| J28 | 125 | 981 | 458 | 1 | *125* |
| J29 | 981 | 392 | 125 | 0 | 981 |
| J30 | 458 | 392 | 125 | 1 | *458* |
| | | | | Correct responses 14 | |

In the tetrad test, four samples are presented, two pairs of the same samples. In other words, two samples are presented two times. The panelist receives four samples and is asked to match the pairs on the basis of their perceived similarities (e.g., "Welcome to this sensory session. Today you will receive four samples. Please taste all the samples and then group them into two groups based on their similarity"). Changing the instructions to "Identify the two most similar samples" induces a different strategy and leads to lower performance. Minor variations in the instructions are likely to cause significant variations in performance (Castura et al. 2018).

In the case of sugar reduction, this test is useful because it eliminates the biases of the triangular test in which a full-sugar sample followed by a reduced-sugar sample will be easier to identify. Recall that the order of presentation in a triangular test can affect the results of the test (Lee et al. 2009). In the case of a tetrad test, the order of presentation seems more adequate for the study of reduced-sugar samples. It is common to select a difference test versus a similarity test in projects concerning sugar reduction. A low percentage sugar reduction (around 15% reduction) should not cause any big differences between the full-sugar versus the reduced-sugar samples, especially because other flavoring technologies are used to reduce the impact of the differences when sugar is reduced. However, with differences higher than 15%, an exact sensory match of the products cannot be achieved by using a similarity test approach because the samples simply might not be the same. In the industry, the most common approach is to use a difference test because the objective is for the consumers not to perceive significant differences across samples (full sugar versus reduced) but rather to perceive the samples as equal.

In our example, 19 trained sensory panelists participate in the study. They all taste four samples that come from two different products: a full-sugar strawberry yogurt and a prototype that consists of a reduced-sugar strawberry yogurt (−45% sugar). The samples are served in plastic coded cups at 8 °C. The codes of the samples used during the study are shown in Figure 10.6. In a tetrad test, there can be only six possible orders of presentation. For example, for product 1 (A) and product 2 (B), the possible orders are as follows:

AABB, ABAB, ABBA, BBAA, BABA, BAAB

In Table 10.4, the order of presentation of each sample is specified and balanced. The last column shows the answers of the panelists. In this case, 9 of 19 panelists correctly identified the pairs of samples.

The results are treated by using the guessing model method with binomial power and a traditional significance level of 5%. With 9 correct responses among 19 panelists, the analysis

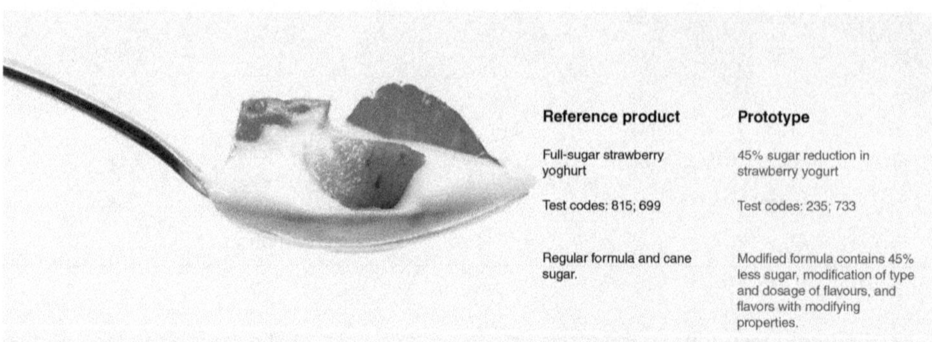


**Figure 10.6** Comparison of full-sugar versus reduced-sugar strawberry yogurt. Test codes used are presented for each sample. *Source:* ajcespedes/Pixabay.

**Table 10.4** Details of the tetrad test: order of presentation by panelists (A1–A19) and answers obtained after the sensory test.

| | Sample 1 | Sample 2 | Sample 3 | Sample 4 | Incorrect/correct |
|---|---|---|---|---|---|
| A1 | 699 | 815 | 235 | 733 | 0 |
| A2 | 699 | 733 | 815 | 235 | 0 |
| A3 | 815 | 733 | 699 | 235 | 1 |
| A4 | 235 | 815 | 699 | 733 | 0 |
| A5 | 815 | 235 | 733 | 699 | 0 |
| A6 | 815 | 699 | 235 | 733 | 1 |
| A7 | 699 | 235 | 815 | 733 | 0 |
| A8 | 733 | 699 | 235 | 815 | 0 |
| A9 | 815 | 699 | 733 | 235 | 1 |
| A10 | 733 | 815 | 235 | 699 | 1 |
| A11 | 733 | 235 | 815 | 699 | 1 |
| A12 | 815 | 733 | 235 | 699 | 0 |
| A13 | 235 | 815 | 733 | 699 | 1 |
| A14 | 699 | 235 | 733 | 815 | 0 |
| A15 | 815 | 235 | 699 | 733 | 0 |
| A16 | 235 | 733 | 699 | 815 | 1 |
| A17 | 699 | 815 | 733 | 235 | 0 |
| A18 | 733 | 699 | 815 | 235 | 1 |
| A19 | 699 | 733 | 235 | 815 | 1 |

can be done by using sensory software such as R or XLSTAT. The software automatically calculates that the $p$ value of the tetrad test with nine correct responses is 0.146. The proportion of discriminators is 0.211, and the power of the test is 0.245. As the $p$ value is more than $\alpha$ (0.05%), we conclude that there are no differences across our samples; therefore, the strawberry yogurt prototype is not perceived to be different from the full-sugar yogurt. In a project such as this, the usual next step is to define the business impact; in other words, if the sales volume or brand importance is high (high product sales or key iconic brand), the next steps can be validating the consumer step of focusing on liking. It is not common to perform a discriminative test with consumers to validate a sugar reduction project (at least this is the experience of the second authors working on many sugar reduction projects with different industries); it is more common to ask consumers about their liking or purchase intent. A study with a minimum of 120–150 consumers is usually done with the two sample tests, either monadically (consumer only tests one sample) or sequential monadically (consumer tests both samples). The action standard is fixed as parity in overall liking and purchase intent (no significant differences across samples). In general, when a discriminative test with a trained panel shows no significant difference across samples, it is expected that consumers (even heavy users of the brand) will not evaluate the liking of the samples differently. Therefore, for less important brands, when a discriminative test shows no difference, the projects stop there and the samples can be considered validated for the sugar reduction project; only for key brands does a final consumer liking validation take place.

#### 10.2.3.3 Salt Reduction in Sausages in Pilot Plant/Plant Samples

As in the case of sugar reduction, salt reduction is a trending topic in the food and flavor industries with new regulations that add nutritional scores or specific warning labels placed at the front of a food package (Arrúa et al. 2017). When a set of samples that a technical team has developed is being evaluated, the first problem that arises is the number of samples produced. In the food industry, because of the speed of reformulations, it is not often the case that samples are evaluated from an experimental design but rather from a set of back-and-forth reformulations based on the technical expertise of the team. Therefore, the number of samples that must be evaluated by the sensory lab is quite high. In addition, having a sufficient number of samples for all tests is a common problem, as the sensory tests often require a large number of samples that are not always available for the studies.

When salt-reduced samples are evaluated, there is often sensory fatigue because of the high number of samples: Imagine making four triangular tests of soups (reduced in salt) in one session. The panelists would in this case receive 12 samples (1 triangular test with 3 samples, multiplied by 4 tests = 12). Evaluating 12 soups correctly is a complex task, as the food matrix is complex, containing multiple ingredients such as fat, spices, and, of course, salt, which can increase sensory fatigue and performance. In this case, the sensory scientist must decide to make the sensory test more efficient, optimize the number of sessions and tests, and decrease the sensory fatigue of the panelists. For our example of salt reduction, our experienced sensory scientist decides to first screen the set of samples due to be presented each day and ideally to present the reference sample only once (imagine the technical team developing prototypes that have only a limited number of references to be used for development, along with the sensory test). The ideal test is, therefore, an R-index test.

The R-index test should ideally be called test A-not A with response reliability (Dacremont and Sauvageot 2019) and is a different type of discriminative test (compared with the tetrad or triangular test). In the R-index test, the panelists must determine whether two samples are from the same product. The panelists must, therefore, decide when the perceptual difference is large enough to declare that the samples come from two different products. This critical perceptual distance is called the decision criterion. It may vary from participant to participant and from session to session for the same participant. In the R-index test, the panelists receive one sample coded as the reference and a set of samples, usually between three and four, from which the references are often used as another sample (to test the discrimination of the panelists). The common instruction to the panelists is as follows:

*Welcome to this session. Today you are performing an A-not A and R-index test. You will receive three samples to test and a reference. Please test the sample marked as reference first and then the other samples. For each sample, you need to evaluate whether it is similar to the reference or not.*

*-Sample 359 evaluation-*

*For sample 359, please indicate whether*

- *It is the reference; I am sure* — *noted S*
- *It is the reference, but I am not sure* — *noted S?*
- *It is not the reference, but I am not sure* — *noted N?*
- *It is not the reference; I am sure* — *noted N*

The wording of the four options can slightly change, but the concept of each response should be the same. The analysis can be performed with regular sensory software or in R or even by hand by using the following formula from O'Mahony (1986):

| | | S | S? | N? | N | |
|---|---|---|---|---|---|---|
| Samples presented | $S$ | $a$ | $b$ | $c$ | $d$ | Total = $(a+b+c+d) = n_S$ |
| | $N$ | $e$ | $f$ | $g$ | $h$ | Total = $(e+f+g+h) = n_N$ |

$$R = \frac{a(f+g+h) + b(g+h) + ch + \frac{1}{2}(ae+bf+cg+dh)}{n_S n_N}$$

The R-index value is then multiplied by 100 to obtain the percentage of R and compared to the critical value by using an R index table for a specific α, which is usually 5% (Bi and O'Mahony 2007).

To illustrate the R-index test with a savory example, Figure 10.7 shows a case study of vegan sausages reduced in salt by 15% and 17%. The objective of the study is to know at what percentage of reduction the panelists perceive the samples to be different. The two samples are presented to the panelists following the traditional R-index method: one sample coded as the reference and the other samples coded with three-digit numbers. For the evaluation, the panelists respond to the same four options: *S, S?, N?, N*.

Twenty-seven panelists evaluated the sausage samples. Table 10.5 shows the results of the evaluation. The R-index equation was used to calculate the results: 58.6% for B1 and 54.9% for B2. Using the significance tables from Bi and O'Mahony (2007) for 27 panelists and an $\alpha$ of 5%, the critical value is 65.35%; therefore, as the calculated R-index of B1 and B2 is lower than the critical value of 65.35%, we conclude that samples B1 and B2 are not significantly different, and the panelists do not detect differences in the salt-reduced vegan sausages. As both samples achieve the action standard of no significant differences versus the full-salt sample, the next step can be to have both samples undergo a consumer test and

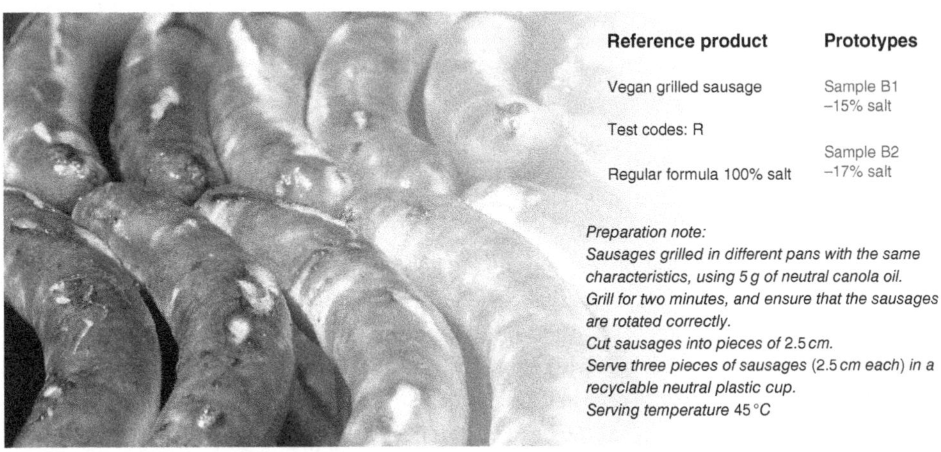


**Figure 10.7** Samples used for the R-index test of sausages with reduced salt content. Preparation and serving conditions are also specified. *Source:* webandi/Pixabay.

**Table 10.5** Results of the R-index test for samples B1 and B2.

| | S | S? | N? | N |
|---|---|---|---|---|
| Reference | 5 | 13 | 5 | 4 |
| B1 | 4 | 9 | 8 | 6 |
| B2 | 4 | 10 | 11 | 2 |

measure liking and purchase intent across the three samples. If no differences are found in liking across the three samples, most probably the sample with the highest percentage of salt reduction can be launched into the market.

## 10.3 Discrimination Testing for Fragrances

### 10.3.1 Product – Stimuli – Fragrance

In the fragrance industry, direct evaluation of fragrance oil, similar to that for flavors, is normally done in alcoholic or propylene glycol solutions by perfumers or fragrance development managers and using smelling strips. Most of the time, however, discrimination tests are performed on finished products.

In perfumery, it is important to think about how consumers will handle or use the products; therefore, the conditions and the moments in which consumers are in contact with the products and smell the fragrances define the conditions of evaluation of the finished products.

Fragrances are consequently often evaluated at different stages, each stage representing one of these key moments. As an example, Table 10.6 displays the key stages of evaluation for laundry products, including detergents and fabric conditioners.

Fragrances can also be evaluated at different time points from the moment the product is applied or used until the end of its usage, that is, until the fragrance lasts. This is the case, for example, for fine fragrance or deodorant products. Fine fragrances are generally evaluated at "fresh," "after 2 hours of application," "after 4 hours of application," "after 6 hours of application," and sometimes "after 8 hours of application." Deodorant time points are evaluated at "fresh," "6 hours after application," and "24 hours after application." The choice of the time points tested is related to consumer expectations of fragrance duration. As each of these moments is key to where consumers interact with the fragrance, it is recommended that discrimination tests be run at each of the defined stages or moments.

Per project and based on objectives, several discrimination tests can, therefore, be required. In such situations, it is important to limit the number of tests per session, as panelists can become saturated by having too many products to smell in one sitting. Note that from one product category to another, the type of stages or the number of time points varies. To cover all the stages and all the product categories, many types of support can be used to evaluate fragrances in finished products. Table 10.7 summarizes the most common supports used in perfumery per category.

**Table 10.6** Evaluation stages that are usually used for laundry products.

| Stage | Key moments for consumers | |
|---|---|---|
| Neat | Panelists smell the neat product in a bottle or jar to mimic consumer behaviors when they open the products | 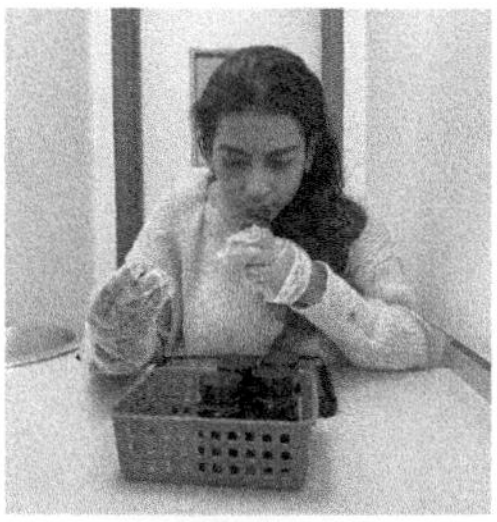 |
| Wet | Panelists smell on a wet cloth to mimic the moment when a consumer opens their washing machine and smells the washed load<br>Note: Wet clothes are wrapped in aluminum foil to keep the cloth humid |  |
| Blooming in sensory cabins | Panelists smell through the hatch to mimic the consumer experience when the laundry is line dried in a room and the consumer enters and smells the fragrance in the air | 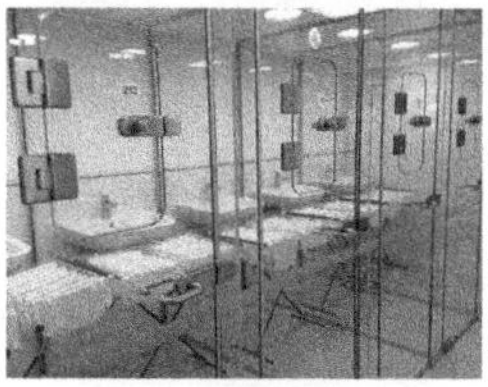 |
| Dry | Panelists smell the dry cloth to mimic the consumer experience of handling the laundry once dried for ironing or storage<br>This evaluation can be done at different moments (after 3 days, 7 days, or even after 8 weeks) |  |

Image from DSM-Firmenich, 2021.

## 10.3.2 Applications

In the perfumery industry, discrimination testing can be used for different purposes. At a factory level or during the development phase of a fragrance oil, discrimination tests can be used for day-to-day quality control or to evaluate the impact of raw material substitution, a process change, shelf life, or batch-to-batch variations.

When final products are evaluated, it is common to use discriminative tests when there is a change in the final product that can modify the fragrance profile, such as a change in base or a fragrance reformulation to reduce the cost or comply with legislation. Currently, Cosmetics Regulation EU 1223/2009 lists 26 raw materials (Table 10.8) as being of concern, called "allergens"; consumers should be aware of their presence in a cosmetics product (soap, lotion, shampoo, hair conditioner, etc.).

**Table 10.7** Different types of support used in perfumery.

| Product | Type of support | | |
|---|---|---|---|
| Laundry care | Liquid format   | Cotton towels 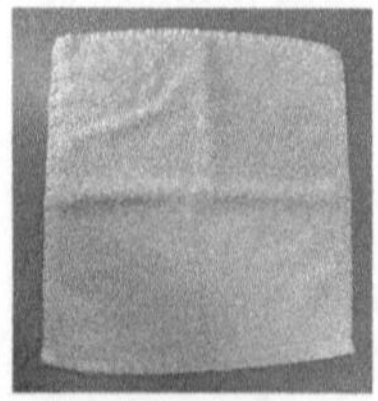 | |
| | Powder format  | Polyester swatch 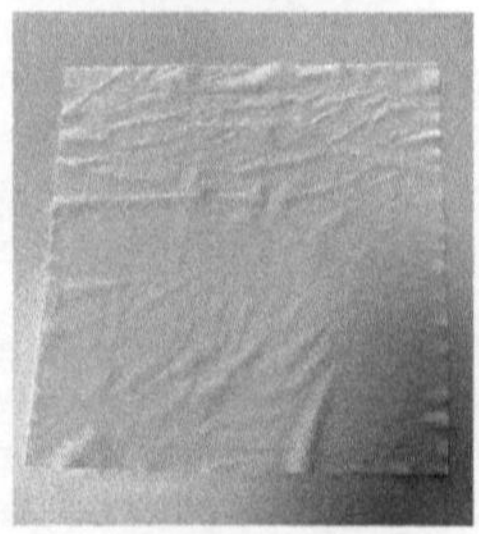 | |
| Hair care | In squeeze bottles   | 10g Caucasian hair 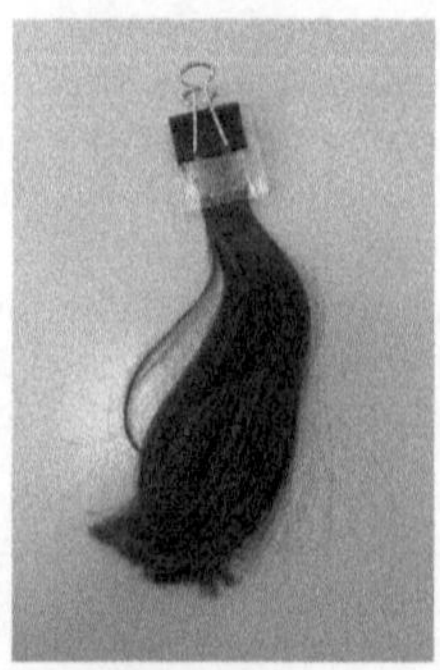 | 20g Asian hair 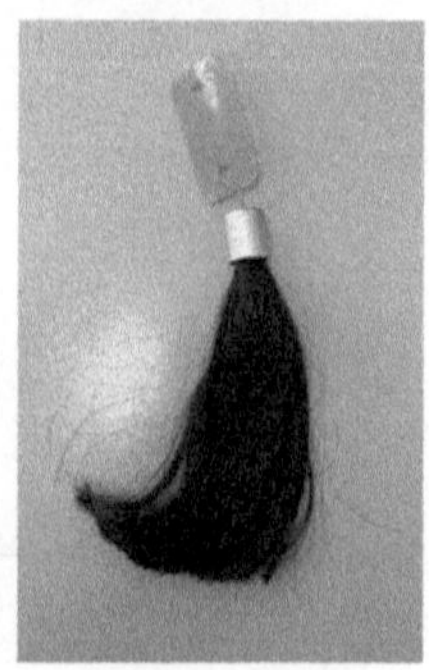 |
| Surface care | Glass brown bottle   | Glass tiles  | |

**Table 10.7** (Continued)

| Product | Type of support |
|---|---|
| Fine fragrance | Glass slides  |
| Deodorant | Large blotters 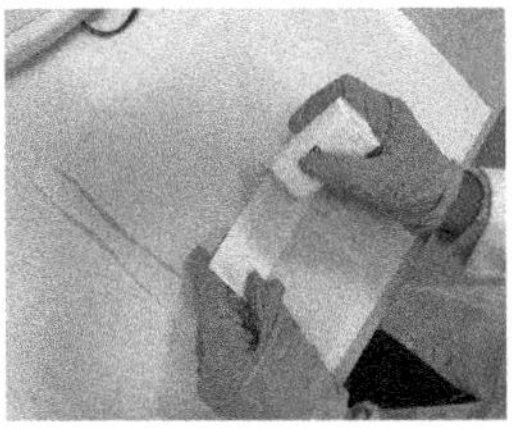 |

Image from DSM-Firmenich, 2021.
*Note:* For other product categories, such as for air care or toilet care, the evaluation supports are the sensory cabins.

These materials can be either restricted from being included in the formula or even banned. Therefore, when a raw ingredient is added to the allergens list, it can have an impact on the perfumery industry, as it often means that fragrances on the market must be reformulated to comply with new regulations.

Cosmetics Regulation EU 1223/2009 also prohibits the use of substances classified as carcinogenic, mutagenic, or toxic to reproduction (CMR). In August 2020, several ingredients were classified as CMR substances, notably the allergen butylphenyl methylpropional, also known as lilial, which has a "floral, muguet, cyclamen, and watery" note, and is a well-known fragrance ingredient commonly found in cosmetic products. This regulation became effective from March 2022. Because of this change, most of the discrimination test projects in the fragrance industry over the past years have focused on the removal of lilial from fragrance formulas.

When final products are evaluated, one other common objective of using discrimination tests is to show better fragrance performance. Fragrances can contain technologies such as capsules that help to prolong the odor of the fragrance over time. For example, in laundry, capsules from softener products are deposited during the washing process into textile fibers. Once the clothes are dried, fragrance is released later as the capsules break with a light touch of the fabric. Throughout the years, the perfumery industry has been continuously working to improve the performance of their capsules while developing a more sustainable delivery system.

In these contexts, discrimination tests are used to show that progress has been made, and tests are, therefore, focusing on showing the differences. Such a test objective allows more

**Table 10.8** List of 26 fragrance allergens from European Union Cosmetics Regulation 1223/2209.

| CAS # | Description |
|---|---|
| 127-51-5 | Alpha-isomethyl ionone |
| 122-40-7 | Amyl cinnamal |
| 101-85-9 | Amylcinnamyl alcohol |
| 105-13-5 | Anise alcohol |
| 100-51-6 | Benzyl alcohol |
| 120-51-4 | Benzyl benzoate |
| 103-41-3 | Benzyl cinnamate |
| 118-58-1 | Benzyl salicylate |
| 80-54-6 | Butylphenyl methylpropional |
| 104-55-2 | Cinnamal |
| 104-54-1 | Cinnamyl alcohol |
| 5392-40-5 | Citral |
| 106-22-9 | Citronellol |
| 91-64-5 | Coumarin |
| 97-53-0 | Eugenol |
| 4602-84-0 | Farnesol |
| 106-24-1 | Geraniol |
| 101-86-0 | Hexyl cinnamal |
| 107-75-5 | Hydroxycitronellal |
| 31906-04-4 | Hydroxyisohexyl 3-cyclohexene carboxaldehyde |
| 97-54-1 | Isoeugenol |
| 5989-27-5 | Limonene (D-Limonene) |
| 78-70-6 | Linalool |
| 111-12-6 | Methyl 2-octynoate |
| 90028-68-5 | Evernia prunastri (oak moss) |
| 90028-67-4 | Evernia furfuracea (tree moss) |

flexibility in the choice of the type of discrimination test, as it does not require as many panelists as when the aim is to demonstrate similarity between two fragrances. The 3-AFC test (cf. Section 10.3.3.2), a very powerful test, can be in these cases a method of choice (Dacremont and Sauvageot 2019).

## 10.3.3 Case Studies

### 10.3.3.1 Lilial Removal in Shampooing Products

As mentioned in the previous chapter, due to new legislation, the fragrance industry in previous years has been asked to remove the ingredient "lilial" from fragrances on the market. As products are already on the market, it is important that new lilial-free developments remain faithful to the original version in hedonics and strength; that is, it is key that the fragrance profile and performance remain the same, as the objective is to do a silent change, in other words, to replace the fragrance on the market without consumers noticing it. To validate the change, the discrimination tests used are triangular tests.

Lilial is highly used in shampooing/conditioner formulations, in some cases representing up to 10% of the fragrance. To reflect the moments when consumers are in contact with the fragrance of a shampoo or a conditioner, the validation of the change is usually done at the stages that highlight the different moments of product usage. These stages are displayed in Table 10.9.

To validate the change, three triangular test sessions are run to cover the three stages, each with a minimum of 42 panelists. The evaluations at the "neat" and "dry hair swatch" stages, the latter prepared 6 or 24 hours before the test, are done in traditional sensory booths (Figure 10.2), whereas the evaluation at the "in use" stage is done in specific cabins that contain a sink (Figure 10.3). In the case of an evaluation at the "in use" stage, the preparation of the hair swatches is done by the same sensory assistant to ensure that it is done consistently and in a standard way; groups of six panelists are organized to minimize sample preparation.

**Table 10.9** Different stages of evaluation of shampoo and conditioner.

| Stage | Shampoo (SH) | Conditioner (CDT) |
|---|---|---|
| Neat   | ✓ | ✓ |
| In use (on hair swatch) | → *Panelists' evaluation @ foaming (2)* → *Panelists' evaluation @ rinsed (wet) (5)* | → *Panelists' evaluation @ after applying (2)* → *Panelists' evaluation @ rinsed (wet) (5)* |
| 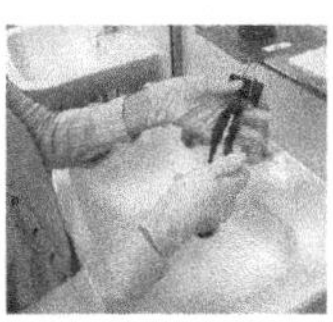 (1) SH/CDT |  (2) Foaming/or applying CDT | 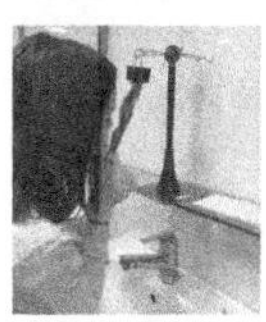 (3) Evaluation  (4) Rinsing 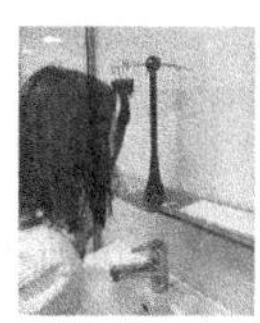 (5) Rinsed |
| Dry – after 6 hours/24 hours | ✓  | ✓  |

At each stage of the evaluation, the order of presentation is balanced among samples and panelists. Each panelist receives three coded samples, two of which are from the same sample and a third from a different sample. The panelists' task is to identify the different sample by smell. In this example, the aim is to silently change the fragrance and therefore to examine the similarity between fragrances, the null hypothesis ($H_0$) being "the products are different." The sensitivity of the test is defined by the values of $\alpha$, $\beta$, and $P_d$ (proportion of distinguishers). The risk alpha ($\alpha$), when considering similarity, is often higher than 5%, as it is not a great concern to falsely conclude that a difference exists when one does not.

In the present case study, it was decided to choose a risk $\alpha$ of 10%, a maximum tolerable proportion of distinguishers of 30% and only a $\beta = 10\%$ chance of failing to detect that level of distinguishers. It is important to acknowledge that this test cannot prove that the two fragrances are identical, but it is possible to ensure that only a reasonably small proportion of the population will be able to detect a difference by selecting a small $P_d$ and risk levels that are acceptable for the business. The results obtained at each stage are presented in Table 10.10. Even though three of four triangular tests show that the fragrances can be considered similar at the selected risk level, at foaming, the fragrances are perceived to be different. As this is one of the most key moments of usage for consumers, the perfumer needs to further rework the fragrance.

#### 10.3.3.2 Validation of New Capsules in a Fabric Softener

The objective in this case is to validate the performance of newly developed capsules and to determine whether the performance of these new capsules is better than that of the current capsules. Note that the different capsules contain the same encapsulated fragrance.

**Current sample**
Free oil X+ Capsules A
Codes 112 and 284

**Candidate with new capsules**
Free oil X+ Capsules B
Code 574

In this context, discrimination tests are used to understand whether the sample with capsules is effectively perceived as different. In preparation for the test, "cotton towels" (Table 10.7) are washed with the softeners in combination with an unperfumed detergent the day before the test, and then the towels are line dried in a controlled drying room (22 °C, 60%RH) for 24 hours. Before the test, the "cotton towels" are placed in coded baskets.

**Table 10.10** Results from the triangular test of the fragrances evaluated at different stages.

| Stage | Correct answers | P value | The fragrances can be considered... |
|---|---|---|---|
| Neat | 17/42 | 0.0649 | Significantly similar |
| In use SH foaming | 20/42 | 0.2779 (NS) | Not significantly similar (different) |
| In use SH rinsed | 14/42 | 0.0071 | Significantly similar |
| Dry six hours | 15/42 | 0.0163 | Significantly similar |

**Table 10.11** Example of sample presentation order for an AFC test.

| Panelist | Sample 1 | Sample 2 | Sample 3 | Assessor's answer | Correct answer |
|---|---|---|---|---|---|
| J1 | 112 | 574 | 284 | 1 | 574 |
| J2 | 574 | 112 | 284 | 1 | 574 |
| J3 | 284 | 112 | 574 | 0 | 574 |
| J4 | 112 | 574 | 284 | 1 | 574 |
| J5 | 574 | 112 | 284 | 1 | 574 |
| J6 | 284 | 112 | 574 | 1 | 574 |
| J7 | 112 | 574 | 284 | 1 | 574 |
| J8 | 574 | 112 | 284 | 0 | 574 |
| J9 | 284 | 112 | 574 | 0 | 574 |
| J10 | 112 | 574 | 284 | 1 | 574 |
| J11 | 574 | 112 | 284 | 1 | 574 |
| J12 | 284 | 112 | 574 | 1 | 574 |
| J13 | 112 | 574 | 284 | 0 | 574 |
| J14 | 574 | 112 | 284 | 1 | 574 |
| J15 | 284 | 112 | 574 | 0 | 574 |
| | | | | Correct responses: 10 | |

In the present study, the 3-AFC test is selected, as our objective is to show that the new capsules deliver significantly more intense fragrance, that is, to examine a directional difference. Compared with a triangular test, the 3-AFC test is statistically more powerful; that is, we are more likely to find a statistically significant difference with a 3-AFC test than with a triangular test. This also means that at an equivalent power, this test does not require as many people as for a triangular test. For example, in a triangular test, to examine a difference, as per standard ISO 4120, when considering an $\alpha$ risk of 5%, a $\beta$ risk of 10%, and a $P_d$ (proportion of distinguishers) of 50% or a $d'$ of 2.3 ($d'$ difference between products, Thurstonian model), 20 panelists are required, whereas for a 3-AFC test, only 5 panelists are needed.

As explained in the previous chapter, the sensitivity of the test is defined by the values of $\alpha$, $\beta$, and $P_d$. The risk $\alpha$, when examining difference, is often equal to 5%, as in this context it is a great concern to falsely conclude that a difference exists; for example, in our case, we could launch worse performing capsules on the market. In contrast, we can be more willing to take a greater risk of missing a difference ($\beta$ risk) that does exist, as we can always further improve the performance of our capsules. Hence, this is why, for this case, it was decided to select a $\beta$ risk of 10% with a $P_d$ of 50% (i.e., 50% of the panelists can detect a difference between the samples in the test), which is equivalent to a $d'$ of 2.3 ($d'$ size of difference between products, Thurstonian model). However, to better determine whether the new capsule type has potential or not, 15 panelists are asked to participate in the test, which means that smaller differences can be captured if there are any ($d'$ and $P_d$ are then reduced).

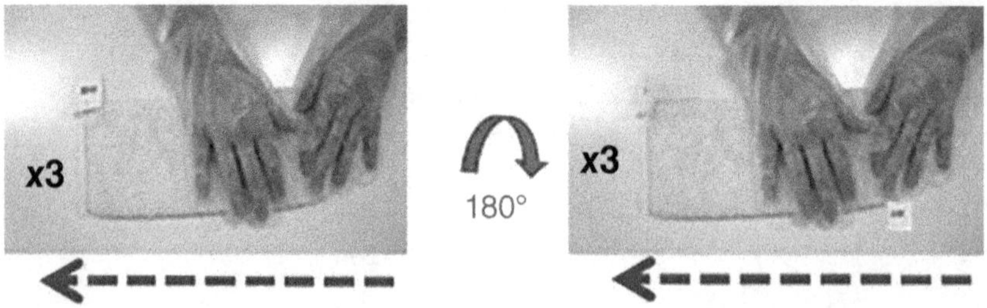


**Figure 10.8** Panelist is asked to rub the towel three times from left to right, turn the towel, rub the towel from left to right, smell the towel, and remember its strength before taking the next sample (Image from DSM-Firmenich, 2022).

On the day of the test, the panelists receive three coded samples, two of which are from the same sample (A, current capsules) and a third sample (B, new capsules), which has more of a specified attribute. Note that to validate the use of the 3-AFC procedure, prior evidence as to whether sample (B) has more of the specified attribute than sample (A) is required. This will also determine the type of triads to use for the test. In our case study, the panelists are asked to select the sample that is stronger in overall fragrance intensity. For sample (B), made with the new capsules, being the sample expected to be "the strongest" in intensity, the only possible triads to use for the test are AAB, ABA, and BAA. To evaluate the performance of the capsules, the panelists need to break them and are asked to rub each towel, as explained in Figure 10.8.

In this example, 10 correct responses were obtained by the panelists, which is equivalent to a 0.667 proportion of correct responses. The guessing probability for this test is 1/3, which is similar to that for the triangular test. The results with the 10 correct responses have a $p$ value of 0.0085, and as the $\alpha$ was fixed at 5%, we can conclude that there are significant differences between the two capsules, the new capsules delivering significantly more fragrance than the current capsules.

## 10.4 Bringing Social Responsibility into Discrimination Testing for Flavors and Fragrances

### 10.4.1 Blind and Visually Impaired Panelists

The use of blind sensory panelists is not new and is probably more common than expected. Performing different sensory tests with blind panelists is actually an easy task, and multiple examples can be found in the scientific literature. The authors actively encourage the use of visually impaired persons as panelists, as it is our duty as sensory scientists to bring diversity into our sensory panels. It is estimated that one billion persons (15% of the world's population) experience some form of disability. Persons with disabilities, on average as a group, are more likely to experience adverse socioeconomic outcomes than persons without disabilities (The World Bank 2022).

The second author reported how a group of blind sensory panelists performed triangular tests of chili samples (Gomez-Corona et al. 2020) in an article that also encouraged the use

**Figure 10.9** Discrimination testing of flavor applications with blind panelists. Left: Sample deliveries to panelist's house. Center: Video-conference sensory session in which all panelists place their smartphone in front of them so that the panel leader (middle of the image with red headphones) can see their evaluations and order of samples. Right: Sensory blind panelists indicate which sample is perceived as different in a triangular test. The panel leader can manually process the results in a file (image from dsm-firmenich, 2022).

of blind panelists. While there is a common belief that blind persons outperform sighted persons in discriminate or descriptive sensory tasks, it has been demonstrated that blind persons do not outperform their sighted counterparts (Mucci et al. 2005; Majchrzak et al. 2017; Sorokowska et al. 2018; Gomez-Corona et al. 2020).

The first two authors of this chapter work regularly with blind panelists and perform all types of sensory discrimination (and descriptive) tests. In the case of flavor applications, the authors recommend doing exactly the same as in a test with sighted panelists: respect the order of the samples, always give the reference first (when using a reference-based method, e.g., R-index), and evaluate from left to right and top to bottom. Blind panelists do not need any special or additional attention. Figures 10.9 and 10.10 show an example of a discrimination task with respectively blind and visual impaired panelists. The images have a flavor application, in which panelists perform a triangular test of different tasting solutions at the panelists' homes during COVID-19 restrictions in Mexico City in 2020. The session was done by using a video-conference system, and the correct order of the samples was confirmed with a visual display (for the panel leader). Panelists were asked to place their phones in front of them, so that the panel leader could see that all the panelists tested their samples in correct order.

### 10.4.2 Panelists in Vulnerable Social Conditions

A sensory scientist should move away from the stereotyped idea that all panelists in the industry are a group of middle-income, middle-age housewives that participate in a sensory panel during their free time during the mornings or the idea that all panelists at a university are students available for a training session. We must think that there are differences across panels and panelists and that one difference can be their vulnerable condition. Although income may be the simplest way to categorize persons as low income, we know that poverty affects billions of humans worldwide, and it is estimated that 1.3 billion humans are multidimensionally poor (UN News 2019). Low-income persons face multiple factors that shape the quality of their lives and make them a vulnerable group, including not only physical deprivation and pain (hunger, deficient healthcare, sanitary conditions, and abuse), exclusion (relationships and community), marginalization, and anxiety and

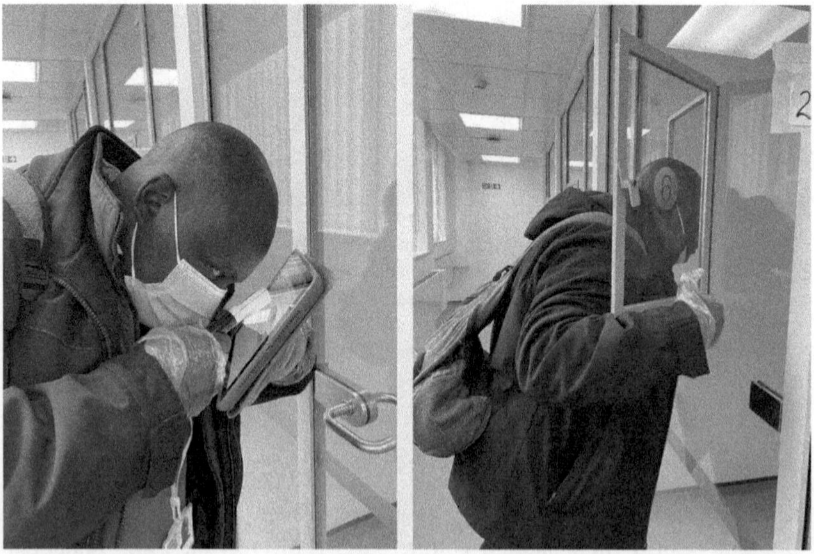

**Figure 10.10** Visually impaired panelists using a magnifying glass to read instructions on the iPad and then smell the fragrance through the hatch to evaluate the intensity of the fragrance diffused in the cabins.

fears about the future but also poor health, poor education (including digital knowledge) and quality of work, and the threat of violence (Gomez-Corona 2020).

Evaluating with low-income persons is common in certain countries, where specific products are targeted for this population; therefore, the discriminative test is in the form of a consumer discriminative evaluation and rarely in the form of a formal sensory panel. Among the previously mentioned factors affecting vulnerable persons, five are most important when vulnerable persons are included in a sensory evaluation session (focusing on discrimination testing): deprivation (hunger), education level, health condition (e.g., diabetes or high blood pressure), digital exclusion, and sanitary conditions (Table 10.12). Of these variables, hunger is the most critical, and we should never perform a sensory evaluation with a person under this extreme vulnerable condition. The sensory scientist should adhere to a high code of ethics and behave with respect and humanity if a person with this condition is identified in the group of panelists or consumers.

For the other vulnerable conditions in Table 10.12, it is possible to perform discriminative tests. For example, in consumers or panelists with a low education level, it is possible that they do not have a sufficient literacy level to answer via a traditional paper questionnaire (not to mention the use of a tablet or computer). To overcome this barrier, the easiest way to perform the task is verbally. Simply explain the task to the person and verbally collect the response. The same applies to persons under a digital exclusion condition. Note that in this case, the person can be of higher income but digitally excluded (e.g., older persons unable to use a smartphone or tablet). On the other hand, when dealing with a special health condition, it is important to always screen consumers and panelists to avoid common chronic diseases such as diabetes, hypertension, or food allergies. However, sometimes people with a specific health condition are the target for a test. For example, in the

**Table 10.12** Effect of vulnerable panelist conditions in discriminative tests.

| Vulnerable condition | Effect of factor in discriminative test | Recommendation |
|---|---|---|
| Deprivation – hunger | Avoid evaluation and search for means to support persons in this vulnerable condition | |
| Education level | Literacy level to read questionnaires<br>Use of direct and simple language<br>Bias of persons external to the community | Perform individual interviews<br>Use the support of local persons to perform the test |
| Digital exclusion | Unable to use a tablet, smartphone<br>Access to Internet connection | Perform individual interviews |
| Health condition | Diabetes is higher in vulnerable persons<br>Hypertension is higher in vulnerable persons<br>Specific medical situation is important for the study (e.g., anosmia, dysphagia) | Always screen persons for health conditions<br>Use medical protocols for tasting |
| Sanitary conditions | Poor personal hygiene – clothes contaminated with unpleasant odors | Ask everyone to wear a lab coat |

development of special nutrition products, it is desirable to test the products in people with the condition (e.g., anosmia, dysphagia). Although this may fall under a consumer research protocol, it does not exclude the possibility of performing discriminative tests with them. In that case, it is important to adhere to medical protocols, but further explanation falls outside the scope of this chapter.

## 10.5 What the Future Might Be in the Flavor and Fragrance Industries

The future of the perfume and flavor industries will depend on increasing consumer demand for sustainable, natural, healthy, and emotionally positive products, as well as on the implementation of sustainable development policies based on environmental, social, and economic pillars. Furthermore, many of the most ground-breaking developments in the F&F industries will come from the realm of artificial intelligence (AI).

With continued technological advancements, and new innovative applications of those technologies, perfumers and flavorists are now able to create unique digital fragrances.

### 10.5.1 Adapting to New Social and Environmental Challenges in the Industry

The next challenges will arrive with sustainability of raw materials and their traceability, as well as societal and economic responsibilities, supported by AI. More especially, AI-augmented creation enables the development of uniquely tailored perfumery and taste solutions with increased speed to market in consideration of the new raw material constraints and complexities. Several top F&F developers have designed and implemented their own in-house AI systems, as well as digital reference directories for use by both consumers and industry professionals.

More particularly, some F&F industries have already developed a formula generator based on AI, which augments the creation of both fragrances and flavors. This system connects the initial composition with data, such as specific applications and olfactive descriptors (scent and taste), helping creators to understand which ingredients to use in products, including shampoo, soap, body spray, beverages, dairy, meat analogs, and more. This provides a base from which creators can build bespoke creations for customers.

Behind the creation of fragrances and flavors with the use of AI, their reformulation and the use of more and more new sustainable raw materials will increasingly need sensory validation. To achieve that, AI will be the key to success to guide and predict the sensory impact of such compounds in the final formulation and client product.

### 10.5.2 What Will Be the Place of Machine Learning and Artificial Intelligence in Discrimination Testing? What Might a Future Panelist Look Like?

AI is intelligence demonstrated by machines, as opposed to natural intelligence displayed by animals, including humans. AI researchers have tried many different approaches and discarded them since the founding of AI, including simulating the brain, modeling human problem solving, using formal logic, creating large databases of knowledge, and imitating animal behavior. In the first decades of the 21st century, highly mathematical statistical machine learning (ML) has dominated the field, and this technique has proved successful, helping to solve many challenging problems throughout industry and academia. ML is the study of computer algorithms that can automatically improve through experience and by using data. ML algorithms build a model based on sample data, known as training data, to make predictions or decisions without being explicitly programed to do so.

On the other hand, discrimination testing with panels is the classically used tool to assess potentially perceived significant differences between products by subjects or consumers. Discriminative tests are time-consuming and need important resources. Optimizing this testing in terms of product formulation and number and type of subjects will be key to delivering the right product on time. One way to achieve this goal will be to develop sensory and subject perception databases or sensitivity databases associated with ML.

First, the sensory product database, connecting the initial composition with data, such as specific applications and olfactive descriptors (scent and taste), could be used to map sensory differences and similarities between products. This would allow us to compare formulas globally and to predict a perceptive distance ($d'$) by using ML to select the most relevant discriminative tests to perform. Performing the discriminative tests would then allow us to validate and fine-tune this information. To be sure, discriminative test results must be systematically standardized and translated into perceptive distances between products, depending on the ingredients, formulas, and associated bases.

AI and ML could also support the better understanding of subject perception. Discriminative tests could be used to establish a sensitivity map for each subject or panelist. Individual results obtained for discriminative tests could be used to build "subject sensitivity" databases to perceive differences between products, depending on the ingredients, formulas, and associated bases. We could then imagine adding subject sensitivity for specific products as a variable to, for example, optimize the number of subjects who detect a significant difference.

### 10.5.3 The Future Panelist Will Be "Connected"

Discrimination testing will continue to see changes with developments in technology, in particular, the application of handheld and wearable devices. Wearable technology allows tracking and recording of many personal health metrics (footsteps, sleeping habits, etc.) and health information (level of fatigue, arousal state, satiety, diet, etc.). All this information could paint a picture of the personal state of the panelist. Furthermore, knowing a person's genetic history, cultural background, fitness level, and other metrics could allow identification of panelists who may have different types of performance (Findlay and Findlay 2017). Of course, the sensitivity of sharing personal information in this way should not be overlooked.

Panelists will use these technologies to record their answers to discriminative tests in a simple manner. Furthermore, innovative technologies could also allow the controlled and remote delivery of stimuli, meaning that depending on the subject's answer, a relevant product could be delivered according to the test objective (minimizing a perceived difference between two beverages, for example). This interactivity between stimuli-subject answers will allow optimization of the number of samples to be evaluated. But more important, discriminative tests will also be performed in a more "ecological" way with consumers, who could do the tests in a meaningful context and during key moments of product consumption. For example, the future panelist will receive products at home and be asked to consume them during usual moments. Everything could be registered online and recorded in databases by using dedicated applications: the context, the explicit answers, and why not some implicit answers such as reaction times.

This will certainly help answer the questions. Difference testing can be performed in a way that ensures a difference will be found. The more powerful the test, the larger the number of assessors with specific training and the more likely a statistically significant difference can be found. The obvious question is, "Does this difference matter for the consumer?" Receiving more information from consumers will help perfumers and flavorists decide what is important and what sensory properties should be tested. Finally, the future of discrimination testing is faster, more accurate, and more oriented to differences that are meaningful to consumers.

## 10.6 Conclusions and Final Considerations

The information contained in this chapter shows how discriminative tests are used in the F&F industry. Some aspects of evaluating flavors in different matrices, or the different moments and places to evaluate fragrances, are unique and may differ from what is done in other industries. This chapter is practical in the sense of showing different applications and examples of several discriminative tests in both flavor and fragrances; nonetheless, different specific needs may arise, depending on the objectives for the discriminative tests.

Regarding the time frame, the tests mentioned in this chapter can almost all be considered traditional (e.g., triangular test or 2-AFC), although some tests are relatively new (tetrad). Nonetheless, all have been used in a similar way through the years, and the past 20 years have not seen any revolution regarding new methods for discriminative tests. In other words, their use and application have been stable over time.

As was briefly mentioned in the last part of the chapter, new tools are emerging with the objective of bringing ML to the field of discriminative tests. To date, there is no scientific publication that clearly explains the process of using big data or ML techniques to be applied to discriminative tests. Although we know that it is a matter of time before there will be new AI tools for discriminative testing, it is also important to highlight that the sensory scientist must understand and master the different aspects of discriminative tests before bringing new AI tools into the sensory lab.

## References

Ares, G. and Varela, P. (2017). Trained vs. consumer panels for analytical testing: Fueling a long lasting debate in the field. *Food Qual. Prefer.* 61: 79–86.

Arrúa, A., Curutchet, M.R., Rey, N. et al. (2017). Impact of front-of-pack nutrition information and label design on children's choice of two snack foods: comparison of warnings and the traffic-light system. *Appetite* 116: 139–146.

Bi, J. and O'Mahony, M. (2007). Updated and extended table for testing the significance of the R-Index. *J. Sens. Stud.* 22: 713–720.

Burns, S.L., Penfield, M.P., Saxton, A.M., and Luckett, C.R. (2018). Comparison of triangular and tetrad discrimination methodology in an applied manner. *Food Qual. Prefer.* 68: 105–112.

Castura, J.C., King, S.K., and Phipps, K. (2018). How task instructions affect performance on the unspecified tetrad test. *J. Sens. Stud.* 33 (3): e12329.

Chollet, S., Valentin, D., and Abdi, H. (2005). Do trained assessors generalize their knowledge to new stimuli? *Food Qual. Prefer.* 16 (13–23): 9.

Dacremont, C. and Sauvageot, F. (2019). Tests de différenciation. In: *Introduction à l'analyse sensorielle et aux études consommateurs: Nouvelles perspectives* (ed. C. Gomez-Corona and M. Lelièvre-Desmas). XOC Estudio: Mexico City.

Ennis, J.M. (2012). Guiding the switch from triangular testing to tetrad testing. *J. Sens. Stud.* 27: 223–231.

Ennis, J.M. and Jesionka, V. (2011). The power of sensory discrimination methods revisited. *J. Sens. Stud.* 26: 371–382.

Findlay, P.M. and Findlay, C. (2017). Chapter 16 – The future of sensory discrimination testing. In: Rogers, L. (eds). *Discrimination Testing in Sensory Science. A Practical Handbook*, Woodhead Publishing Series in Food Science, Technology and Nutrition, vol. 2017, 347–359. Woodhead Publishing.

Gomez-Corona, C. (2020). Sensory and consumer research for good: a review on social responsibility. *Curr. Opin. Food Sci.* 33: 115–123.

Gomez-Corona, C., Pohlenz, A., Cayeux, I., and Valentin, D. (2020). Panel performance and memory in visually impaired versus sighted panels. *Food Qual. Prefer.* 80: 103807.

Hautus, M.J., van Hout, D., and Lee, H.S. (2009). Variants of A Not-A and 2AFC tests: signal detection theory models. *Food Qual. Prefer.* 20 (3): 222–229.

Ishii, R., Kawaguchi, H., O'Mahony, M., and Rousseau, B. (2007). Relating consumer and trained panels' discriminative sensitivities using vanilla flavored ice cream as a medium. *Food Qual. Prefer.* 18: 89–96.

Ishii, R., O'Mahony, M., and Rousseau, B. (2014). Triangular and tetrad protocols: small sensory differences, resampling, and consumer relevance. *Food Qual. Prefer.* 31: 49–55.

ISO 4120:2021. Sensory analysis — Methodology — Triangular test.

ISO 16657:2006. Sensory analysis — Apparatus — Olive oil tasting glass.

Lee, H.S., van Hout, D., and Hautus, M.J. (2007). Comparison of performance in the A-Not A, 2-AFC, and same–different tests for the flavor discrimination of margarines: the effect of cognitive decision strategies. *Food Qual. Prefer.* 18: 920–928.

Lee, Y.M., Chae, J.E., and Lee, H.S. (2009). Effects of order of tasting in sensory difference tests using apple juice stimuli: development of a new model. *J. Food Sci.* 74 (6): S268–S275.

Majchrzak, D., Eberhard, J., Kalaus, B., and Wagner, K.H. (2017). Do visually impaired people develop superior smell ability? *Perception* 46 (10): 1171–1182.

Mucci, A., Garitta, L., Hough, G., and Sampayo, S. (2005). Comparison of discrimination ability between a panel of blind assessors and a panel of sighted assessors. *Journal of Sensory Studies*, 20(1), 28–34.

O'Mahony, M. (1986). *Sensory Evaluation of Food. Statistical Methods and Procedures.* New York: Dekker.

Rousseau, B. (2015). Sensory discrimination testing and consumer relevance. *Food Qual. Prefer.* 2015 (43): 122–125.

Rousseau, B., Stroh, S., and O'Mahony, M. (2002). Investigating more powerful discrimination tests with consumers: effects of memory and response bias. *Food Qual. Prefer.* 13 (1): 39–45.

Shepard, T.G., Veldhuizen, M.G., and Marks, L.E. (2015). Responses times to gustatory-olfactory flavor mixtures: role of congruence. *Chem. Senses* 40: 565–575.

Sinding, C., Valadier, F., Al-Hassani, V., Feron, G. et al.(2017). New determinants of olfactory habituation. *Scientific Reports*, 7(1), 41047.

Sorokowska, A., Sorokowski, P., Karwowski, M. et al. (2018). Olfactory perception and blindness: a systematic review and meta-analysis. *Psychol. Res.* 83: 1–17.

The World Bank. (2022). Disability inclusion. https://www.worldbank.org/en/topic/disability#:~:text=Results,One%20billion%20people%2C%20or%2015%25%20of%20the%20world's%20population%2C,million%20people%2C%20experience%20significant%20disabilities (accessed September 2022).

UN News (2019). New UN poverty report reveals 'vast inequalities' between countries. https://news.un.org/en/story/2019/07/1042231 (accessed June 2020).

Valentin, D., Dacremont, C., and Cayeux, I. (2010). Does short-term odour memory increase with expertise? An experimental study with perfumers, flavourists, trained panellists and novices. *Flavour Fragr. J.* 26: 408–415.

# 11

# Kids as Sensory Detectives

## Application of Discrimination Testing with Children to Identify If Differences Exist Between Two or More Products

*Cindy Beeren*

*EdiBee, Altea, Spain*
*Whyteleafe Community Hub, Whyteleafe, UK*

## 11.1 Introduction: Sensory Testing by Kids

Kids' foods are a key interest for many governmental and non-governmental bodies, not least UNICEF, often with a key focus to enhance the nutritional quality of these foods to reduce and prevent diseases as well as foods developed for kids with certain allergies (Gutierrez 2020).

The global kids' food and beverage market is an important sector within the total food and beverage market. It is currently estimated to grow with a compound annual growth rate (CAGR) of 6% between 2022 and 2032 (Miller 2023). Within the food and beverage sector major players as well as smaller manufacturers and start-ups are introducing new brands aimed at kids.

Also outside of the food industry we can see a growing trend in products offered to kids. Over-the-counter health products in the UK saw a 4% raise despite a decline in sickness, according to Mintel (Greenslade 2020), kids' cosmetics are set to grow with 6.8% CAGR from 2020 to 2026 (Market Research Future 2021). Moreover, new product launches also included new brands focusing on specific age groups of kids both in the food and non-food industry; examples here include Little Spoon, Fruu Cosmetics and Scrubbington's.

Before these new kids products hit the market, most will be tested using consumer research, where kids will take part in the product testing.

However, when carrying out more objective sensory research, manufacturers would typically use an adult trained sensory panel. For many applications, you could argue that indeed the use of an expert adult panel will more than suffice to carry out testing. These trained sensory panels are designed to be able to detect and describe all aspects of a product's sensory profile. In addition, established adult expert panelists are already trained and depending on the age group of the kids' product to be tested, adults might find it easier to adhere to testing protocols. However, in some cases you might decide it is better to test with kids. Whether this is because you want to know the true perception of a child, or be able to

*Discrimination Testing in Sensory Evaluation*, First Edition. Edited by Lauren Rogers, Joanne Hort, Sarah E. Kemp and Tracey Hollowood.

use descriptions generated by kids in any further work, this chapter is here to support you when you decide to carry out sensory testing with kids, giving practical advice as well as further information about kids vs. adults sensory testing.

In this chapter, we will be reviewing the differences between kids' and adults' perceptions, looking at when and why to use kids for sensory testing and supporting you choosing the right methods and protocols when you do use kids for sensory testing.

## 11.2 Kids' vs. Adults' Sensitivities

The term sensitivities describes the ability of kids and adults to perceive different stimuli from different products that we handle, for example, the ability to perceive the sweetness in an apple, or the ability to see a gloss on toothpaste. Although much more research has been done into adult sensitivities (especially to sweet and bitter taste), from research we also know that kids' perception is different to that of adults. To start off, our sensitivities change as we grow; the key differences between kids and adults are described in this section. Sensitivities towards different tastes and smells have also shown to have a significant impact on the foods we consume, possibly even more so for kids than for adults. Coulthard and Blissett (2009) show that fruit and vegetable consumption of kids usually is in line with fruit and vegetable consumption of the mother, however, kids who were sensitive to taste and smell stimuli showed lower consumption of these foods. This certainly indicates that sensitivities play a large impact on food liking and are thus important to be well understood when developing kids' products.

Sensitivities can also differ for our different senses; let's have a look at each of the different senses and the key differences between adult and kids' sensitivities.

### 11.2.1 Smell

Kids' ability to pick up on different smells is comparable with adults, however adults are better able to describe the odors. The association of the different smells with the words is particularly harder for kids under six whilst for us adults we quickly associate smells with previous occasions and experiences (Lumeng et al. 2005). As kids grow older, more associations are stored in their brain, and associating smells with previous experiences and occasions becomes a more natural thing to do.

### 11.2.2 Taste

Taste and smell often get confused, and probably rightly so, as around 95% of taste disturbances are actually caused by a loss of smell (Sobek and Jagielski 2022) – think of colds, medications, ear infections etc. when we suddenly notice we cannot "taste" as much. As we eat our food, different chemicals are released into our nasal cavity and we thus start perceiving (smelling) them. When we eat a jam sandwich and we perceive a strawberry taste, in sensory terms it would be referred to as a strawberry flavor or retronasal smell. Orthonasal smell is the traditional smell by sniffing, whilst retronasal smell is perceived from the mouth.

In the sensory world, when we talk about taste, we refer to just the basic tastes perceived using our taste papillae in our mouth (on our tongue/palate and throat): sweet, salt, sour, bitter, and umami. Kids' taste thresholds in comparison with adult taste thresholds has been researched quite a bit, with conflicting results. Several papers seem to suggest that kids' taste sensitivity is either similar or poorer than that of adults.

However, we may want to bear in mind that some of the differences in results may actually come from the test protocols themselves. For example, kids' reading ability, how comfortable they felt doing the test, length of test, and a likelihood to confirm what adults say, may all influence the results of kids tests (Vennerød et al. 2017).

Popper and Kroll (2011) analyzed that differences found can be due to differences in taste perception, cognitive ability (in the test), or response strategy (of the test). Popper and Kroll conclude that differences in perception between adults and kids is mostly observed in more complex taste stimuli (mixtures rather than single taste aqueous solutions), suggesting that mental processing plays a role in the different performance. They suggest that this in turn indicates that kids may not notice changes in product formulation to the same degree as adults.

### 11.2.3 Vision

Kids' vision also differs from adult vision. It appears that the largest difference is actually the way that our brain processes images (Pappas 2010). Kids under 12 separate visual information, whilst adults process different visual cues into one set of information. This makes it easier for adults to see depths and textures, and easier for kids to spot differences on different elements of the visual input. The lenses on younger kids' eyes are also clearer, letting in more light. As we age, the lenses accumulate protein, and by the time adults are in their 40s, the light getting into our eyes is a bit more dim (Star Tribune 2018). These findings support the general knowledge that appearance of products is a more important factor to overall consumption and liking of the product for kids. Kids are also likely to study the appearance of a food more prior to eating the food, compared to adults.

### 11.2.4 Touch

Although sensitivities in texture perception (touch) is generally less explored, the texture of food products is especially important to kids. Research by Werthmann et al. (2015) shows that the texture of yoghurt (not taste or color) led to a reduction of food intake and acceptance of the yoghurt by kids, similar to that of fussy eating.

Kids tend to reject textures that are difficult to process in the mouth, and prefer soft and uniform food, whilst adults tend to have a preference for harder textures. As kids grow older, a shift can be seen towards a preference for harder and also lumpier foods. (Szczesniak 2002; Laureati et al. 2020). It is also known that kids refuse certain foods based on their texture, and the exact preparation method for example for vegetables may have a big impact on a kid accepting or rejecting the food.

The question that arises with regards to sensitivity to touch is whether the bias for certain textures of foods by kids can be (partly) explained by a stronger sensitivity for the sense of touch (and thus texture), which may or may not be explained by not fully developed oral processing properties.

Also for non-foods, kids are assumed to have more sensitive skin, examples where this would affect kids are an increased sensitivity to the feel of clothing, such as the seams in socks near the toes, or the texture of certain materials (wool, tights, etc.), or even the hardness of the spokes of hair brushes. Surfaces that may be perceived as smoother by adults could be perceived as rougher by kids.

### 11.2.5 Hearing

It is probably no surprise to any parent or carer that kids' hearing is different from adults. In general kids tend to be more sensitive to sounds as they do not have the lifetime of exposure to noise, plus also our hearing diminishes naturally with age.

From a simple frequency science fair experiment by Kevin Krick (2011), it shows that kids found it easier to hear between 15,000 and 20,000 Hz than adults, at lower frequencies (up to 5000 Hz), no difference was found, although the differences at this higher frequency is unlikely to make a difference at food consumption, one might want to bear it in mind for some particular products – think for example about appliance buzzers and alarms.

Interestingly, it appears that adults focus on expected sounds, allowing them to identify sounds better, whilst young kids listen to broader frequencies, thus hearing unexpected frequencies equally well to expected frequencies (Werner 2007). Although there is little research done on the difference in perception of sounds from food, one could assume that adults have a clearer expectation of the sound a food should make, whether this is the fizz on opening a carbonated beverage, the snap when breaking a piece of carrot or the crunch when biting into a crisp. A similar reasoning would be valid for non-foods. Adults would focus on expected sounds, such as the click that a closure "should" make, the "pchit"-sound of a spraying can or the click-clacking sound of shoes walking.

### 11.2.6 Sensory Fusion

As already indicated in some of the modalities above, adults tend to have certain expectations when using their senses. As adults, we have learned to fuse the information of our different senses to create one picture. Combining all this sensory information, may lead to an inability to look at the separate pieces of information coming from each of our senses (Nardini et al. 2010). Kids – depending on their age and experiences – tend to separate their perception for each sense and can thus focus better on individual stimuli.

## 11.3 Kids vs. Adults' Preferences

We all know that our preferences change over time. Our taste buds are developed when we are still in the womb. At only 8 weeks of pregnancy taste cells start to appear and at around 15 weeks into pregnancy, fetuses develop their taste buds and can actually start to taste. Studies have shown that the fetus drinks more inside the womb when the mother has just eaten something sweet. In a recent study by Ustun et al. (2022), fetuses at 32–36 weeks gestation showed facial expressions associated with laughter when the mother had taken a carrot powder capsule (carrot being usually associated with sweetness), and a "cry-face" expression when kale powder capsule (kale usually associated with bitterness) was taken.

Research has also show that flavors consumed during pregnancy and/or consumed during breastfeeding increase liking of the food during weaning (Mennella et al. 2001) and we thus learn to like foods in the womb and during very early infancy, setting the foundation for continued food preference development, then build onto this by social, environmental and biological factors (Ventura and Worobey 2013). Although a lot of the development for our food preferences occurs during early childhood, it continues throughout our childhood, adolescence and adulthood (Nicklaus et al. 2004).

Kids like greater intensities of sweetness and saltiness (Desor et al. 1975; Liem 2017), and adults are more tolerant towards sour and bitter tastes. One of the reasons for the innate likes and dislikes of these tastes may be biological; a liking towards energy rich and protein dense (sweet and salty) foods, and an aversion towards potential poisonous or toxic (bitter and sour) products.

As we get older, we learn to like certain foods, some because we like the effect it has on our body – such as coffee and alcoholic products – and many with repeated exposure and social influences.

Our liking is also dependent on our current diet, the diets of our peers and availability of foods and those we see and hear about. Some examples of those factors:

- If we have just eaten some chocolate, we might not rate strawberries as sweet and tasty as we might otherwise.
- If we are on a specific diet such as not eating any bread, we might start to crave it after some time.
- If we see our friends enjoying paella we create positive associations and are more likely to order it ourselves.
- If we see our local vegetable market displaying romanesco for a period of time, we may get intrigued and want to try it.
- If a pear and ginger jam has clear and attractive sensory descriptors displayed in a advertisement, it might just make our mouth water.
- If our parents often purchase, display and/or consume beetroot, we are more likely to enjoy beetroot at some point in our lives too.

For both adults and kids, our food choices are actually more driven by food dislikes than our food likes. We may either eat it or not eat food that we like but food we dislike, we would only rarely eat (Keller et al. 2022). With kids being more likely to be identified as "picky eaters," a larger amount of food dislikes and lower variety in intake can be assumed. Picky eating is common and usually only whilst we get used to a new food, though there are also some persistent picky eaters (Toyama and Agras 2016) both in childhood and adulthood.

## 11.4 Food Neophobia

Food neophobia and picky eating are two main forms of kids' food rejections and a reduced variable diet (Lafraire et al. 2016).

Food neophobia affects the willingness to try new foods, and it is best described as the avoidance or fear of novel foods, whilst picky eating is defined as the rejection of a larger number of foods that may both be familiar and novel to the child.

Although anyone can be affected by food neophobia, it typically develops at a very young age and may be inherited from the parents (Faith et al. 2013). Kids who have been eating fine up to the age of about two years, may suddenly show an aversion for any unknown foods. The degree of food neophobia differs from child to child, with a greater degree of food neophobia usually leading to a narrower variety of particularly fruits and vegetables in their diet, but also a higher energy intake from discretionary foods in young children (Perry et al. 2015).

In a research setting, when confronted with a new food, neophobic kids may be less likely to try and test that particular food.

Kids (and adults) can be tested on their degree of food neophobia using food neophobia scales (Pliner and Hobden 1992; Pliner 1994; Loewen and Pliner 2000).

Various articles in literature have shown that repeated exposure to foods reduces the fear and novelty of food and increases the liking towards previously disliked food (Birch and Marlin 1982; Wardle et al. 2003; Lakkakula et al. 2010). Even having certain foods in the home, and/or being enjoyed by family members and close friends, may have a positive impact on the child.

Other ways to encourage kids to eat new foods is by making a food "less novel." This can, for example, be done by:

- Provision of positive sensory descriptors of food.
- Provision of the new food alongside a well-liked food.
- Providing background information about the food, for example how it is traditionally eaten, the nutritional value of the product.
- Providing a "taster plate" with a pea-sized amount of any new foods to try first.
- Trying the food at a friend's home, by whom that novel food is already liked.
- Encouraging kids to just look or smell a product prior to asking them to taste a product will also help them build up to tasting a novel food product.
- Presenting the food in an attractive and fun manner.

During testing with kids, researchers rarely screen kids on their degree of food neophobia. It is however common practice to exclude kids with a strong aversion to the specific food being tested. When testing with kids, especially when using a generally lesser liked or less common/novel food, you might want to consider including some of the above examples in the test: providing a good description of the products to be evaluated, describing who would usually eat it, when and possibly how, will ease a child into trying the new food category.

## 11.5 So Why Use Kids?

Understanding differences between kids' and adults' perceptions in more detail, it might have given you some thoughts on when you would use kids rather than adults for testing. The differences observed between kids and adults makes it more interesting to work directly with children for successful product development (Nicklaus 2015). When optimizing products, the best formulation for a product aimed at kids is likely to be different from a product formulation aimed at adults. Any tests surrounding the preference of products is thus (where possible) logically done by the appropriate age group. There are many examples in literature of using kids to measure preference to different products (Vereecken et al. 2010; Chanadang et al. 2018; Praet et al. 2023 – as examples).

But when should we consider using kids for more objective sensory testing? When looking through literature, you will find this subject highly controversial. Objective sensory testing often requires more complex tests to be carried out, which in turn requires concentration and better cognitive abilities of the panelists performing the tests (Garcia 2011). As with any research carried out, carefully considering objectives and listing out requirements of testing is important – the latter maybe even more so when testing with kids.

Here are three examples when testing with kids could be beneficial.

### 11.5.1 Appropriate Sensitivity

As described above, kids and adults may perceive foods differently. If we assume, as some literature suggests, that adults are able to identify small differences better than kids, this would suggest that if a sensitive adult panel would not taste the difference, kids would not either. In some cases, commercially, this adult trained panel might be too strict. For different reasons (such as health (Belc et al. 2019), marketing, availability, cost, ethical, or legal reasons), a food manufacturer may wish to change an ingredient in their product – for example, reduce the amount of sugar in a sandwich spread aimed at children. A difference found by an adult panel between the new and the current spread may indicate it not to be appropriate to reduce the amount of sugar in the spread, however, a panel of the right target group, kids aged 6–10, may show that the kids do not taste the difference, and the manufacturer could go ahead with reducing the amount of sugar in their spread.

Also, as described earlier, whilst adults may be better in identifying differences when combining all characteristics of a product together (sensory fusion), kids may hone in on very specific areas better, for example, just the smell. For example, as adults we have learned that certain aromas are associated with certain tastes e.g., caramel aroma is associated with sweetness, but younger kids may not have these associations yet, and may not perceive a difference in taste, even though they would notice the difference in smell.

### 11.5.2 Understanding the Sensory Perceptions as Perceived by Kids

Although as described earlier, kids may not always describe their perceptions as accurately as adults can, during any sensory testing with kids' products it is most definitely useful to have kids' own descriptions. It often indicates which characteristics are most dominant and also most important to them and descriptions obtained are useful in any communications about the products to kids. Sensory claims are an important marketing tool to use on products, it increases familiarity with the product and creates subconscious triggers about a product (Beeren 2016).

As kids grow older, perceptions and descriptions will become more aligned to adult descriptions and more complex in nature and again, using kids of the right age will be helpful in marketing communications about the product.

### 11.5.3 Appearance

We eat with our eyes. As adults we have learned certain associations with the appearance of foods (Zampini et al. 2007). A red color is perceived sweeter, a darker color more intense. This effect can even be seen in trained sensory panels. However, this effect is not seen as much with kids (Moskowitz 1994). Thus if a difference test on a kids' product was required, whereby the focus was on taste/flavor of a product, but a difference was present in appearance, a kids' panel would be an appropriate choice.

Conversely, appearance often plays a major part of kids' decision making around food – liking scores and likelihood of consumption are often strongly affected by it. Research around the use of free choice profiling by kids and adults showed that kids' chosen attributes to describe different juices were more related to the appearance and color of the products, whilst adults were more objective and technical (Blay and Beeren 2007). In a study by Jansen et al. (2010), fruit intake by 4- to 7-year-old kids was measured whilst presenting the fruit in a more visually appealing manner alongside the fruit's regular presentation. Results showed that appealing visual presentation had a strong effect on consumption of the fruit.

If appearance is of a critical consideration to a food or non-food product, it would be interesting to obtain kids' objective views.

## 11.6 Ethical Considerations

As you are considering working with kids, the ethical considerations are obviously vital for any test. Foremost, any risks to testing with kids should be considered, including the risk that the environment could bring as well as the products to be evaluated. Usually, companies will have independent quality officers and ethics committees who would review and or work together with the researchers to create a detailed risk assessment. Tests may also be reviewed by external ethical committees. Prior to a first test, contact with such a committee should always be sought. The committee itself will indicate, based on the test and national requirements, whether a further review and formal approval should take place (Nicklaus 2015).

During the test, always ensure kids are supervised by either their own parent/guardian or another responsible adult. This responsible adult may require a criminal record/children barred test depending on the country you are operating in. In different countries, there may also be different legislation that could affect research with kids - in terms of supervision, items to test, participation, incentives/payment, etc.

For any test with kids, kids should not be forced to test/eat anything that they do not want to. Test participation should be completely voluntary, with the kids providing consent to participate in the test prior to the start and having the option to leave the test if they wish to do so. In addition, consent should be sought from a parent/guardian prior to the test, usually a consent form with information on the test and test product will be provided to the parent/guardian, and signed by them prior to participation. Please see Figure 11.1 below as an example of a consent form. Consent forms should be adapted to the test, the country's legal national requirements and reviewed by the company's legal team.

Usually, kids with allergies and intolerances cannot participate in testing, due to the risk this may bring to the child. However, there may be exceptions, and of course panels are also organized specifically for individuals with allergens. For example, a gluten-free bread panel

**Consent form**
**Cereal X Product Development**

Dear parent/carer,

Thank you for considering your child's participation on our research study. We are conducting this study to improve our cereal products.

During our test, your child's safety is very important to us. Unfortunately, children with allergies cannot participate for their own safety. Any information about your child will not be shared by us to any third parties and will be kept in accordance with GDPR legislation. Please tick the following box(es) relevant to your child:

- ☐ My child **has** allergies or intolerances.
- ☐ My child **does not** have any allergies or intolerances
- ☐ My child does have dietary restrictions or preferences, namely:........................................................

The products tested are current market products found in country X, and variations thereof.
The products and test is of a confidential nature to us. If you agree for your child to participate, you and your child may not disclose any information to any third party. Taking pictures of the products or children during testing is also not allowed.

By signing this document you agree for your child to participate in this test, to acknowledge that you have read, understood and will adhere to the information. For multiple children, please complete 1 form per child.

Researcher name, Company
Date.

I ......................................consent for my child .....................to participate in the research study involving cereals.
I declare that I do not know of any reasons of health or diet which preclude my child from this programme, and I am happy for my child to participate. I also understand that participation is voluntary and that my child may withdraw from this study without prejudice.

Signature......................................................................... Date............................

**Figure 11.1** Example consent form.

can be organized for individuals with gluten allergy/intolerances. For this occasion, all areas including kitchen, testing, storage, and waiting areas should not store any products containing the allergen (gluten) and should be deep cleaned prior to the test and product storage and preparation. Additional risk assessments should be carried out to ensure no cross contamination can happen and that testing is safe for any individuals participating in the test.

For consumer testing, the Market Research Society created comprehensive guidelines for testing with children (MRS 2012, 2020). Much of this guidance is very useful for sensory discrimination testing too. A link can be found in the reference list of this chapter.

For testing of foods, the IFST's sensory group has created guidelines for ethical and professional practices with some practical advice on ethical considerations for testing with kids and adults (IFST 2020). A link to this document can also be found in the reference list.

## 11.7 Environment for Testing with Kids

Testing with kids can be done in different environments, each with their own benefits and limitations. The environment you choose will likely be dependent on the availability of the environment – for example, do you have a dedicated sensory environment available to you? The timing and duration of the testing, the products to be evaluated and the budget available are also important factors.

**Dedicated sensory environments or external "halls"** will give more control to the testing area, and may remove some bias. However, putting kids in "foreign" environments may lead to kids feeling uncomfortable and ultimately risk faulty test results. Aside from a company's owned sensory testing area, you could consider hiring a testing hall from a sensory provider, or hiring a village hall, space in a hotel, etc. These latter spaces are often quite neutral, spacious and with availability of a separate kitchen area to prepare for testing. Some tips to put children more at ease in these environments are listed at the end of this section.

**School, clubs, or home** may be more comfortable for kids and they may feel more relaxed completing tests here. Clearly in such environments, there may be more distraction for the kids, which could lead to bias or again invalid test results. When using external environments such as a school or club, be certain to agree which area will be used. Go and see it and agree as to how flexible they are with some furniture being moved around, certain products being tested and of course the timings of the test and other uses of the test-room, and any adjacent rooms at that same time.

**Products that require preparation**, such as cooking, diluting, and assembling are ideally carried out in an environment where the researcher is in charge of the preparation. You may obtain too many variables if these products are prepared at home and/or by (too many) different individuals. The more preparation required, the more space and potentially equipment you will require. You will need to ensure that each product tested is prepared under the exact same conditions. When you are testing 10 different pan-fried sausages, you would need quite a bit of hob space, which will drastically limit your available options.

**Products that require certain storage conditions** for products that require a certain temperature to store, the recommendation is very similar to products that require preparation. You do not want to introduce variables due to some half melted ice-creams, or chocolate that has bloomed due to wrong storage conditions.

**For testing that requires additional time**, you could consider multiple small tests. If a test is longer in duration, it is not advisable to have a very long setting with kids. The kids may get bored and the quality of the results obtained are likely to be affected. Unless testing can be made part of a curriculum, schools and clubs are unlikely to want testing carried out in their time. Several after school or club sessions could be good options as kids will already be there. If kids would need to come to a specific dedicated area over several appointments, drop outs are very likely, and higher incentives are almost certainly required for full participation. Home testing would be the easiest choice for multiple testing sessions. For each of these options incentives should be considered to encourage test completion over multiple sessions. For any sessions over 30 minutes, consider doing breaks after each 15 minutes, a quick jump around and change of scenery will make a huge difference in their focus on the test when they return.

Whichever environment is chosen, always try to remove any factors that may result in poor test conditions and biased results.

Tips to create optimal testing space:

- Create a welcoming reception. First impressions count, make a child feel welcome and important upon arrival of any external testing space. Introduce researchers/staff to the kids, explain the setup and let them know who to go to for any help.
- Younger children (under 5) may feel uncomfortable to be left without their parent/ carer with them; consider having the parent/carer observe testing to help them feel safe.
- Create fun waiting and testing areas. Research has shown that better results are achieved when testing and waiting areas are fun for kids (Moskowitz 1985). You do want to make the actual testing area quite neutral so as to not to distract, but close surrounding areas, and particular areas that kids wait in, walk through and stay in between testing should be appropriately furnished. Think about coloring papers, little games, some books or magazines or games on tablets. It is important for kids not to be bored prior to a test, their likelihood of active participation during the test will be affected.
- Brief kids in smaller groups. It will be difficult for some kids to focus on research protocol explanations. They may be anxious about the test, and are likely in an unfamiliar environment. When explaining the setup to kids, it will be easier and more likely taken in if done with fewer kids. The smaller the group the better, but try to limit to a maximum of around 30 (similar to a school class set-up).
- Do not refer to the taste research as a test. "Test" may have negative connotations and sound scary to children. You could consider referring to it as research, quiz, investigation, or a detective game – see next bullet.
- Make the actual test into a game or story. Everyone engages better when it is a more fun environment. Researchers from Nofima and the Norwegian University of life sciences took this to a new level when they tested kids on their sensitivities with taste detective games, asking them to solve different taste mysteries (Ervina et al. 2020). Urbick (2011), suggests using fairy tales such as Goldilocks and the three bears or mimicking computer games for older kids as examples to engage with the kids. Kids are more excited to participate and more engaged with testing when the testing is carried out as a game. Dropout rates during testing have also shown to decrease when using these approaches.
- Ask kids to complete the test by themselves. Kids may have a tendency to talk with their friends during testing. Let the kids know that you are keen to have their personal answers and perceptions.
- Let the kids know that there is no right or wrong answer. Sensory testing will feel like a testing scenario, which will make some kids anxious to give the "right" answers. Let kids know that each person has a different perception and for the games/tests to work well, the researchers need the perceptions from quite a few kids.

Aside from the above tips geared towards testing with kids, normal sensory environment conditions and practices should be adhered to as much as possible. Full details can be found in ISO8589:2007 (ISO 2007), general guidance for the design of test rooms.

## 11.8 Sensory Screening

When carrying out preference testing, usually kids are not screened for their ability to taste, smell and verbalize themselves. Instead, researchers take a large subset of their target group and aim to have a good representation of their customer base.

This is different from discrimination testing. When carrying out discrimination testing, we usually apply a much smaller group of panelists who are capable of finding sensorial differences (if they exist) and describing these in an appropriate manner. We assume that if this group could not find a difference between samples, the general target group for the product will not either.

Depending on the test objective, duration and number of tests, as a sensory researcher you should consider whether or not to screen and train your kids' panel. This section will look at the specific screening and training considerations for testing with kids.

It is anticipated that general sensory screening practices are known by the reader. For further reading on these, different ISO standards are available, most notable ISO8586:2023 (ISO 2023).

### 11.8.1 Why Screen Kids for Testing?

Just like when using adults for sensory testing, when carrying out sensory tests with kids, you may want to ensure or at least know the different sensitivities your panelists have for certain tastes, general or specific odors, visual stimuli, textural, and sound perception. For most sensory panels extensive taste screening is carried out, alongside semi-extensive smell and appearance testing and some vocabulary assessments. Texture and sound are often the poorer cousins of the other senses during traditional sensory screening sessions.

For the particular purpose of testing with kids and whether or not to screen, the researcher should first ask themselves about the purpose of the test. If there is a clear focus on a certain element (e.g., sugar reduction in a food) clearly sensitivity testing towards sweet and other basic tastes would be important to carry out. However, even when it is not as clear cut, when using kids for multiple or more extensive testing, a simple screening test is likely to prove valuable when analyzing the test results. When outlying scores are received from certain child assessors, the researchers could look back at the screening results. Is a particular (in)sensitivity a likely cause of a different result?

Also, when testing with smaller groups, as is often the case for sensory testing, there is a higher risk of taking a group of non-average sensitivities amongst the group. You could happen to be unfortunate and your test result could provide false information.

If your test set up allows, consider doing a screening session. Another benefit – aside from obtaining the actual data – is that kids will see the task as more important and also learn from the experience itself, each having the potential of enhancing the motivation for the actual sensory test.

The next sections provide some ideas on the actual screening testing that may be applied.

### 11.8.2 Basic Taste Screening

Basic taste screening is the most common type of screening. Basic taste screening is certainly worth doing with kids too if you are measuring any differences in sweet, salt, sour, bitter or umami in products.

During this screening, the kids are provided with solutions of water mixed with the different tastants (sugar, salt, an acid [e.g., citric acid], caffeine or quinine and msg) plus water and the kids are asked to identify the taste of the solution. The solutions can be presented in different intensities, starting with lower intensities to measure kids' sensitivity levels, or just one level.

It is important to familiarize kids to the different tastes prior to the test, as they would otherwise not know how to describe the taste, rather than not taste it. For younger kids the number of solutions should be minimized as concentration on the test for longer periods may be harder for this group.

Kids do tend to like a basic taste screening if it is brought in a fun, detective setting. It may set the scene for further testing too.

### 11.8.3 Odor Descriptive Screening

When doing odor screening, it is near impossible to screen for all possible odors associated with the products due for testing. A close secondary objective of odor screening is also to test a panelist's ability to describe different perceptions. In a panel, you do want panelists who are capable of giving a clear description. Usually (most) odors presented in a screening test will be ones familiar to the panelist. If possible, use some odors that are relevant to the actual test too. General odors that could be considered include vanilla, cinnamon, mint, lemon, green grass, woody, and rose/floral.

Kids over the age of six could smell different odors on perfume strips and pretend to be "perfumers." They can first individually write down the smell and then in a small group setting discuss if they like it, where they have smelled it before and if they would use it for anything.

### 11.8.4 Visual Screening

There are different visual screening tests that can be performed. Probably the most commonly used in the sensory industry are the Ishihara test and Farnsworth Munsell 100 hue test. Working with kids, the Ishihara test would be the easier of the two to complete. With this test you will easily be able to identify whether kids are color blind. The test is easy to perform, and kids often do like it too.

### 11.8.5 Texture Descriptive Screening

We perceive the texture of our food (mainly) by our sense of touch. For food, this tends to be predominantly with our fingers/hands and in our mouth. Within this, we perceive surface responses – for example, whether something is smooth or grainy and we perceive deep responses – by putting pressure we can feel if something is hard or tough for example. Throughout the eating experience, we can perceive the food's texture and the texture of the food changes. Think here about the whole process, from when the kids see the product on their plate ("yuck that looks slimy," or as a friend's kid once said "Grandma, that looks 'volting'") to the feeling left after they have finished and swallowed the food ("this feels tingly in my throat").

For non-food, texture is also often a critical criteria for a successful product, in particular consider how products are held whilst holding or feel when in use. For products used by (smaller) kids, smaller more sensitive hands and skin may perceive textures differently from adults. With texture screening, individual kid's sensitivity can be measured prior to the test.

For the texture screening process one or more of the following could be considered;

- Texture discrimination: choose a relevant product to the actual discrimination test. Can kids distinguish different textures within that product group? For example, if the discrimination test is on fat reduced chocolate, prepare or buy some different chocolates and ask kids which one is smoothest, and which one melts the quickest in the mouth.
- Texture descriptive ability: use this test if it is important to the test to receive clear texture descriptions – for example, for the use of product development, marketing, or Quality Control. Ask kids to describe textures of different foods/products. Examples could be:
  - Orange: juicy, jelly parts.
  - Pear: gritty, juicy.
  - Sugar: crystals, coarse.
  - Cornflakes: crispy, crunchy.
  - Plush fabric: furry, soft.
  - Play dough: squidgy, soft, rubbery.
  - Rock: hard, rough, gritty.
- Recognition of differences in texture, for example rank different yoghurts in thickness, or super fun for kids, different slimes in softness.
- Efficiency in chewing could be measured by chewing and spitting out two colored sweets. Place 0.3 mm of two different colored sweets on top of each other, ask kids to place it on their tongue and bite with their molars five times, then spit out and see how well the colors are mixed. Repeat (with new sweets!) with 10 chews.
- Oral spatial acuity can be measured by placing icing letters or numbers on the tongue and recognizing the letter – be mindful that children know the letters or numbers well, or by using grating squares of 1–2 cm with 0.5/0.75 mm grooves placed in horizontal or vertical direction on the tongue (Lee et al. 2022).

Be mindful of kids' ages and tooth development. Chewing performance will be different when kids are missing some teeth and biting forces are lower for kids under 6 (Almotairy et al. 2018). When hard or sticky products are being evaluated, you may wish to screen, or at least be aware of kids missing certain teeth.

Kids' biting forces will also change lots as they grow up. A food considered hard by a younger child may be perceived less hard by an older child.

Sound is also usually associated with texture. Think of the fizz of carbonated drinks, snap of chocolate or the crunch of an apple. Where sound is important, kids will often love to be involved in a little sound-test. Think for example of measuring the length of a sound (putting hand up when it starts and down when it stops) describing the sound certain products make and finding out which speaker or direction a sound comes from. A little group session after this sound screening with a few kids to discuss what they heard and if sound is important is a great way to introduce them to be mindful of sounds during further testing too.

## 11.9 Training Session

In the section after this, the training methods appropriate for kids are discussed. However, whichever test you decide to use with kids, some training will be inevitable. As discussed earlier, trying to keep the training simple and fun is much more likely to be a success. Try to have hands-on elements to keep kids engaged too.

### 11.9.1 Training in Different Methods

Similar to an adult panel, some discrimination methods are much harder than others to train in. Where possible try to keep things simple. A visual test run/warm up on very clearly different products (e.g., a red vs. a green apple) is a simple way to explain the test protocols to kids. For slightly older children (9 years +), a bruised and intact apple were successfully used by Velazquez-Mendoza et al. 2022. Often after establishing a general difference, testing includes one or more questions on the sensory characteristics, additional training may be required to learn key characteristics. Try to find some completely different products suitable for kids to introduce these sensory characteristics.

Aside from discrimination testing, simplified descriptive testing (Murray et al. 2001) can also be performed by kids to identify differences between two or more products. For younger kids, free choice profiling (Lachnit et al. 2003), sorting (Cartier et al. 2006) or CATA (Rocha et al. 2021) may be more suitable. For each descriptive method, kids will need to be taught any test protocols. For more traditional profiling methods, such as Quantitative Descriptive Analysis (QDA) based profiling (Bleibaum et al. 2020), which is often applied since the data analysis may be simpler or already set up within an organization, more extensive training on the attributes will be required. QDA based methods would be more suitable for older kids (minimum age 8–9 years). Similarly, ranking based on specific attributes other than appearance may be more suitable for a slightly older age group. From about 12 years onwards, similar techniques can be used as those with adult panels.

Be mindful that younger kids tend to rate products more towards the extremes of the scales, whilst adults and older kids use the middle parts of the scale – to leave open the option that there might be a product that might be even more or less intense.

Kids base a lot of their sensory perception on the appearance of food. Focusing first on appearance prior to any other sensory attributes might help them feel that this element has been looked at. Nevertheless – if there is a difference in appearance – taste and texture may be perceived differently based on the appearance. This perception will equally be true for the product in the market, and thus this bias during the testing could be considered part of obtaining a true reflection of kids' perception of the product. If however, the main objective of a test was to assess the aroma, taste/flavor or texture/feel of a product, any appearance differences may be obscured by using red lighting and/or dark serving pots. In this case, no appearance questions should be asked, as that would draw kids' attention to it.

When carrying out the training sessions, ideally you want to do this in person in smaller groups. Groups of 5–10 kids are likely to interact with the researcher and each other, plus it is easier to check if all kids understand the test method and any protocols. If you are unable to train in person, consider video materials instead of written instructions. A video will be more engaging and is likely easier to follow. Follow this up with a few key points in writing to emphasize certain protocols.

When training sessions are only short, refer to them as something like a "warm-up" or a "practice" session. This will come across less formal and more inviting to participate in. It will also help the children settle in before the actual test.

Examples of training sheets can be seen in Figures 11.2, 11.3, and 11.4.

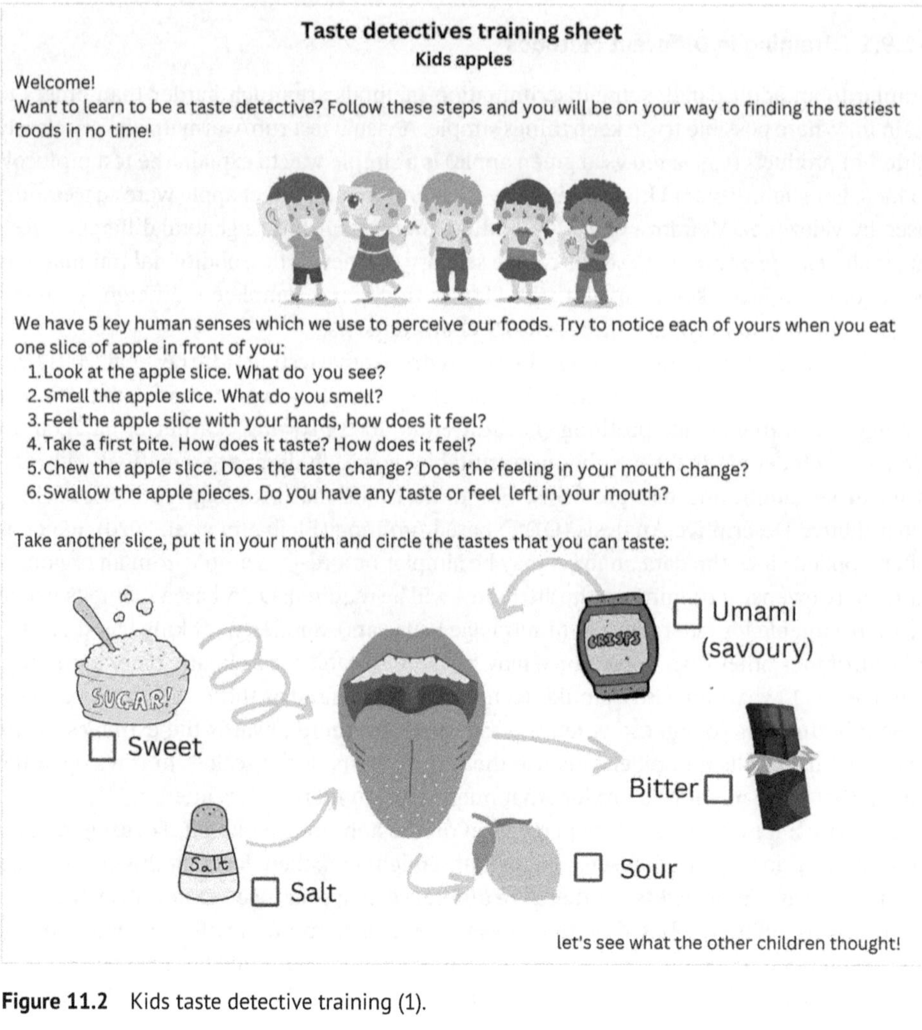

**Taste detectives training sheet**

**Kids apples**

Welcome!
Want to learn to be a taste detective? Follow these steps and you will be on your way to finding the tastiest foods in no time!

We have 5 key human senses which we use to perceive our foods. Try to notice each of yours when you eat one slice of apple in front of you:

1. Look at the apple slice. What do you see?
2. Smell the apple slice. What do you smell?
3. Feel the apple slice with your hands, how does it feel?
4. Take a first bite. How does it taste? How does it feel?
5. Chew the apple slice. Does the taste change? Does the feeling in your mouth change?
6. Swallow the apple pieces. Do you have any taste or feel left in your mouth?

Take another slice, put it in your mouth and circle the tastes that you can taste:



let's see what the other children thought!

**Figure 11.2** Kids taste detective training (1).

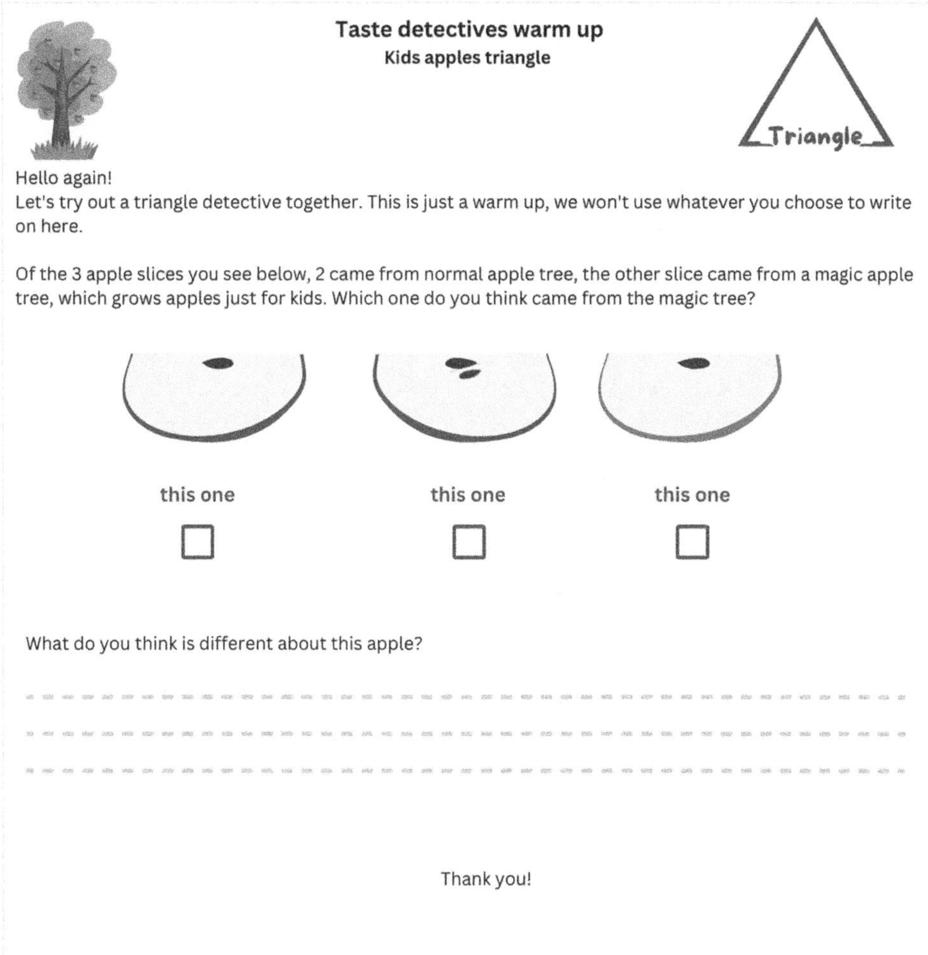


**Figure 11.3** Kids taste detective training (2).

## 11.10 Test Methods

Guinard (2001 – table 1) and ASTM (2021 – table 1) have published useful reviews for different skills and behaviors that children generally present at different age groups. Guinard's summary takes into account kids' cognitive abilities from infants to teenagers, and identifies factors to be considered when designing sensory tests. From this review, other reviews (e.g., Thomas and Murray 1980; Velazquez-Mendoza et al. 2022) and personal experience, the following overview will give an indication of tests and potential adaptations per age group. Of course, each kid is different and capabilities differ from one kid to another. As such, this overview should be seen as a guide; for any age group, ensure additional panel host(s) is (/are) available during testing to support kids as required.

**Taste detectives training sheet**
**Kids apples**

Ready for another taste detective training? Let's learn about different ways to describe apples to help you solve this mystery of the magic kids apples!

Taste your apple slice and have a look at all the descriptions below. Tick the ones that you think it tastes like. And don't worry. We will not use these results. It is just part of your detective training!

- ☐ Red skin, bright red colour
- ☐ Glossy skin, smooth shiny surface
- ☐ Light yellow flesh, inside is light yellow
- ☐ With a crispy sound when you bite in it
- ☐ Sweet like sugar
- ☐ Sour like a lemon
- ☐ Floral like a flowers
- ☐ Mealy, with a soft feeling with you bite in it
- ☐ Astringent feeling, like the inside of my cheeks want to come in
- ☐ With pieces of skin I can feel in my mouth
- ☐ Green like grass or green apples
- ☐ Taste of honey
- ☐ Juicy taste like apple juice
- ☐ Fruity taste like different fruits
- ☐ Sweet apple taste after I have swallowed the apple slice

Wanna know what the other taste detectives in learning thought? Let's have a look together.

**Figure 11.4** Kids taste detective training (3).

- *From teenage (12–15) onwards*, kids display very similar cognitive skills and behaviors as adults, and are assumed to participate in similar tests to those developed for adults, without the need for many adjustments. Keeping kids of these ages motivated throughout the length of the test is probably the key factor in a successful test.
- In the *pre-teen years (8–12 years)*, kids are pretty competent in most cognitive and skills behaviors to complete most "adult" tests, however, keeping interest is a key area of focus for the researcher. As attention span may not be quite as great as that of the slightly older age groups, sessions should be kept shorter or very engaging to ensure test results are not biased by a child getting bored of a test. Language used should be appropriate for the age group, avoiding the use of complex wording. Other than that, most sensory tests can be successfully completed by this age group. Self-administering tests should be mostly fine for this age group too. Discrimination tests that can be considered include:
  - Triangle Test (Meilgaard et al. 2006).
  - Paired Comparison (Meilgaard et al. 2006).
  - Same/Difference tests (ASTM 2018).

- Tetrad (Garcia et al. 2012).
- A-not-A-test (Velazquez-Mendoza et al. 2022).

If the researcher chooses to use the "A-not-A test," a recent study by Velazquez-Mendoza et al. (2022) may be of interest. The researchers found that this age group was very capable of performing this type of test, however, framing the A-not-A test based on liking showed higher ability to discriminate the samples and more accurate results. Garcia et al. (2012), compared the tetrad test with the triangle test using over 400 kids aged 6–11, their conclusions show in favor of tetrad testing, this test being more powerful.

- At the *early readers stage (5–8 years)*, we expect some more assistance to be required from sensory scientists. As children at this age may also be keen to get adult approval, objective guidance is vital. Any individual with a vested interest in a specific outcome of the test may not be suitable to guide the kids in their testing. At this age group, as kids only just have started reading and their attention span is still more limited, simplification of tests is important (e.g., visual scales). Paired Comparison (Meilgaard et al. 2006) (e.g., which one is sweeter, which one is darker) or Same/Different tests (ASTM 2018) (is this product the same as that one or different?) are appropriate tests for this age group. Some scaling of intensity of attributes could be included, but keep sample sets to smaller sizes. With a little bit of visual training, a Triangle Test (Meilgaard et al. 2006) could be completed, increased guidance is suggested for the lower age group (5/6).
- *Pre-school/school starters (3–5 years)*, any reading or writing skills at this age group tends to be very basic or just beginning, and thus testing would need to be conducted verbally at a 1:1 or small group level. Kids at this age level might find discrimination tests more difficult. A sorting test would be recommended. Simple paired comparison tests or ranking of a few samples could be carried out on attributes that are based around the appearance of a product or possibly some simpler attributes like odor strength, etc. Keep tests simple; Liem et al. (2004), showed that ranking of sweet solutions on sweetness with 4- and 5-year-old children was difficult, however, ranking the same samples on preference gave acceptable results.
- *Toddlers (18 months–3 years)* are generally not able to read or write yet and would not be able to complete many tests as yet. If testing products specifically to this age group, observational tests are likely most suitable (e.g., Moding and Fries 2020). Verbal liking tests can of course be conducted too, or towards the end of this age group, children can be asked for some words to describe a product. Bear in mind that these descriptions are most likely pretty subjective and not necessarily very useful for product development purposes. Investigations into performing discrimination tests with this age-group has previously indicated that these type of tests are not appropriate for this age group (Kimmel et al. 1994) although paired comparison has been seen to be less problematic for kids aged 2–3 years (Kimmel et al. 1994).
- Infant (birth – 18 months), generally very few sensory science tests are carried out with this age group, for obvious reasons. Any products for this age group are best tested by the parent/guardian. Be aware that bias may occur when a parent/carer is asked to make judgements of products for a child, the person may answer based on personal experiences rather than for the infant, or may give a response that puts their child in a better

light to the researcher (ASTM 2021). Providing products to the infant first and describing/recording facial expressions may help overcome some of the bias. Observational studies in general can be carried out for this age group too, although these tend to be more around product acceptability. Use of video material is useful when made by parent/carer and reviewed by expert behavioral researchers.

### 11.10.1 Language

Depending on the age and other personal factors, language/verbal skills used and understood by kids will vary. This will lead to potential misunderstanding of the test procedures, test questions and the sensory attributes evaluated during the test.

Attributes such as sour and bitter are often confused (also by adults). For any attribute where there may be a possible confusion, labelled reference samples should be provided alongside with an age appropriate verbal and written description to teach the kids these attributes and avoid confusion and error during the actual test.

Pay special attention to any language used during testing. Test questionnaires with a few kids before going to a larger scale, and ask for feedback during the test to ensure kids understand the test instructions.

A "warm-up" or "example" test prior to the actual experiment is recommended for any new protocols or new tests for the kids. These could be done in a visual way, to not overwhelm other senses. For example, if kids are doing a triangle test for the first time, show three dinosaurs, one T-rex and two stegosauruses and ask them which one is the odd one out. This example can be adapted to age appropriate items/characters – from Disney characters to different planets.

Three digit codes are commonly used to label samples in the sensory industry, these codes are there to reduce bias, as codes used should have no association with anything, whilst letters or single digits may have certain associations. These codes are randomly generated, and known three digit numbers, such as 007 and 123, should be avoided. Younger kids might find it harder to use three digit codes. Rather than using any code, or label that kids might just like the look of or associate with something, try not to use any label and instead create a place on the paper where you can actually add the test products. See Figure 11.5 for an example.

**Figure 11.5** Example of test paper without use of sample labels.

### 11.10.2 Scales

For certain tests, scales are used to measure attribute intensity or a measure of difference.

As adults, we are very used to scales and scoring. This is different for kids, especially younger kids who have not been exposed to much testing yet. Proper explanations of scales used in any tests is critical for sensory testing. As outlined previously, if possible try to have engaging training to keep kids interested.

Although often used for consumer acceptability, pictorial scales such as facial scales, have been shown to bias the results. From research studies on hedonic (liking) scales, it was concluded that pictorial facial scales perform worse than child oriented verbal scales (Kroll 1990). Some kids may choose a part of the scale because they like that picture more, rather than accurately reflecting the appropriate point on the scale for the product. Zeinstra et al. (2009), concluded that facial scales are good for identifying dislikes, but not to measure different levels of food acceptance in kids aged 5–13 years.

In discrimination testing, scales are sometimes used to identify a size of difference (measured with statistics such as R-Index (Society of Sensory Professionals 2021; Chapter 7 in this book)), or also to identify a strength of a certain sensory characteristic. End of scales usually have descriptions such a none to very to indicate the intensity of certain sensory attributes. Especially for younger children, a more engaging end of scale – such as "super light" to "super dark" or "can't taste any sweet taste" to "this tastes super sweet" can be considered.

When possible, add pictures or photos on parts of the scale to differentiate samples reliably – for example for some appearance attributes such as color intensity. As an alternative, to gather information on sensory profile without the need for scales, CATA can be used for children 8–12 years of age (Rocha et al. 2021). In this test kids choose (check) sensory attributes that they feel are describing the product from a larger list of attributes. The authors conclude that children can adequately discriminate food sensory profiles using CATA methodology with kids-generated attributes.

## 11.11 Final Thoughts

Some final, but important thoughts when you do hopefully decide to carry out sensory testing with kids.

### 11.11.1 Motivation and Incentives

There are many reasons why kids may want to participate in sensory testing. Sometimes it is the fun of the test itself, in other cases it may be the incentive that initially draws kids into the test.

Usually all sensory testing, with adults or kids, some form of incentive or payment is made. Shorter tests with much liked products usually require less of an incentive for kids to participate than attending multiple sessions testing Brussels sprouts. If testing during school or club time, schools or clubs themselves would likely expect or appreciate some form of incentive of reimbursement. In this situation, kids do not need to be paid/given incentives.

For home testing, especially where products are handed out at school or brought home, and are liked by the kids, incentives can be very little to none. Especially where there are products left over, or that can be reused (such as non-food items like shampoo or toys), the product itself can often be plenty of incentive to participate.

When not much budget is available, do ensure to keep testing short, as people do expect to be compensated for frequently attending sessions or longer sittings. Fundamentally, the incentive has to be enough to encourage the kids and their parent/guardian and/or the organization where the test is held, to participate in the testing and cover their time as well as any expenses from travelling to the test location. Yet, the incentive should not be too high either as this could potentially skew test data – for example, kids may feel a certain pressure to perform well and be worthy of the incentive (Patterson and Beeren 2011). Incentives could be monetary (cash or vouchers), and/or in the form of (gift) items that kids can keep. Check with your finance/legal teams as you may need to obtain social security numbers and other personal information from kids and their guardians when providing any incentives (not just cash).

Incentives may bring kids to a test, but motivation is thereafter most important to complete a test and keep them focused.

According to Urbick (2011), the most important key driver for kids of any age is control. Control to be in charge of themselves and their immediate environment. In their lives, kids are told a lot what to do, their striving for control often manifests in foods and other items that they can choose. Where possible within the test set-up, try to give the kids some control. Maybe it is as simple as adding the milk to the cereal or even just letting them choose where to sit for the test.

To keep motivation up, keep all elements of the test engaging, have regular breaks and interaction with the kids.

Both motivation and incentives will lower the amount of drop-outs from no-shows and incomplete tests. Do always recruit extra kids as some drop-outs are always expected.

Motivation to participate in testing also does not end at the end of the test. Kids love hearing the outcomes and how they have helped to shape a product or campaign!

### 11.11.2 Health and Sensory Test

Performing sensory tests in a fun way with kids is an excellent way to introduce new and healthy foods to kids, and is perfect to combine at schools and clubs in combination with some nutrition education. Action for Healthy Kids has created a page with ideas for hosting taste tests, as an avenue to explore new foods and learn about health benefits of foods (Action for Healthy Kids n.d.).

### 11.11.3 Future Kids Sensory Detectives

With more academic research and methods being trialed with kids, the kids' market continuing to grow, and sensory tests being carried out more practically and less rigidly to traditional methods, we are likely to see a growth in the number of discrimination tests carried out with kids. Moreover, involving kids will be a great benefit to them to learn about food and other products and to learn how new products are brought to market.

The way we test and develop kids' products may further evolve by even more involvement of kids going forward. Involving kids in every aspect of product development and also to review sensory (discrimination) testing to ensure it is done in an age and generation appropriate manner. As adult sensory researchers we may just need to be "down with the kids" to get most out of our research.

## References

Action for Healthy Kids (n.d.). Host a taste test. https://www.actionforhealthykids.org/activity/host-a-taste-test/ (accessed 18 January 2023).

Almotairy, N., Kumar, A., Trulsson, M., and Grigoriadis, A. (2018). Development of the jaw sensorimotor control and chewing – a systematic review. *Physiol. Behav.* 194: 456–465. https://doi.org/10.1016/j.physbeh.2018.06.037.

ASTM (2018). Standard test method for same–different test. https://www.astm.org/e2139-05r18.html (accessed 20 January 2023).

ASTM (2021). *Standard Guide for Sensory Evaluation of Products by Children and Minors*. ASTM.

Beeren, C. (2016). Putting sensory science into the marketing mix how sensory research can strengthen marketing strategies. https://www.leatherheadfood.com/wp-content/uploads/2016/08/White-Paper-NEW-COVER-Putting-Sensory-Science-into-the-Marketing-Mix.pdf (accessed 18 January 2023).

Belc, N., Smeu, I., Macri, A. et al. (2019). Reformulating foods to meet current scientific knowledge about salt, sugar and fats. *Trends Food Sci. Technol.* 84: 25–28. https://doi.org/10.1016/j.tifs.2018.11.002.

Birch, L.L. and Marlin, D.W. (1982). I don't like it; I never tried it: effects of exposure on two-year-old children's food preferences. *Appetite* 3 (4): 353–360.

Blay, M.Y. and Beeren, C.J.M. (2007). *Adults Versus Children: Differences in Perception of Food Sensory Attributes. Research Report*, 916. Leatherhead Food International.

Bleibaum, R., Clara, T., and Thomas, H. (2020). *Chapter 3|Quantitative Descriptive Analysis*. ASTM. https://www.astm.org/mnl1320170005.html (accessed 1 February 2023).

Cartier, R., Rytz, A., Lecomte, A. et al. (2006). Sorting procedure as an alternative to quantitative descriptive analysis to obtain a product sensory map. *Food Qual. Prefer.* 17 (7–8): 562–571.

Chanadang, S., Chambers, E.I., Kayanda, R. et al. (2018). Novel fortified blended foods: preference testing with infants and young children in Tanzania and descriptive sensory analysis. *J. Food Sci.* 83 (9): 2343–2350.

Coulthard, H. and Blissett, J. (2009). Fruit and vegetable consumption in children and their mothers. Moderating effects of child sensory sensitivity. *Appetite* 52 (2): 410–415.

Desor, J., Greene, L., and Maller, O. (1975). Preferences for sweet and salty in 9- to 15-year-old and adult humans. *Science* 190 (4215): 686–687.

Ervina, E., Berget, I., and Almli, V.L. (2020). Investigating the relationships between basic tastes sensitivities, fattiness sensitivity, and food liking in 11-year-old children. *Foods* 9 (9): 1315.

Faith, M.S., Heo, M., Keller, K.L., and Pietrobelli, A. (2013). Child food neophobia is heritable, associated with less compliant eating, and moderates familial resemblance for BMI. *Obesity* 21 (8): 1650–1655.

Garcia, K.M. (2011). Sensory discrimination testing with children. LSU Doctoral Dissertations. p. 3871. https://digitalcommons.lsu.edu/gradschool_dissertations/3871 (accessed 16 January 2023).

Garcia, K., Ennis, J.M., and Prinyawiwatkul, W. (2012). A large-scale experimental comparison of the tetrad and triangle tests in children. *J. Sens. Stud.* 27 (4): 217–222.

Greenslade, E. (2020). *UK Children's Healthcare Market Report*. London, UK: Mintel.

Guinard, J.-X. (2001). Sensory and consumer testing with children. *Trends Food Sci. Technol.* 11(8): 273–283.

Gutierrez, T.J. (2020). Food manufacturing for babies and children: editor's perspective. In: *Food Science, Technology and Nutrition for Babies and Children* (ed. T.J. Gutierrez), 1–2. Cham: Springer.

IFST (2020). IFST guidelines for ethical and professional practices for the sensory analysis of foods. IFST.org. https://www.ifst.org/sites/default/files/IFST%20Guidelines%20for%20Ethical%20and%20Professional%20Pr (accessed 28 September 2022).

ISO8586:2023 (2023). *Sensory Analysis – General Guidelines for the Selection, Training and Monitoring of Selected Assessors and Expert Sensory Assessors*. Geneva: International Organization for Standardization. https://www.iso.org/standard/36385.html.

ISO8589:2007 (2007). *Sensory Analysis – General Guidance for the Design of Test Rooms*. Geneva: International Organization for Standardization. https://www.iso.org/standard/36385.html.

Jansen E, Mulkens S, Jansen A. How to promote fruit consumption in children. Visual appeal versus restriction. *Appetite.* 2010 54(3):599–602. https://doi.org/10.1016/j.appet.2010.02.012.

Keller, K.L., Shehan, C., Cravener, T. et al. (2022). Do children really eat what they like? Relationships between liking and intake across laboratory test-meals. *Appetite* 2022 (172): 105946. https://doi.org/10.1016/j.appet.2022.105946.

Kimmel, S.A., Sigman-Grant, M., and Guinard, J.-X. (1994). Sensory testing with young children. *Food Technol.* 48: 92–99.

Krick, K.B. (2011). Kids vs. adults: frequency perception. California State Science Fair. https://csef.usc.edu/History/2011/Projects/J0711.pdf (accessed 16 January 2023).

Kroll, B.J. (1990). Evaluating rating scales for sensory testing with children. *Food Technol.* 44: 78–86.

Lachnit, M., Busch-Stockfisch, M., Kunert, J., and Krahl, T. (2003). Suitability of free choice profiling for assessment of orange-based carbonated soft-drinks. *Food Qual. Prefer.* 14 (4): 257–263. https://doi.org/10.1016/s0950-3293(02)00082-4.

Lafraire J, Rioux C, Giboreau A, Picard D. Food rejections in children: Cognitive and social/environmental factors involved in food neophobia and picky/fussy eating behavior. *Appetite.* 2016;96:347–357. https://doi.org/10.1016/j.appet.2015.09.008.

Lakkakula, A., Geaghan, J., Zanovec, M. et al. (2010). Repeated taste exposure increases liking for vegetables by low-income elementary school children. *Appetite* 55 (2): 226–231.

Laureati, M., Sandvik, P., Almli, V.L. et al. (2020). Individual differences in texture preferences among European Children: development and validation of the Child Food Texture Preference Questionnaire (CFTPO). *Food Qual. Pref.* 80 103828

Lee, J., Russell, C.G., Mohebbi, M., and Keast, R. (2022). Grating orientation task: a screening tool for determination of oral tactile acuity in children. *Food Qual. Prefer.* 95: 104365. https://doi.org/10.1016/j.foodqual.2021.104365.

Liem, D. (2017). Infants' and children's salt taste perception and liking: a review. *Nutrients* 9 (9): 1011.

Liem, D.G., Mars, M., and de Graaf, C. (2004). Consistency of sensory testing with 4- and 5-year-old children. *Food Qual. Prefer.* 15: 541–548.

Loewen, R. and Pliner, P. (2000). The Food Situations Questionnaire: a measure of children's willingness to try novel foods in stimulating and non-stimulating situations. *Appetite* 35 (3): 239–250.

Lumeng, J.C., Zuckerman, M.D., Cardinal, T., and Kaciroti, N. (2005). The association between flavor labeling and flavor recall ability in children. *Chem. Senses* 30 (7): 565–574.

Market Research Future (2021). Children's cosmetics market to procure USD 1,795.15 million revenue by 2026 at substantial CAGR of 6.79%, predicts market research future (MRFR). GlobeNewswire News Room. https://www.globenewswire.com/news-release/2021/02/25/2182540/0/en/Children-s-Cosmetics-Market-to-Procure-USD-1-795-15-Million-Revenue-by-2026-at-Substantial-CAGR-of-6-79-Predicts-Market-Research-Future-MRFR.html (accessed 20 September 2021).

Meilgaard, M., Civille, G., and Carr, B. (2006). *Sensory Evaluation Techniques*, 4e. CRC Press.

Mennella, J.A., Jagnow, C.P., and Beauchamp, G.K. (2001). Prenatal and postnatal flavor learning by human infants. *Pediatrics* 107 (6): e88–e88.

Miller (2023). Kids Food and Beverages Market Size, Share, Growth Future Trends Analysis Research Report to 2033. https://www.linkedin.com/pulse/kids-food-beverages-market-size-share-growth-future-trends-miller/ (accessed 2 February 2023).

Moding, K.J. and Fries, L.R. (2020). Feeding practices demonstrated by parents of toddlers: an observational analysis of breakfast, lunch, dinner, and snacks. *Appetite* 155: 104825.

Moskowitz, H.R. (1985). Product testing with children. In: *New direction for product testing and sensory analysis of foods*, 371. Westport, CT: Food and Nutrition Press Inc.

Moskowitz, H.R. (1994). 'Children vs. adults' in food concepts and products. In: *Just-in-Time Development*, 293–332. Trumbull, CT: Food & Nutrition Press.

MRS. (2012). MRS guidelines for research with children and young people (September 2014). MRS.org.uk. https://www.mrs.org.uk/pdf/2014-09-01Children%20and%20Young%20People%20Research%20Guidelines.pdf (accessed 28 September 2022).

MRS (2020). MRS guideline: conducting data collection activities with children. https://www.mrs.org.uk/pdf/MRS-Guideline-conducting-data-collection-activities-with-children.pdf (accessed 2 February 2023).

Murray, J.M., Delahunty, C.M., and Baxter, I.A. (2001). Descriptive sensory analysis: past, present and future. *Food Res. Int.* 34 (6): 461–471.

Nardini, M., Bedford, R., and Mareschal, D. (2010). Fusion of visual cues is not mandatory in children. *Proc. Natl. Acad. Sci.* 107 (39): 17041–17046.

Nicklaus, S. (2015). 22 – Sensory testing in new product development: working with children. In: *Rapid Sensory Profiling Techniques* (ed. J. Delarue, J.B. Lawlor, and M. Rogeaux). Woodhead Publishing.

Nicklaus, S., Boggio, V., Chabanet, C., and Issanchou, S. (2004). A prospective study of food preferences in childhood. *Food Qual. Prefer.* 15 (7–8): 805–818.

Pappas, S. (2010). Your brain processes images differently when you're a kid. livescience.com. https://www.livescience.com/10753-brain-processes-images-differently-kid.html#:~:text=Adults%20and%20children%20really%20don (accessed 20 September 2021).

Patterson, N.J. and Beeren, C.J.M. (2011). Case studies of consumer testing of food products using children. In: *Developing Children's Food Products* (ed. D. Kilcast and F. Angus), 188–203. Cambridge: Woodhead Publishing.

Perry, R.A., Mallan, K.M., Koo, J. et al. (2015). Food neophobia and its association with diet quality and weight in children aged 24 months: a cross sectional study. *Int. J. Behav. Nutr. Phys. Act.* 12 (1): 13. https://doi.org/10.1186/s12966-015-0184-6.

Pliner, P. (1994). Development of measures of food neophobia in children. *Appetite* 23 (2): 147–163.

Pliner, P. and Hobden, K. (1992). Development of a scale to measure the trait of food neophobia in humans. *Appetite* 19 (2): 105–120.

Popper, R. and Kroll, J.J. (2011). Consumer testing of food products using children. In: *Developing Children's Food Products* (ed. D. Kilcast and F. Angus), 163–187. Cambridge: Woodhead Publishing.

Praet, E., Baeza-Álvarez, J., De Veer, D. et al. (2023). Bottle with a message: the role of story writing as an engagement tool to explore children's perceptions of marine plastic litter. *Mar. Pollut. Bull.* 186: 114457.

Rocha, C., Ribeiro, J.C., Lima, R. et al. (2021). Application of the CATA methodology with children: qualitative approach on ballot development and product characterization of innovative products. *Food Qual. Prefer.* 88: 104083.

Sobek, G. and Jagielski, P. (2022). The number of fungiform papillae, taste sensitivity and smell functions of children aged 11–15. *Nutrients* 14 (13): 2578.

Society of Sensory Professionals (2021). Society of sensory professionals. https://www.sensorysociety.org/knowledge/sspwiki/Pages/R-index%20in%20Sensory%20Evaluation.aspx (accessed 2 February 2023).

Star Tribune (2018). Children's eyes let in more light than adults' so go dark to help them sleep. https://www.startribune.com/children-s-eyes-let-in-more-light-than-adults-so-go-dark-to-help-them-sleep/476337673/ (accessed 20 September 2021).

Szczesniak, A.S. (2002). Texture is a sensory property. *Food Qual. Prefer.* 13 (4): 215–225.

Thomas, M.A. and Murray, F.S. (1980). Taste perception in young children. *Food Technol.* 34 (3): 38–41.

Toyama, H. and Agras, W.S. (2016). A test to identify persistent picky eaters. *Eat. Behav.* 23: 66–69.

Urbick, B. (2011). Working with children and adolescents for food product development. In: *Developing Children's Food Products* (ed. D. Kilcast and F. Angus), 204–228. Cambridge: Woodhead Publishing.

Ustun, B., Reissland, N., Covey, J. et al. (2022). Flavor sensing in utero and emerging discriminative behaviors in the human fetus. *Psychol. Sci.* 33 (10): 1651–1663.

Velazquez-Mendoza, A.L., Vidal, L., Varela, P., and Ares, G. (2022). Can children use the A–not-A test? *J. Sens. Stud.* 37 (4): 1–5.

Vennerød, F.F.F., Hersleth, M., Nicklaus, S., and Almli, V.L. (2017). The magic water test. An affective paired comparison approach to evaluate taste sensitivity in pre-schoolers. *Food Qual. Prefer.* 58: 61–70.

Ventura, A.K. and Worobey, J. (2013). Early influences on the development of food preferences. *Curr. Biol.* 23 (9): R401–R408.

Vereecken, C.A., Vandervorst, S., Nicklas, T. et al. (2010). Test–retest reliability and comparison of children's reports with parents' reports of young children's fruit and vegetable preferences. *Appetite* 55 (3): 574–581.

Wardle, J., Cooke, L.J., Gibson, E.L. et al. (2003). Increasing children's acceptance of vegetables; a randomized trial of parent-led exposure. *Appetite* 40 (2): 155–162.

Werner, L. (2007). What do children hear? How auditory maturation affects speech perception. *The ASHA Leader*. https://doi.org/10.1044/leader.FTR1.12042007.6 (accessed 24 May 2022).

Werthmann, J., Jansen, A., Havermans, R. et al. (2015). Bits and pieces. Food texture influences food acceptance in young children. *Appetite* 84: 181–187.

Zampini, M., Sanabria, D., Phillips, N., and Spence, C. (2007). The multisensory perception of flavor: assessing the influence of color cues on flavor discrimination responses. *Food Qual. Prefer.* 18 (7): 975–984.

Zeinstra, G.G., Koelen, M.A., Colindres, D. et al. (2009). Facial expressions in school-aged children are a good indicator of 'dislikes', but not of 'likes'. *Food Qual. Prefer.* 20 (8): 620–624.

# 12

# Expanding Attribute-Specific Difference Tests with Multisample Paired Comparison Paradigms

*Curtis R. Luckett*

*Department of Food Science, University of Tennessee, Knoxville, TN, USA*
*Department of Food Science, SCiFi Foods, San Leandro, CA, USA*

In discrimination testing, overall or unspecified discrimination methods seem to be more popular. Unspecified discrimination methods are concerned with detecting the presence of a difference not the nature of a difference (i.e., what attributes are different across the products). However, attribute-specific difference tests can be a valuable tool to test the nature of a sensory difference. In attribute-specific difference tests, several sensory tasks can be used, and all these tasks pit two samples against each other, while broader objectives often have an interest in more than two samples. Examples would include forced choice tasks in which participants are asked to choose which sample is sweeter or sorting tasks where participants are asked to sort samples based on similarity in bitterness intensity. These attribute-specific tasks tend to pair two products and attempt to decide which of these products are the same or different for a particular attribute. However, it is often the case that there are more than two products of interest. In the case of having more than two products of interest, alternative data analysis procedures can be used. This chapter will address attribute-specific discrimination and highlight alternative analysis methods for directional discrimination tasks that can be used in multisample paradigms. These methods are also helpful in incomplete block designs that often accompany multisample discrimination paradigms.

## 12.1 Introduction

Attribute-specific discrimination tests are valuable to sensory practitioners who wish to test product differences on a specified attribute. Attribute-specific discrimination tests are also referred to as directional difference test or specified difference test. These tests follow a methodology and statistical analysis approach that is much the same as general discrimination tasks. However, there are key factors to account for when transitioning from a general

*Discrimination Testing in Sensory Evaluation*, First Edition. Edited by Lauren Rogers, Joanne Hort, Sarah E. Kemp and Tracey Hollowood.

**Table 12.1** List of methods and analysis methods to be covered in this chapter.

| Sensory methods | Data analyses |
|---|---|
| n-AFC | Exact binomial test |
| Unspecified tetrad | Pearson's chi-squared |
| | Thurstonian modeling |
| | Mosteller's extension of Thurstone |
| | Bradley-Terry models |
| | Elo models |

difference test to an attribute-specific paradigm. First, it is important to ensure that there is no ambiguity in the participants' understanding of the attribute of interest. More plainly, some level of training is needed to confirm all participants have a precise and accurate understanding of the attribute to be used to discriminate the samples. Second, while the tasks seem similar, underlying differences in decision rules and subsequent psychometric functions coincide with notable changes in the statistical power for a given sensory difference. Decision rules refer to the process of breaking down each test into two mutually exclusive and exhaustive regions (Green and Swets 1966). Characterizing the fashion in which a participant makes their decision in discrimination tasks is key to linking the percentage of correct answers and the size of the difference between the samples. This relationship, known as a psychometric function, is key for comparing the sensitivity and statistical power across different discrimination tasks (Ennis et al. 2014). Decision rules are typically dictated for a discrimination task, meaning that test instructions can alter the decision rule of a particular discrimination test. For example, the power benefits the tetrad test has over the triangle test are not in the change of guessing correctly (both have the probability of guessing correctly = 1/3). The increased power of the tetrad test comes from the decision rule given by the instruction to group the samples. When participants do not use the grouping strategy, the increase in power of the tetrad goes away (Rousseau and Ennis 2013). The instructions and unspecified tests such as triangles and unspecified tetrads are less powerful than their specified relatives, 3-AFC and specified tetrads, respectively. Along similar lines, you would expect more correct answers for a specified test (3-AFC) than an unspecified test (triangle) for participants of equal sensitivity (Frijters 1979; O'Mahony et al. 1994). Third, the sensory practitioner must be sure that only differences in only one attribute are present or only one attribute is of interest in terms of larger objectives. For example, reformulation with increased cocoa may increase the bitterness and chocolate flavor. In this case, the sensory practitioner must chose an unspecified difference or be sure that only one of the two attributes is of part of the larger reformulation objective.

Before continuing, the author would like to align on the terminology in describing sensory discrimination tests and methods. Going forward, products will be used to describe the items of interest in the testing paradigm, usually variations on the same theme (Rogers 2017). For example, a simple situation where we want to compare an original soup formulation to a reduced sodium formulation. We will use the term samples to describe the number of items the participant will be asked to choose from. Continuing our example with the two soup products, a tetrad test would involve four samples, two of each product (i.e., control and reduced sodium). If we were to choose a triangle test for our soup discrimination test, we would have three samples, which could consist of two controls and one reduced sodium or one control and two reduced sodium.

Several methods can be applied for directional difference testing (Table 12.1), for example, specified tetrads, where participants see multiple samples and make judgments based on a singular attribute of the product. A specified tetrad on sweetness would ask the participants to group the samples based on sweetness intensity – instead of similarity like in an unspecified triangle. Like the similarities between specified and unspecified tetrad tests, the attribute-specific sibling of the widely popular triangle test is the 3-AFC (alternative forced choice) test. While 3-AFC tests are popular, AFC tests can take the form any $n$-AFC arrangements, where $n$ (the amount of samples involved in the test) can be any number greater than 1. Like all discrimination tests, the statistical power stemming from the likelihood of a participant guessing correctly is often a guiding factor in choosing the number of samples in an AFC testing paradigm. Other factors such as products having strong carryover effects or products that fatigue the palate are also at play. The balance of the statistical power and not fatiguing the palate of participants tends to lead most sensory practitioners to choose 2-, 3-, or 4-AFCs.

While the simplest of the options presented here, 2-AFCs (paired comparisons) are also a valuable tool to accomplish many sensory objectives (Yang and Ng 2017). In relation to discrimination testing, the paired comparison is often used in assessing attribute-specific directional differences. These tests have the obvious benefit of not just identifying a perceivable difference between samples, but also the modus in which they differ. In a traditional analysis of paired comparisons, the analysis typically focuses on: *how many times could people notice a difference between products A and B?* From this point, a Thurstonian or through a guessing/percent-distinguisher approach (e.g., binomial or chi-squared distributions) can be used to analyze the data. Most would choose the Thurstonian approach because of the added benefit of a measure of magnitude of difference between the two products ($d'$). Additionally, the Thurstonian approach accounts for the aforementioned differences in performance of certain discrimination tests (Meilgaard et al. 2006). As a reminder, different discrimination tasks have different decision rules, which is directly related to the psychometric function that accompanies each test. These differences tend to lead to unspecified versions of discrimination tasks being less powerful than their specified counterparts (Frijters 1979; O'Mahony et al. 1994).

Uniquely, methodological decisions around implementations of attribute-specific discrimination tests are often opposed by not only unspecified difference tests but also some sort of rating paradigm. There are numerous resources on choosing a discrimination test (Meilgaard et al. 2006; Ford 2017); however, this chapter will mainly focus on the appropriateness of attribute-specific discrimination tasks to particular situations. Just like all sensory methodology, the test method must match the objective of the test. One area that we will demonstrate is that attribute-specific discrimination tests can be used to garner information not only around the product but also the consumers who participate in the test. Discrimination tests have often been thought of as solely sources of information regarding products (Adjei 2017). Typically, consumer relevant information is thought to come through affective testing. While affective testing can take the form of a preference or a rating task, rating tasks can also be used to assess differences in particular attribute intensities. These rating paradigms would have participants rate the samples of interest for the strength of specific attributes, and those ratings could be compared across samples through an analysis of variance or *t*-test. Examples of these rating paradigms typically involve trained panelists conducting traditional descriptive analysis or degree of difference testing. Furthermore, methods with less trained participants such as magnitude estimation could also be used. While rating attribute intensity is acceptable, humans tend to be more sensitive comparing stimuli than rating. Much of the benefit of attribute-specific difference testing is the fact that it does not require graded responses but instead breaks down the test into the simplest possible decision units (Gridgeman 1955). This allows sensitive sensory work to be done with untrained participants, and with certain data analysis strategies, the size of the difference can still be estimated. The major downside of using attribute-specific degree of difference testing is the relative nature of all resulting data. Discrimination tasks are comparison tasks, and therefore, what is being compared directly influences the data. With rating tasks, this is not the case. Additionally, a rating paradigm can easily be designed for a single participant to assess several attributes. Lastly, a sequential monadic rating paradigm would require less of each product.

The crux of the decision is weighing the availability of product, drawbacks of relative data, the number of attributes of interest, and the variance introduced by asking participants to make graded judgments.

Let us look at an example of a situation where an attribute-specific discrimination test can be used. Additionally, we will work through several different ways to analyze the data.

### 12.1.1 Simple Examples

**The background**: Supporting product reformulations are part of life for many sensory professionals. Reformulations often represent a good opportunity to employ discrimination tasks. As mentioned earlier, it is advisable to use an unspecified test if the nature of any difference is not known. However, in the case of aroma-active ingredients, which are typically used in miniscule amounts, the reformulated product is not likely to change beyond the flavor note of interest. When one is reasonably sure the reformulated product is only possibly different a single dimension, the attribute-specific difference test presents a good option due to the increase in statistical power.

**The situation**: Natural vanillin is an extremely expensive ingredient. To stay cost-competitive with our competitors, we would like to slightly reduce the amount of vanillin in our all-natural confectionary product. However, we will need to verify that consumers of our product do not notice the decreased vanilla flavor.

**Objective:** Test if reducing the amount of vanilla flavor in the confection reduces the perceived vanilla flavor.

**Test structure**: The test sample, which is our confection with reduced vanilla flavor, will be compared for vanilla flavor strength to the control sample. In this scenario, we have a modest budget, coupled with a product that has little carryover. Additionally, we are confident that the participants would not be fatigued if they are required to taste several samples. After weighing the factors at play, our team decides on a 3-AFC, which gives us a more statistical power per participant and is straightforward in terms of participant expectations. More explicitly, the 3-AFC is not as reliant as a specified tetrad on ensuring that the participants employ the correct cognitive strategy (O'Mahony et al. 1994). This is not to say that unspecified tetrads are not of use, but in this case, our participants have no experience performing tetrads and our team wants to err on the side of caution. In weighing the project's budget with statistical parameters related to type I and type II errors, we settle on a sample size of 66 participants. This number gives us a 5% type I risk and a 5% type II risk, when the proportion of distinguishers ($P_d$) is 0.3 (Schlich 1993). The power of any experiment is dependent on the size the effect of interest, and in percent discriminator models, the $P_d$ is typically used, which corresponds the maximum percentage of discriminators that is acceptable for the products to be classified as interchangeable. Each of our 66 participants will receive three samples and be asked to select the sample they find to have the strongest flavor.

**Interpreting results**: As mentioned earlier, we have several data analysis options. We will use three methods: two guessing/percent discriminator methods – Pearson's chi-square and binomial approach, along with a Thurstonian approach. As mentioned earlier, the Thurstonian approach would be preferred because it acknowledges the specific decision rule of various discrimination tasks. Additionally, Thurstonian models give an estimate of the size of the difference. The estimate of the size of the difference is denoted by d′. As a quick refresher, the Thurstonian approach assumes that the sensations elicited by the samples are normally distributed with equal variance. In practice, this involves translating the proportion of hits and false alarms to $z$-scores, where hits are defined as correctly identifying a different sample and false alarms are defined as participants identifying a difference that does not exist. The $d'$ is the distance between the two $z$-scores; the larger the $d'$, the larger the perceptual difference. However, this chapter shares guessing/percent discriminator methods, as they are still commonly found in the sensory program of many organizations. The first percent discriminator method, the binomial test, characterizes our results in terms of their likelihood of being due to chance. The binomial test stems from the binomial distribution which is the basis of the test. The chi-squared test, which uses the chi-squared distribution, is used to compare an observed outcome to an expected outcome. In terms of sensory discrimination tests, this equates to testing to see if your observed frequency of correct answers differs than what would be expected by chance. It also should be noted here that the results of a discrimination task can also be analyzed by comparing the results to the areas under a normal distribution (Stone and Sidel 1978); however, this will not be covered in this chapter.

Both guessing and percent discriminator methods can be quickly executed in the widely accessible Microsoft Excel or Google Sheets, reducing the chance of a simple computational error. The *CHISQ.TEST* (Excel) or *CHITEST* (Sheets) function provides a $p$ value from the number of observed judgments for each sample, combined with how many answers for each value you expect due to chance. The binomial approach requires a simple formula in conjunction with the *BINOMDIST* function (both Excel and Sheets use this function name).

**Solution**: We find that 27 of the 66 participants found the control sample to be stronger in terms of vanilla flavor (Table 12.2). Both Pearson's chi-square and binomial approach will be testing to see if our number of votes for the control (27) is greater than what you expect by chance, assuming that there is no difference in vanilla flavor strength. In our 3-AFC task, there is a one-third chance of simply guessing correctly, meaning that out of 66 participants we would expect 22 correct answers by pure chance. Pearson's chi-square approach will directly compare our observed results (27 votes for the control and 39 votes for the test sample) to what we would expect by chance (22 votes for the control and 44 votes for the test product). For our chi-square analysis, we can simply use the *CHISQ.TEST* or *CHITEST* function and included the necessary values for each key judgment.

*Microsoft Excel*

= CHISQ.TEST (Range of Observed Judgments, Range of Expected Judgments)

*Google Sheets*

= CHITEST (Range of Observed Judgments, Range of Expected Judgments)

For data arranged in the following manner:

**Table 12.2** Example data for a 3-AFC test on vanilla flavor intensity.

| | Control | Test |
|---|---|---|
| Observed | 27 | 39 |
| Expected | 22 | 44 |

The correct formula would be =CHISQ.TEST(B2:C2,B3:C3), where cells B2 and C2 are the observed results and B3 and C3 are the results expected by chance. The *p* value given in this case is 0.1917. Therefore, we could conclude that there is a 19.17% chance we would observe data like we collected if there was no perceivable difference in vanilla flavor strength.

Additionally, we could use an exact binomial approach. In Excel or Google Sheets, we would use the *BINOMDIST* function where our results cell would have the following formula:

$$= 1 - \text{BINOMDIST}(C_c, T, P_c, \text{TRUE})$$

where $C_c$ = votes for the control – 1, $T$ = total judgments, and $P_c$ = probability of guessing correctly. The TRUE comes from our desire to look at cumulative probability under the binomial distribution. We are interested in testing if the probability of the votes for the control is equal to or fewer than expected by chance. If we did not specify a cumulative distribution, we would only be assessing the probability of getting exactly the number of votes for the control. This thought dovetails with our definition of a *p*-value, where we estimate the probability of getting data at least as extreme as what we have observed.

For those interested in using R, this same method can be used by employing the *pbinom* function that is in base R code (Team 2020). Again, where $C_c$ = votes for the control – 1, $T$ = total judgments, and $P_c$ = probability of guessing correctly.

```
1-pbinom(Cc, T, Pc)
```

In our case, we would use the following values:

$$C_c = 26 \text{ (Number of votes for control} - 1)$$

$$T = 66$$

$$P_c = 1/3$$

This calculation would give us a $p$ value of 0.1209, which can be interpreted as there is a 12.09% of observing the data we did if there was no perceivable difference in vanilla flavor strength.

Lastly, there is the Thurstonian option of analyzing the data. Using the sensR package in R, the *discrim* function provides a simple platform for this Thurstonian analysis (Brockhoff and Linander 2017). The *discrim* function can use more information, but if we simply provide the function with the number of votes for the control, the number of total votes, and the discrimination method used, we can get our results. So that would be discrim(votes for control, total votes, method = "threeAFC"). Below is the code and output from the sensR package.

```
discrim(27,66, method = "threeAFC")

    Estimates for the threeAFC discrimination protocol with
    27 correct answers in 60 trials. One-sided p-value and 95
    % two-sided confidence intervals are based on the 'exact'
    binomial test.

             Estimate Std. Error  Lower  Upper
    pc         0.4091    0.06052 0.3333 0.5371
    pd         0.1136    0.09078 0.0000 0.3056
    d-prime    0.2592    0.21142 0.0000 0.6771

    Result of difference test:
    'exact' binomial test:  p-value = 0.1209
    Alternative hypothesis: d-prime is greater than 0
```

The $p$ value given in this case is 0.1209, which is based on the same binomial test we ran earlier. However, we do have the benefit of the $d'$ value, which is a direct measure of the size of the difference between the samples. Like previously, we could conclude that there is a 12.09% chance we would observe data like we did if there was no perceivable difference in vanilla flavor strength.

As you can see, the $p$ values differ slightly across the three analysis methods, which is due to assuming different underlying distributions. Taken together, there is weak evidence that people do notice the reduced vanilla flavor quantity in the products.

At this point, one may wonder which method should be chosen for testing to see if there is a significant difference among products in terms of the attribute of interest. There is no easy answer here, but in practice, both the chi-square test and the binomial test should roughly agree with each other. In the author's opinion, the benefits and criticisms of each test do not apply to attribute-specific discrimination testing.

More specifically since the output of a sensory discrimination task will always be measured as correct or incorrect, both methods seem appropriate, and the author is unaware of any convincing arguments for either test when the data are structured as such.

### 12.1.2 Going Beyond a Pair of Products

The options above are statistically rigorous and accepted as valid; however, if your interests go beyond two products, these methods begin to lose utility. Mainly, there is the lack of a transitive component in the analysis. These pairwise methods of data analyses do not use the valuable information from discrimination tasks involving one of the samples, but not both. For example, if I have three products of interest (A, B, and C), knowing how products A and B performed against product C can help us answer questions about products A and B, even before directly comparing them. If the sensory judgments do not break transitivity (e.g., a participant rates product A higher than product B, B higher than product C, but C higher than A), this information is helpful in discrimination tasks. Unfortunately, this information from *common opponents* has no place in the common methods of discrimination analysis. Second, there is the issue of compounding of error rates when doing multiple comparisons of sample pairs. If one sets alpha (false-positive rate) to 0.05 or 5% and analyzes 10 pairs of samples, regardless of the analysis method, the false-positive rate grows well past the stated 5%. Of course, a correction of alpha can be made to account for these multiple comparisons, but this would still lead to reductions in statistical power through inflating the likelihood of a false negative (Holm 1979), and in discrimination tasks, a false negative is often the more detrimental of the two possible errors.

The circumstance that products are compared against each other in pairs does not necessitate strictly separate pairwise data analyses. Outside of the current suite of discrimination data analysis methods, there are several frameworks that are common in other fields of inquiry; coincidently, some of the methods were developed out of sensory science before finding more footing in other fields. For example, Bradley-Terry models were first used to quantitate food sensory judgments (Bradley and Terry 1952; Terry et al. 1952; Hopkins 1954; Gridgeman 1955), but they are used sparsely in current research on human food judgments. It should also be mentioned that when we assess different statistical frameworks for suitability in these scenarios, it is also important to ensure the *matched* nature of data which would stem from a series of paired comparisons. Treating each sensory evaluation as independent leads to incorrect estimates of variance (Agresti 2018).

When we look for methods that would suit the analysis of several samples, which were all compared in a pairwise manner, several examples stand out. The most pressing need when analyzing directional difference tasks across a multisample product set is the ability to handle incomplete block designs, as the number of comparisons needed increases geometrically according to the equation below.

$$\textit{Number of comparisons} = \frac{n(n-1)}{2}$$

where $n$ is the number of samples in the sample set. From basic calculations, it becomes easy to see the benefit of designing discrimination studies with multiple samples using incomplete block designs. For example, a sample set of 7 products has 21 combinations of sample pairs, well beyond what is possible in a single sensory analysis session.

One of the most obvious uses of these analysis frameworks is sport. In many popular sports, two teams or individuals compete and a winner is determined. Typically, not every team/player would face each other; therefore, these methods are designed around incomplete block designs. One arena where these methods have caught on is chess, where metric hungry competitors and fans have devised many measures of skill through analyzing paired comparisons (e.g., chess matches). While stated earlier, the paired nature of the comparisons does not necessitate a strictly paired approach to data analysis, and the methods used in sports are well designed to integrate all the comparisons a sample is involved in since not all team/players face each other regularly.

There are many benefits to integrating the performance of a sample across all comparisons it was involved in. From a practical standpoint, an incomplete block design can be utilized without much of a downside. These designs are useful when the samples are fatiguing, there are simply too many possible pairs, or the samples have unique characteristics such a skin product or other personal care products. Additionally, these methods could also be used to assess attributes across cultures or age groups, where scale usage differences are documented.

## 12.2 Practical Considerations

Any decision on methodology also has practical considerations. Unlike most applications presented in this book, the decision to adopt a multisample paired comparison design is often set in opposition to a sequential monadic-rating paradigm. Therefore, in discussing this topic, it has to be mentioned that choosing a comparison paradigm over rating paradigm has several practical drawbacks. The first is fatigue, products that are heavily fatiguing for participants should likely be assessed in a rating paradigm. Furthermore, the size of the sample set should be taken into account when determining if a comparison paradigm is a good idea. As shown earlier, the number of possible paired comparisons increases geometrically with the number of samples in the sample set (see Table 12.3). Lastly, if the amount of sample is limited, one should forgo a comparison strategy. All three of the practical considerations can be minimized by incomplete block designs, where participants do not see every possible pair of samples. All of the methods in this chapter are well suited for analyzing incomplete block comparison designs.

**Table 12.3** The number of combinations needed for a complete block design in discrimination testing.

| Number of products | 2 | 3 | 4 | 5 | 6 | 7 | 8 | 9 | 10 |
|---|---|---|---|---|---|---|---|---|---|
| Combinations of product pairs for a complete block design | 1 | 3 | 6 | 10 | 15 | 21 | 28 | 36 | 45 |

## 12.3 Mosteller's Extension of Thurstone

The first method that will be discussed is the Mosteller method of analyzing paired comparisons. This method was introduced by Frederick Mosteller as an extension of Roger Thurstone's model of comparative judgment (Thurstone 1927). At its core, the Thurstonian model suggests that when stimuli are judged by a participant, each stimulus elicits a sensation that exists on a continuum. Thurstone provided five cases for his model, with each case having unique assumptions. Case V is most commonly discussed in terms of sensory discrimination. In Thurstone's Case V, the variance around the sensation is uncorrelated between stimuli and uniformly distributed. In a series of three papers, Mosteller laid out the least squares solution to Thurstone's Case V (Mosteller 1951a), how unequal standard deviations across samples would manifest in Thurstone's law (Mosteller 1951b) and a test of significance for Thurstone's Case V (Mosteller 1951c). The end results of Mosteller's work on this topic is the framework for which multiple samples can be placed on the same discriminable continuum and goodness of fit can be assessed. The sensation being placed on a continuum coupled with the sensations tracking on a normal distribution creates the case where the data are imperfect. For example, see Figure 12.1 where four stimuli are placed on a continuum, where the mean sensation is the midpoint of each distribution, denoted $S_i$. The shaded density represents the variation around the mean sensation. Since in this model the sensation attributed to each stimulus is not simply a rank order, but has variance around it, a certain proportion of time a participant makes a judgment ($X$) that $X_i < X_j$, even when $S_i > S_j$. This phenomenon is highlighted in the below figure, where $X_4$ represents a judgment of $S_4$ and $X_3$ represents a judgment of $S_3$.

As you can see in this example, $X_3$ was judged to be stronger than $X_4$, even when $S_3 < S_4$. This phenomenon, described as fallible data by Mosteller, leads to the proposed model accounting for the probability of this type of result, not just the average sensation given by each stimulus ($S_x$). Furthermore, Mosteller also deviated from Thurstone's Case V, in which Thurstone assumes the products on the continuum to be uncorrelated. Using these assumptions, Mosteller takes a least squares approach to the paired comparisons that places numerous samples on the same continuum and their location can be ascertained using a $S_i - S_j$ matrix as we will do in the Case Study III.

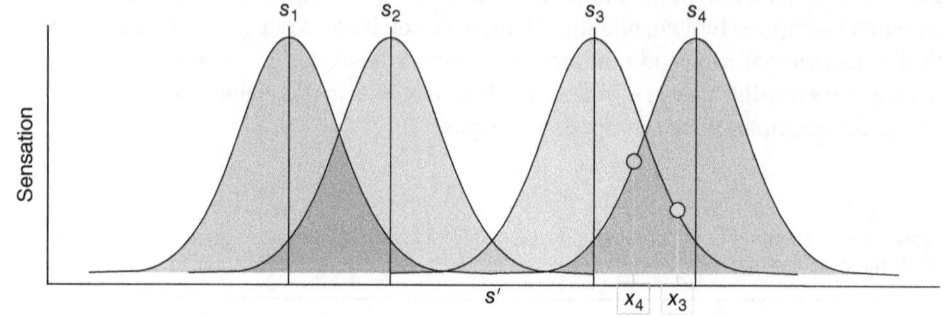


**Figure 12.1** Example of a Mosteller's multisample extension of Thurstone.

Unfortunately, these innovations have not been implemented into modern day statistical software, limiting the utility of the method compared to methods that have more accessible software applications. However, through using the functions in Microsoft Excel or R, the method can still be used to implement to a reasonable degree of effectiveness. A more detailed guide to performing Mosteller's method in Microsoft Excel and R is at the end of this chapter.

## 12.4 Bradley-Terry Models

Like Thurstone's model, the Bradley-Terry model assumes perception exists on continuum but differs in the expected distribution of the sensation. As mentioned earlier, the Bradley-Terry model was initially used in food preference testing (Gridgeman 1955) and I will use preference to initially describe the model. In general, Bradley-Terry models use a maximum likelihood function on previous comparisons to build a logistic model without an intercept that gives the probability of a sample being preferred in future comparisons.

$$\log\left(\frac{\Pi_{ij}}{\Pi_{ji}}\right) = \beta_i - \beta_j \text{ or the generalized form of } logit\left(\Pi_{ij}\right) = \beta_i - \beta_j$$

where $\Pi_{ij}$ represents the probability of sample $i$ being preferred to sample $j$ in a future paired preference test and $\beta_i$, $\beta_j$ are sample parameters. These probabilities are known as Bradley-Terry ratings, or scores, and are logit transformed odds ratios. The logit transformation is commonly used to linearize sigmoid functions, such as probability. Linearizing probability functions, such as odds ratios, allow us to use probability functions outside of the constraint of their 0, 1 limits. While the parallels to logistic regression are strong, the binomial distribution does not account for the numerous factors that could influence the probability estimates (Agresti 2018).

The model estimates the $\Pi_{ij}$ value through

$$\widehat{\Pi_{ij}} = \frac{e^{\left(\widehat{\beta_i}-\widehat{\beta_j}\right)}}{1+e^{\left(\widehat{\beta_i}-\widehat{\beta_j}\right)}}$$

In terms of sensory attribute-specific direction differences, a Bradley-Terry rating is the ratio of the attribute intensity between samples.

A key differentiator between the Bradley-Terry models and others presented here is the extent at which Bradley-Terry models have been expanded and built into common software applications. For example, the leveraging of Bradley-Terry models as a generalized linear model that can be easily used to see how product characteristics can influence the outcome of a paired comparison (Turner and Firth 2012). Furthermore, additional software can be used to cluster participants on their discrimination responses (Séménou 2015), which could be used for identifying subgroups of a population that are more/less sensitive to an attribute and calculate the Bradley-Terry ratings for each cluster.

## 12.5 Elo and mElo

From the world of chess, the Elo rating can be implemented in these multisample paradigms as well (Elo 1978). In the Elo method, each player (or in our case product) is assigned a rating, and the fact that not all chess players would compete against each other is considered. When two products are compared, the product chosen to be stronger in the attribute of interest gets its rating adjusted upward, and the product chosen to be weaker in the attribute of interest gets its rating changed down. The extent of the adjustment is directly related to how unexpected the result is based on the prior sample ratings:

*When higher rated product is preferred:*

$$Winner\ Rating_{new} = Winner\ Rating_{old} + (1-p) \times k \tag{12.1}$$

$$Loser\ Rating_{new} = Loser\ Rating_{old} - (1-p) \times k \tag{12.2}$$

*When lower rated product is preferred:*

$$Winner\ Rating_{new} = Winner\ Rating_{old} + p \times k \tag{12.3}$$

$$Loser\ Rating_{new} = Loser\ Rating_{old} - p \times k \tag{12.4}$$

where $p$ is the probability of the higher rated product being preferred and $k$ is a constant. From the equations above, it is evident that $k$ is a significant determinate of how much the scores change in the short term, yet only a minor influence over the ratings when numerous comparisons are made (Albers and de Vries 2001; Neumann et al. 2011). Additionally, in some sensory studies, the number of comparisons may be quite small leading to the sequence of testing influencing the final Elo values. To address this issue, the sequence of comparisons is randomized and then averaged allowing for stable estimates (Neumann 2015). One hundred to 1000 randomizations have shown similar results (Clark et al. 2018).

One major difference between the Elo or mElo method and the other three methods presented in this chapter is variance. For simplicity, the Elo method did not consider that a player performs within a range of their true ability. This lack of variance is a known shortcoming of the Elo rating system and has since been updated to include measures of variance (Glickman 1999). Because Elo ratings do not have variance, there is not a rigorous statistical method to compare sample ratings. Therefore, Elo methods are likely best employed when the situation does not necessitate comparing individual samples. Of the four methods presented here, mElo has only been applied in a sensory setting for preference.

## 12.6 Friedman-Style Rank Analysis

Interestingly, while all these methods mentioned in this chapter deal explicitly with paired comparisons, Friedman-style rank analysis has been used to assess a series of paired comparisons. Since 2-AFC tests are the simplest form of ranking, rank sums can be calculated for each sample and used to construct reliable measures of an attribute strength (Civille and Carr 2015; Luckett et al. 2020). The Friedman-style rank analysis was originally proposed in 1985 (Carr 1985) and later described in detail in the widely used sensory textbooks (Meilgaard et al. 2006; Yang and Ng 2017). Therefore, while it remains a viable option, this chapter will not cover this method in detail.

## 12.7 Considerations on Choosing an Appropriate Method

This chapter has introduced four methods to be used in a multisample attribute-specific testing paradigm, and the reader probably wants to know which one is best. Unfortunately, an easy answer to this question does not exist. There is little publicly available data on the performance of these methods against each other. A study on perceived physical strength of humans found that Bradley-Terry models and mElo performed comparably to Likert scale ratings. On food, another study showed mElo ratings from paired preference tests correlated slightly better to 9-point hedonic ratings than Bradley-Terry, Friedman-style rank analysis, or Mosteller's extension of Thurstone (Luckett et al. 2020). However, this relationship was dependent on the food product, and all four methods performed reasonably well. That being said, there are clear benefits of specific methods that are mainly based on the software options available to implement these methods. Primarily, some of these methods lack any current software to be conveniently implemented. As mentioned earlier, Bradley-Terry models are a feature of several R packages and XLSTAT-Sensory. Conversely, the author is not aware of any statistical software that offers the Mosteller method for analyzing paired comparisons. Of special note here is the development of the BradleyTerry2 (Turner and Firth 2012) R packages, which provides a framework to investigate the effect of explanatory variables on product performance. Additionally, the CompR package for R is useful for calculating Bradley-Terry scores, but also clustering participants from their unique response patterns.

As mentioned earlier, humans tend to be very sensitive in comparison paradigms, which could lead to samples that are chosen extremely infrequently as the stronger of the two samples for the attribute of interest. More specifically, without the noise in data from asking participants to make graded responses, it could be expected that there would be very little votes the control. Therefore, if a slight but noticeable difference between samples is observed by the participant pool, the magnitude of the difference will be calculated to be very large due to the infrequent votes for the control. When there are few observations for each pair of samples, empty or sparsely populated cells, either through small sample sizes or clear differences in the attribute of interest, these methods should be used with a grain of salt. The robustness of these models to small cell frequencies has not been tested. However, in cases of large sensory differences, these methods will likely exacerbate the difference in ratings extracted from the paired comparisons. Therefore, when a sample is chosen very infrequently, it is better to exclude if from the sample set all together.

**Case Study I: Spreadability of a Sun Cream**

**The situation**: A manufacturer of sun cream is interested in increasing the spreadability of their sun cream. The R & D team has devised three prototypes, all which take a different approach to increasing spreadability.

**Objective**: Test sunscreen samples for spreadability.

**Test structure**: Each of the five samples (three prototypes plus the current formulation and a key competitor) will be compared for spreadability by an internal panel of 50 employees. The newly formulated samples have been coded SplA, SplB, and SplC by the R & D team, which are included in the data set along with our Current formulation and our Competitor formulation. After a discussion of timing and resources, it was determined that an incomplete block design will be used, where each panelist judges 4 of the 10 possible pairs of samples.

**Interpreting results**: For this scenario, let us use the Bradley-Terry method because of the generalized linear model (glm) options presented in the BradleyTerry2 R package. Of course, this method will also give us the Bradley-Terry ratings, which will be a unitless measure of spreadability.

**Solution**: Fitting a Bradley-Terry model in the BradleyTerry2 package requires us to have the data as a list of the sample judged to be more spreadable (winners) and the sample judged to be less spreadable (losers) for each pair. Each row is a single 2-AFC trial. Table 12.4 provides an example of the first several rows of such data.

First, this example will use three R packages. After installing, be sure to load the packages with the following code:

```
library(BradleyTerry2)
library(dplyr)
library(ggplot2)
```

The data are available as an R data file here: https://github.com/crluckett attribute_difference_testing. As explained above, this dataset is simply a list of paired comparisons where the most and least spreadable samples from each pair are listed. The *readRDS()* function is used to call in our dataset, which is a .rds file. These files are R-specific data storage file.

```
sun <- readRDS(file = "sun.rds")
```

To meet the BradleyTerry2 data expectations, we need to add a column outlining which sample was found to be most spreadable, which can be done by creating a

**Table 12.4** Example of the data used for this example.

| Participant ID | More spreadable | Less spreadable |
|---|---|---|
| 1 | SplA | SplC |
| 1 | SplA | SplD |
| 1 | SplB | SplC |
| 1 | SplD | SplB |

column of 1s as evidenced by the result object in the below code. After creating the result object, a Bradley-Terry model can be fit using BradleyTerry2's BTm function. This function uses the result, the winner, the loser, and a data statement. In this example, we also add the *refcat* statement to ensure that the other samples are being compared to our current sun cream formulation.

```
result <- rep(1, nrow(sun$comparisons))
model <- BTm(result, More.Spreadable, Less.Spreadable,
refcat = "Current",    data = sun)
```

Now we have created our data, we can look at the results using the *summary* function in Rs base code.

```
summary(model)
```

The below summary is the core of the output, with minor formatting changes for publication purposes. The estimate is the Bradley-Terry score, which in our case is the unitless measure of spreadability of the samples. Additionally, you can see that each score was tested against our current formulation for a statistical difference. From these data, we can see that all three prototypes (SplA, SplB, SplC) are significantly more spreadable than our current formulation (i.e., Current) if we define significant as $p<0.05$. In this output, the samples being used as a reference are not displayed in the results table.

```
              Estimate Std. Error z value Pr(>|z|)
Competitor     0.2143      0.2075    1.033   0.30170
SplA           0.5701      0.2107    2.705   0.00682 **
SplB           0.6729      0.2108    3.191   0.00142 **
SplC           0.5856      0.2123    2.759   0.00580 **
Signif. codes:  0 '***' 0.001 '**' 0.01 '*' 0.05 '.' 0.1 ' ' 1
```

After seeing the above data, one might want to compare other combinations of samples. This can be completed changing the sample designated by the *refcat* statement. If we were interested in seeing if any of our protypes are more spreadable than the competitor's sample, we do the following model.

```
comp.model <- BTm(result,  More.Spreadable, Less.Spreadable,
refcat = "Competitor", data = sun)
```

Which would give us:

```
          Estimate Std. Error z value Pr(>|z|)
Current   -0.2143      0.2075   -1.033    0.3017
SplA       0.3558      0.2081    1.710    0.0873 .
SplB       0.4586      0.2080    2.205    0.0275 *
SplC       0.3713      0.2096    1.771    0.0765 .
Signif. codes:  0 '***' 0.001 '**' 0.01 '*' 0.05 '.' 0.1 ' ' 1
```

From the above output, we can see that only SplB is significantly more spreadable than the competitor's sample (again defining $p<0.05$ as statistically significant). From these analyses, we can conclude that all three protypes of our sun cream are more spreadable than our current formulation. The prototype SplB was the only sun cream that was more spreadable than our competitor's, which makes it the best candidate to compete in the product space where spreadability is of utmost importance to the consumer.

## Case Study II: Sweetness of Zero Sugar Beverage

**The situation**: A soft drink company is looking to reduce the amount of zero sugar formulations in its portfolio. The objective is to see if there are differences in sweetness across the formulations. Additionally, from past work, we are concerned that there will be subpopulations of the consumer-base that are more/less sensitive to specific sweeteners.

**Test structure**: We have five different beverages and each of these samples will be compared for sweetness by 124 consumers. After a discussion of timing and resources, it was determined that a complete block design will be used, where each consumer judges all 10 possible pairs of samples. For each pair, the consumer is tasked with selecting the sample that is most sweet.

**Interpreting results**: For this scenario, let us use the Bradley-Terry method because of the available clustering algorithm in the CompR package. The Bradley-Terry scores from each product represent the magnitude of the attribute of interest. So, for our sample, the higher the Bradley-Terry score, the sweeter the sample is. To assess the issue of heterogeneity, we look to the possible occurrence of clusters of consumers who show unique patterns of sweetness judgments.

To start, we will load our dataset and change it into a paired comparison format that this R package is designed for. CompR has a built-in function (ClassDataPairComp) to do this transformation from a matrix in R. The dataset cola.RDS can be found here with the full code for this example: https://github.com/crluckett/attribute_difference_testing.

The first few rows of these data look like what is shown in Table 12.5, where the cell values are binary values based on if the row sample was found to be sweeter than the column sample.

```
cola.data <- readRDS(file = "cola.rds")
sample.names <- c("SplA",   "SplB", "SplC", "SplD", "SplE")
data_ <- ClassDataPairComp(data, labelprod = sample.names,
labelcons = NULL, labelcrit = NULL)
```

To answer our first question, is there a difference in sweetness among the samples, we can use the *EstimBradley()* function, which will create Bradley-Terry scores for each product. For our case we can say that higher Bradley-Terry scores translate to higher

**Table 12.5** Observed votes for each sample. The cell values are the votes for the row sample being sweeter than the column sample.

| | SplA | SplB | SplC | SplD | SplE |
|---|---|---|---|---|---|
| SplA | 0 | 0 | 0 | 0 | 0 |
| SplB | 1 | 0 | 1 | 0 | 1 |
| SplC | 1 | 0 | 0 | 0 | 1 |
| SplD | 1 | 1 | 1 | 0 | 1 |
| SplE | 1 | 0 | 0 | 0 | 0 |

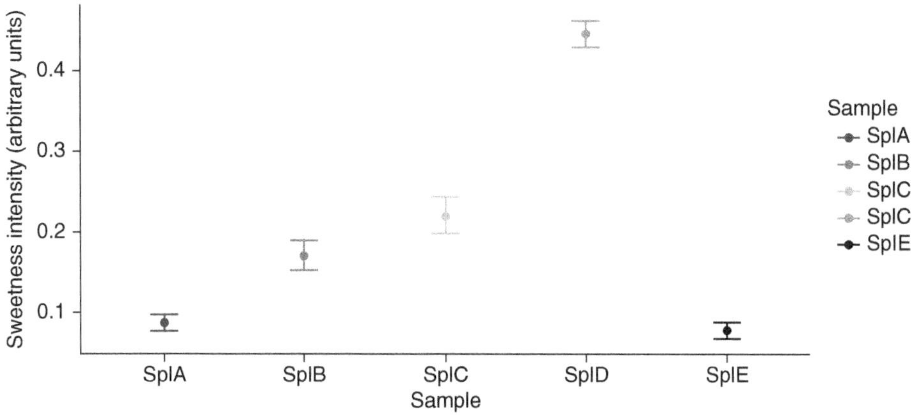


**Figure 12.2** Sweetness intensity across our samples.

sweetness. If we were to plot that data, we can see that sample D (SplD) is considered the highest in sweetness, while samples A and E are lowest in sweetness (Figure 12.2).

However, to address the concern of subgroups of people being more/less sensitive to specific sweeteners, we need to start to look at heterogeneity of the responses. There is more than one way to do this, but since we are using the CompR package, we can use the clustering functionality to assess the optimal number of clusters in our participant pool. The selection of the number of clusters should follow general guidelines found with other consumer sensory studies, where one of the main guidelines is the number of participants in the study. For this study ($n$ = 124), we chose to assess up to five clusters. By adjusting the Tcla value in the EstimBradley() function, you can specify the number of clusters. For example, a model with five clusters can be fit using the following statement:

```
EstimBradley(data_,Constraint=0,Tcla=5,eps=0.001,
eps1=0.001, TestPi=TRUE)
```

As you can see Tcla = x, where x is the number of clusters you want to fit to the data. In this package, we have three parameters to help us assess the optimal number of clusters, all three based on the likelihood function. In general, maximizing likelihood is a common method for finding the model that best fits a specific dataset. The CompR package calculates three such values, Akaike information criterion (AIC), Bayesian information criterion (BIC), and corrected Akaike information criterion (cAIC) by using the *getIC()* function. These three measures have subtle differences which go beyond the scope of this chapter, but the central principle here is that among a finite set of models, the model with the lowest information criterion is optimal. For all three information criterions, the lowest value corresponds with the maximum likelihood. For example, let us use BIC to guide us on the optimal number of clusters. The plot below shows the BIC value across models where the number of clusters ranges from 1 to 5. As you can see in Figure 12.3, the lowest BIC value is found with the model splitting the participants into four clusters. This can be taken as evidence as there is a heterogeneity in sensitivity to sweetness.

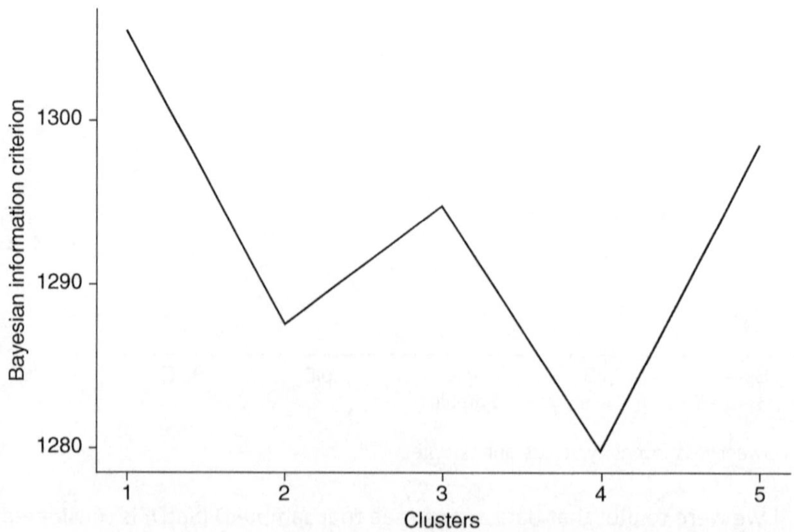


**Figure 12.3** Bayesian information criterion (BIC) considering 1–5 clusters of consumers.

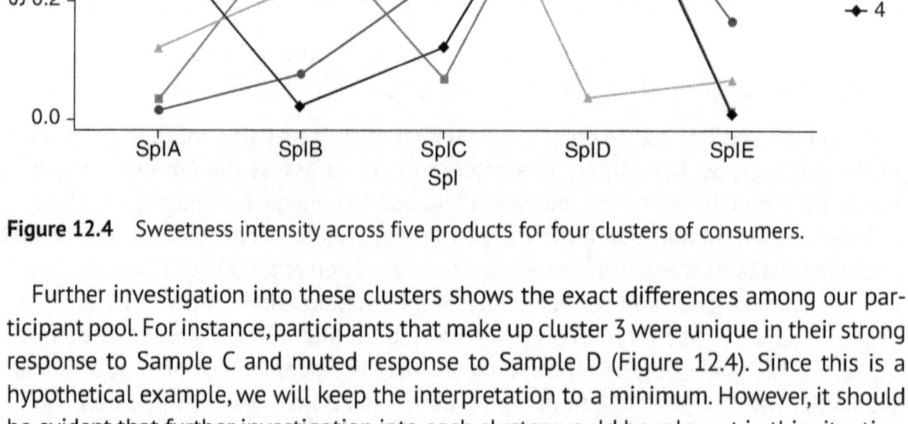


**Figure 12.4** Sweetness intensity across five products for four clusters of consumers.

Further investigation into these clusters shows the exact differences among our participant pool. For instance, participants that make up cluster 3 were unique in their strong response to Sample C and muted response to Sample D (Figure 12.4). Since this is a hypothetical example, we will keep the interpretation to a minimum. However, it should be evident that further investigation into each cluster would be relevant in this situation.

From this analysis, we can conclude that there is a difference among our samples in terms of sweetness. However, there seems to be clusters of consumers that are more/less sensitive to specific sweeteners. While many of the next steps involve decisions beyond the sensory team, but on average, SplD is the sweetest formulation. The demographics and importance of each cluster to the company is a major factor on the decision whether the number of formulations in the portfolio can be reduced.

**Case Study III: Saltiness in a Reformulated Snack Food**

**The situation**: To reduce the sodium content of a savory snack, seven prototypes have been made using a variety of saltiness enhancers on a low-sodium version of a savory snack. We would like to see if these prototypes differ in terms of saltiness.

**Test structure**: We have eight different samples: our current product plus seven prototypes made with various combinations of saltiness enhancers. After a discussion of timing and resources, it was determined that an incomplete block design will be used, where each panelist judges 10 of the possible 28 pairs of samples over the course of 2 sessions. With the resources allotted, we were able to recruit 126 participants, meaning that each pair of products will receive 45 total votes.

**Interpreting results**: For this scenario, we will employ Mosteller's method of paired comparisons because of positive management views on Thurstonian modeling in discrimination testing. More specifically, method choice is always influenced by what is typical in the organization. In our example, the decision was made several years ago to transition to a Thurstonian approach for better estimates of effect size and accounting for the decision rule of each test. After calculating $S'$, we will have the unitless estimate of saltiness across the samples. Additionally, we use a $\chi^2$ distribution to test to ensure that our model correctly fits our data. More specifically, we want to ensure that the use of two sessions does not break the principle of transitivity.

**Solution**: *Data format*: To begin to implement the Mosteller method, it is prudent to start with a Table 12.6 where each sample is a row and a column. Each cell should consist of the proportions of times the column sample was chosen over the row sample. Our results are below, for example, sample 6 was considered saltier than sample 1 44.4% of the time and sample 4 was chosen to be saltier than sample 5, 40.0% of the time.

**Table 12.6** Observed proportions of votes for each sample. The cell values are the proportion of times the column sample was chosen as saltier than the row sample.

| | Spl 1 | Spl 2 | Spl 3 | Spl 4 | Spl 5 | Spl 6 | Spl 7 | Spl 8 |
|---|---|---|---|---|---|---|---|---|
| Spl 1 | — | 0.472 | 0.556 | 0.333 | 0.417 | 0.444 | 0.361 | 0.389 |
| Spl 2 | 0.528 | — | 0.417 | 0.472 | 0.371 | 0.343 | 0.417 | 0.389 |
| Spl 3 | 0.444 | 0.583 | — | 0.457 | 0.389 | 0.333 | 0.306 | 0.278 |
| Spl 4 | 0.667 | 0.528 | 0.543 | — | 0.600 | 0.194 | 0.371 | 0.333 |
| Spl 5 | 0.583 | 0.629 | 0.611 | 0.400 | — | 0.457 | 0.343 | 0.444 |
| Spl 6 | 0.556 | 0.657 | 0.667 | 0.806 | 0.543 | — | 0.571 | 0.444 |
| Spl 7 | 0.639 | 0.583 | 0.694 | 0.629 | 0.657 | 0.429 | — | 0.500 |
| Spl 8 | 0.611 | 0.611 | 0.722 | 0.667 | 0.556 | 0.556 | 0.500 | — |

Next, we need to convert our proportions to their normal derivative using either the *normsinv* function in Microsoft Excel or the *qnorm* function in R (Table 12.7). $S'$ can now be calculated by averaging across all the rows in a column as shown below.

So, from the above, we can see that there are numerical differences in saltiness among the samples (Figure 12.5). The magnitude of the $S'$ value, which can be thought of a corollary to the d′ found in more the more traditional Thurstonian framework, represents the saltiness intensity.

**Table 12.7** Normal derivatives of observed probabilities.

| | Spl 1 | Spl 2 | Spl 3 | Spl 4 | Spl 5 | Spl 6 | Spl 7 | Spl 8 |
|---|---|---|---|---|---|---|---|---|
| Spl 1 | — | −0.070 | 0.140 | −0.431 | −0.210 | −0.140 | −0.355 | −0.282 |
| Spl 2 | 0.070 | — | −0.210 | −0.070 | −0.328 | −0.405 | −0.210 | −0.282 |
| Spl 3 | −0.140 | 0.210 | — | −0.108 | −0.282 | −0.431 | −0.508 | −0.589 |
| Spl 4 | 0.431 | 0.070 | 0.108 | — | 0.253 | −0.862 | −0.328 | −0.431 |
| Spl 5 | 0.210 | 0.328 | 0.282 | −0.253 | — | −0.108 | −0.405 | −0.140 |
| Spl 6 | 0.140 | 0.405 | 0.431 | 0.862 | 0.108 | — | 0.180 | −0.140 |
| Spl 7 | 0.355 | 0.210 | 0.508 | 0.328 | 0.405 | −0.180 | — | 0.000 |
| Spl 8 | 0.282 | 0.282 | 0.589 | 0.431 | 0.140 | 0.140 | 0.000 | — |
| Sum | 1.349 | 1.436 | 1.848 | 0.759 | 0.085 | −1.985 | −1.627 | −1.864 |
| Average ($S'$) | 0.193 | 0.205 | 0.264 | 0.108 | 0.012 | −0.284 | −0.232 | −0.266 |

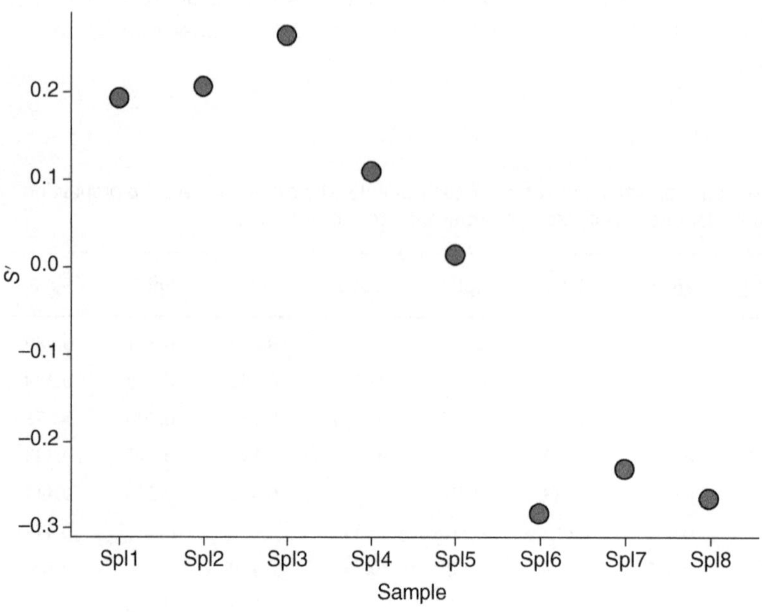


**Figure 12.5** $S'$ values for saltiness across all eight products.

Next, we will seek to determine if our model fits the data through a $\chi^2$ test that compares our fitted values to expected values. The first step to this process involves creating a matrix in which we subtract the $S'_j$ values from the $S'_i$. The results are shown in Table 12.8.

The next step involves converting our cell values back to probabilities which can be done using the *pnorm()* function in R or the NORMDIST function in Excel. The result should look as in Table 12.9.

The final step is to calculate our sum of squares and use that value to test our model for statistical significance. To calculate the sum of squares, we first perform an arc sine transformation of the square root of each value in our $p'$ and $p''$ matrices. The arc sine transformation can be performed using the arcsin function in R or Excel. Be sure to calculate the new values in degrees, as both software use default units of radians. In Excel, this calculation will use the DEGREES function, which will look like this:

= DEGREES (ASIN(x)), where $x$, is the square root of the $p'$ or $p''$ value.

**Table 12.8** Difference between $S'_j$ and $S'_i$ values.

| | Spl 1 | Spl 2 | Spl 3 | Spl 4 | Spl 5 | Spl 6 | Spl 7 | Spl 8 |
|---|---|---|---|---|---|---|---|---|
| Spl 1 | — | | | | | | | |
| Spl 2 | 0.01 | — | | | | | | |
| Spl 3 | 0.07 | 0.06 | — | | | | | |
| Spl 4 | −0.08 | −0.10 | −0.16 | — | | | | |
| Spl 5 | −0.18 | −0.19 | −0.25 | −0.10 | — | | | |
| Spl 6 | −0.48 | −0.49 | −0.55 | −0.39 | −0.30 | — | | |
| Spl 7 | −0.43 | −0.44 | −0.50 | −0.34 | −0.24 | 0.05 | — | |
| Spl 8 | −0.46 | −0.47 | −0.53 | −0.38 | −0.28 | 0.02 | −0.03 | — |

**Table 12.9** Fitted probabilities, where the cell values are the proportion of times the column sample was expected to be saltier than the row sample.

| | Spl 1 | Spl 2 | Spl 3 | Spl 4 | Spl 5 | Spl 6 | Spl 7 | Spl 8 |
|---|---|---|---|---|---|---|---|---|
| Spl 1 | — | | | | | | | |
| Spl 2 | 0.50 | — | | | | | | |
| Spl 3 | 0.53 | 0.52 | — | | | | | |
| Spl 4 | 0.47 | 0.46 | 0.44 | — | | | | |
| Spl 5 | 0.43 | 0.42 | 0.40 | 0.46 | — | | | |
| Spl 6 | 0.32 | 0.31 | 0.29 | 0.35 | 0.38 | — | | |
| Spl 7 | 0.34 | 0.33 | 0.31 | 0.37 | 0.40 | 0.52 | — | |
| Spl 8 | 0.32 | 0.32 | 0.30 | 0.35 | 0.39 | 0.51 | 0.49 | — |

In R:

```
asin(x)*180/pi, where x is the square root of the p′ or
p″ value.
```

After completing the arc sine transformations of the observed probabilities, your data should look like the values found in Table 12.10.

After completing the arc sine transformations of the fitted probabilities, your data should look like the values found in Table 12.11.

Lastly, it should be noted that the same calculations can be made for 3-AFC or other discrimination tasks where the expected proportion of votes for a sample is not 50%. In this case, the arc sine transformation can be forgone (Mosteller 1951c).

The next step is to square the differences in between the arcsine fitted probabilities and the arc-sine observed probabilities (Table 12.12).

If the squares are added up, the value of 5696.805 is obtained, and this is our sum of squares. To calculate our $\chi^2$ value, we divide our sum of squares by 821/observations in each cell. In our case, our $\chi^2$ value is $\chi^2 = 5696.805/(821/45)$ or 312.2488. While our calculated $\chi^2$ value is very high, meaning that there is strong evidence that our fitted values do not align

**Table 12.10** Arc sine treated squares of observed probabilities ($p'$).

| | Spl 1 | Spl 2 | Spl 3 | Spl 4 | Spl 5 | Spl 6 | Spl 7 | Spl 8 |
|---|---|---|---|---|---|---|---|---|
| Spl 1 | — | | | | | | | |
| Spl 2 | 46.605 | — | | | | | | |
| Spl 3 | 41.785 | 49.778 | — | | | | | |
| Spl 4 | 54.756 | 46.605 | 47.467 | — | | | | |
| Spl 5 | 49.778 | 52.476 | 51.413 | 39.232 | — | | | |
| Spl 6 | 48.215 | 54.150 | 54.756 | 63.867 | 47.467 | — | | |
| Spl 7 | 53.070 | 49.778 | 56.415 | 52.476 | 54.150 | 40.918 | — | |
| Spl 8 | 51.413 | 51.413 | 58.180 | 54.756 | 48.215 | 48.215 | 45.000 | — |

**Table 12.11** Arc sine treated squares of fitted probabilities ($p''$).

| | Spl 1 | Spl 2 | Spl 3 | Spl 4 | Spl 5 | Spl 6 | Spl 7 | Spl 8 |
|---|---|---|---|---|---|---|---|---|
| Spl 1 | — | | | | | | | |
| Spl 2 | 45.279 | — | | | | | | |
| Spl 3 | 46.631 | 46.352 | — | | | | | |
| Spl 4 | 43.080 | 42.801 | 41.453 | — | | | | |
| Spl 5 | 40.883 | 40.606 | 39.265 | 42.797 | — | | | |
| Spl 6 | 34.257 | 33.990 | 32.702 | 36.113 | 38.270 | — | | |
| Spl 7 | 35.387 | 35.117 | 33.817 | 37.257 | 39.427 | 46.172 | — | |
| Spl 8 | 34.634 | 34.366 | 33.074 | 36.495 | 38.657 | 45.392 | 44.220 | — |

**Table 12.12** Difference in arc sine treated squared probabilities.

| | Spl 1 | Spl 2 | Spl 3 | Spl 4 | Spl 5 | Spl 6 | Spl 7 | Spl 8 |
|---|---|---|---|---|---|---|---|---|
| Spl 1 | — | — | — | — | — | — | — | — |
| Spl 2 | 1.759 | — | — | — | — | — | — | — |
| Spl 3 | 23.488 | 11.732 | — | — | — | — | — | — |
| Spl 4 | 136.331 | 14.469 | 36.171 | — | — | — | — | — |
| Spl 5 | 79.122 | 140.902 | 147.582 | 12.711 | — | — | — | — |
| Spl 6 | 194.821 | 406.425 | 486.355 | 770.278 | 84.574 | — | — | — |
| Spl 7 | 312.711 | 214.928 | 510.653 | 231.613 | 216.776 | 27.601 | — | — |
| Spl 8 | 281.531 | 290.597 | 630.277 | 333.454 | 91.365 | 7.969 | 0.608 | — |

well with our actual values, we may still want to calculate a *p*-value for our model fit. The *p*-value can be obtained using software or a table, where the degrees of freedom equals the number of observations per cell. In our case, our degrees of freedom are 44 (45 – 1). For those using R, the *p*-value can be calculated using the pchisq() function. In Excel, the *p*-value can be calculated using the CHI.DIST function.

This calculation reveals a *p*-value of $4.1 \times 10^{-42}$, indicating it is extremely unlikely that our model fits the data. In plain terms, the principle of transitivity was not present in our data, or that participants were not consistent in their judgments of saltiness. In this case, we could resort to a series of binary comparisons in saltiness or choose a method not dependent on transitivity (e.g., Friedman rank analysis), as the validity of the individual paired comparisons is not in question.

## 12.8 Future Developments

Attribute-specific difference tests are an important sensory tool. While not as commonly used as their unspecified counterparts, attribute-specific difference tests can be the best option if the nature of the hypothesized difference is known. The utility of attribute-specific difference tests is increased with the addition of a Thurstonian approach to analyzing the data. Additionally, multisample designs add another layer of functionality to attribute-specific difference tests. These multisample designs will only gain more analyses options as the methods discussed here become more user friendly and are integrated into more statistical analysis/sensory software. As mentioned earlier and highlighted by Case Study III, much of the ability to do these types of analyses is dependent on computing accessibility. Maximizing the insights gleaned from choice-based studies is of high interest across many scientific disciplines (Neumann et al. 2011; Clark et al. 2018; Hopper et al. 2019), which will impact sensory evaluation. It seems inevitable that more statistical frameworks are popularized for use with these types of data, including Bayesian methods (Issa Mattos and Martins Silva Ramos 2022). Additionally, as peripheral fields, such as chemosensory psychophysics adopt these methods, sensory science will see the diverse capabilities of choice-based studies such as attribute-specific discrimination tests (Weir et al. 2022).

## References


Adjei, M.Y.B. (2017). Chapter 4—Applications and Limitations of Discrimination Testing. In: *Discrimination Testing in Sensory Science* (ed. L. Rogers), 85–105. Woodhead Publishing. https://doi.org/10.1016/B978-0-08-101009-9.00004-6.

Agresti, A. (2018). *An Introduction to Categorical Data Analysis*. Wiley.

Albers, P.C.H. and de Vries, H. (2001). Elo-rating as a tool in the sequential estimation of dominance strengths. *Anim. Behav. 61* (2): 489–495. https://doi.org/10.1006/anbe.2000.1571.

Bradley, R.A. and Terry, M.E. (1952). Rank analysis of incomplete block designs: I. The method of paired comparisons. *Biometrika 39* (3/4): 324–345. https://doi.org/10.2307/2334029.

Brockhoff, P.B. and Linander, C.B. (2017). Chapter 15 – Analysis of the data using the R package sensR. In: *Discrimination Testing in Sensory Science* (ed. L. Rogers), 303–344. Woodhead Publishing https://doi.org/10.1016/B978-0-08-101009-9.00015-0.

Carr, B.T. (1985). Statistical models for paired-comparison data. *American society for quality control 39th congress transactions* (pp. 295–300). Baltimore, MD: American Society for Quality.

Civille, G.V. and Carr, B.T. (2015). *Sensory Evaluation Techniques*. CRC Press.

Clark, A.P., Howard, K.L., Woods, A.T. et al. (2018). Why rate when you could compare? Using the "EloChoice" package to assess pairwise comparisons of perceived physical strength. *PLoS One 13* (1): e0190393. https://doi.org/10.1371/journal.pone.0190393.

Elo, A. E. (1978). *The rating of chessplayers, past and present*, New York, NY: Arco Pub.

Ennis, J.M., Rousseau, B., and Ennis, D.M. (2014). Sensory difference tests as measurement instruments: a review of recent advances. *J. Sens. Stud. 29* (2): 89–102. https://doi.org/10.1111/joss.12086.

Ford, R.A. (2017). Chapter 3 – Deciding which test to use in discrimination testing. In: *Discrimination Testing in Sensory Science* (ed. L. Rogers), 67–83. Woodhead Publishing. https://doi.org/10.1016/B978-0-08-101009-9.00003-4.

Frijters, J.E.R. (1979). The paradox of discriminatory nondiscriminators resolved. *Chem. Senses 4* (4): 355–358. https://doi.org/10.1093/chemse/4.4.355.

Glickman, M. E. (1999). Parameter estimation in large dynamic paired comparison experiments. *Journal of the Royal Statistical Society: Series C (Applied Statistics), 48* (3): 377–394

Green, D.M. and Swets, J.A. (1966). *Signal Detection Theory and Psychophysics*, vol. 1. New York: Wiley.

Gridgeman, N.T. (1955). The Bradley-Terry probability model and preference tasting. *Biometrics 11* (3): 335–343. https://doi.org/10.2307/3001772.

Holm, S. (1979). A simple sequentially rejective multiple test procedure. *Scand. J. Stat. 6* (2): 65–70.

Hopkins, J.W. (1954). Incomplete block rank analysis: some taste test results. *Biometrics 10* (3): 391–399. https://doi.org/10.2307/3001593.

Hopper, L.M., Egelkamp, C.L., Fidino, M., and Ross, S.R. (2019). An assessment of touchscreens for testing primate food preferences and valuations. *Behav. Res. Methods 51* (2): 639–650.

Issa Mattos, D. and Martins Silva Ramos, É. (2022). Bayesian paired comparison with the bpcs package. *Behav. Res. Methods 54*: 2025–2045.

Luckett, C.R., Burns, S.L., and Jenkinson, L. (2020). Estimates of relative acceptability from paired preference tests. *J. Sens. Stud. 35* (5): e12593. https://doi.org/10.1111/joss.12593.

Meilgaard, M.C., Carr, B.T., and Civille, G.V. (2006). *Sensory Evaluation Techniques*, 4e. CRC Press.

Mosteller, F. (1951a). Remarks on the method of paired comparisons: I. The least squares solution assuming equal standard deviations and equal correlations. *Psychometrika 16* (1): 3–9.

Mosteller, F. (1951b). Remarks on the method of paired comparisons: II. The effect of an aberrant standard deviation when equal standard deviations and equal correlations are assumed. *Psychometrika 16* (2): 203–206. https://doi.org/10.1007/BF02289115.

Mosteller, F. (1951c). Remarks on the method of paired comparisons: III. A test of significance for paired comparisons when equal standard deviations and equal correlations are assumed. *Psychometrika 16* (2): 207–218. https://doi.org/10.1007/BF02289116.

Neumann, C. (2015). EloChoice: Preference rating for visual stimuli based on elo ratings. https://Cran. r-Project.Org/Web/Packages/EloChoice/Index.Html (Accessed March 1, 2023).

Neumann, C., Duboscq, J., Dubuc, C. et al. (2011). Assessing dominance hierarchies: validation and advantages of progressive evaluation with Elo-rating. *Anim. Behav. 82* (4): 911–921. https://doi.org/10.1016/j.anbehav.2011.07.016.

O'Mahony, M., Masuoka, S., and Ishii, R. (1994). A theoretical note on difference tests: models, paradoxes and cognitive strategies. *J. Sens. Stud. 9* (3): 247–272. https://doi.org/10.1111/j.1745-459X.1994.tb00246.x.

RC Team (2020). *R: A Language and Environment for Statistical Computing (Version 4.0.2).* R Foundation for Statistical Computing.

Rogers, L. (2017). *Discrimination Testing in Sensory Science: A Practical Handbook.* Woodhead Publishing.

Rousseau, B. and Ennis, J.M. (2013). Importance of correct instructions in the tetrad test. *J. Sens. Stud. 28* (4): 264–269.

Schlich, P. (1993). Risk tables for discrimination tests. *Food Qual. Prefer. 4* (3): 141–151.

Séménou, M. (2015). CompR: paired comparison data analysis. R package version 1.0.

Stone, H. and Sidel, J.L. (1978). Computing exact probabilities in sensory discrimination tests. *J. Food Sci. 43* (3): 1028–1029.

Terry, M.E., Bradley, R.A., and Davis, L.L. (1952). New designs and techniques for organoleptic testing. *Food Technol. 6* (7): 250–254.

Thurstone, L. L. (1927). A law of comparative judgment. *Psychol. Rev.*, *34*: 273–286.

Turner, H. and Firth, D. (2012). Bradley-Terry models in R: the BradleyTerry2 package. *J. Stat. Softw.* 48 (1): Article 1. https://doi.org/10.18637/jss.v048.i09.

Weir, E. M., Hannum, M. E., Reed, D. R., Joseph, P. V., Munger, S. D., Hayes, J. E., & Gerkin, R. C. (2022). The Adaptive Olfactory Measure of Threshold (ArOMa-T): A rapid test of olfactory function. *MedRxiv*.

Yang, Q. and Ng, M.L. (2017). Chapter 5 – Paired comparison/directional difference test/2-Alternative Forced Choice (2-AFC) test, simple difference test/same–different test. In: *Discrimination Testing in Sensory Science* (ed. L. Rogers), 109–134. Woodhead Publishing. https://doi.org/10.1016/B978-0-08-101009-9.00005-8.

# 13

# Summary

*Sarah E. Kemp[1], Tracey Hollowood[2], Joanne Hort[3] and Lauren Rogers[4]*

[1] *Consultant and formerly Head of Global Sensory and Consumer Guidance, Cadbury Schweppes, UK*
[2] *Formerly Managing Director of Sensory Dimensions Ltd, UK*
[3] *Food Experience and Sensory Testing Laboratory (Feast), Massey University, Palmerston North, New Zealand*
[4] *Freelance Sensory Scientist, UK*

## 13.1 Introduction

In 2022, a survey was conducted for the ASTM to establish the new "Selecting Discrimination Methods" standard. The survey collected responses from 74 participants (refer to Figure 13.1) and the findings indicate that a significant majority (73%) of the participants choose a discrimination method based on its alignment with the test objective, or they consider the alpha and beta risks, rather than relying solely on a single method (9%) as a solution to all discrimination testing requests. The process of selecting the most appropriate test method is not always easy or obvious. For an interesting discussion about the choice of sensory test related to the objective(s) of the research, see Lawless and Heymann (2010, pp. 8–10). There are many different factors that come into play, and this chapter addresses these through references to the other chapters in this book.

It is worth noting that not all discrimination tests are driven by product-related objectives. Sensory discrimination tests find applications in various contexts, including basic and medical research: for example, in COVID-19 research (see Hunter et al. 2023). They are used for determining a panellist's threshold, prescreening individuals for further tests, panellist training and monitoring, or simply gaining insights into human behavior and sensitivity (Sabiniewicz et al. 2021; Garcia-Burgos et al. 2022; Norton et al. 2022; ISO 2023).

## 13.2 Overall Comparison of Methods

Sensory discrimination tests encompass a range of categories, as illustrated in Table 13.1. The predominant approach in these tests involves comparing two products, typically denoted as "A" and "B" (Category 1). For instance, in the tetrad test, although the assessor is presented with four samples, only two distinct products, A and B, are involved. "A" may represent the reference or gold standard product, while "B" could be a modified

*Discrimination Testing in Sensory Evaluation*, First Edition. Edited by Lauren Rogers, Joanne Hort, Sarah E. Kemp and Tracey Hollowood.

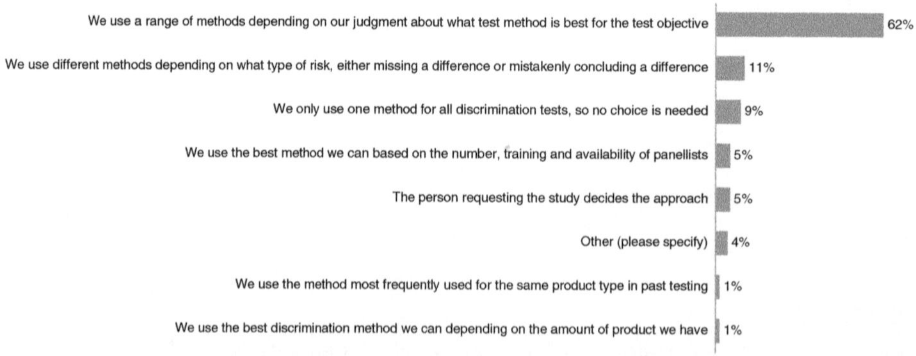


**Figure 13.1** ASTM Survey Q1. Which one of the following best describes how you select a discrimination method?

**Table 13.1** Categories of discrimination tests.

| Category | Example |
|---|---|
| 1. The number of *products* presented to the panellist in one test | The majority of sensory discrimination tests involve two products; however, some tests involve multiple products, e.g. ranking, difference from control, and polyhedral tests |
| 2. The number of *samples* presented to the panellist in one test | From 1 to 12, e.g. 1 sample in some presentations of the A-not A test, through to 12 samples in the six-out-of-twelve test |
| 3. Asymmetrical or symmetrical | The tetrad is an example of a symmetrical test (AA-BB) while the triangle is an example of an asymmetrical test (A-BB or B-AA) |
| 4. Specified or unspecified | Whether the test has a specified attribute (e.g. saltiness or stickiness) or not |
| 5. Reference present or not | If a reference or control sample is identified as such in the test. For example, the duo-trio test has a reference while the ABX test (Greenaway 2017) does not |
| 6. The panellist task | The way in which the assessor is asked to make the judgment |
| 7. If there is a response bias inherent in the test | The differences between subjects in terms of their criterion for stating whether there was a difference or not |
| 8. If a rating scale is used as part of the test | For some tests, an additional rating scale is used |

version, such as a reduced-salt or reduced-fat variation. Each product in the tetrad test is therefore represented by two samples, which are generally sourced from the same batch.

The number of samples[1] employed in these tests (Category 2) can vary significantly. Some tests utilize as few as one sample, as seen in some iterations of the A-not A test, or two samples

1 Samples refer to the quantity of items categorized as "A" and "B" provided to the panellists. These items can include glasses of drinks, plates of biscuits, pots containing odors, laundry swatches, hair switches, and similar articles. In the context of the triangle and duo-trio tests, the number of samples presented to the panellists would be three. For the two-out-of-five test, the number of samples would be five.

for the same-different test (Yang and Ng 2017). Other tests employ five samples, as exemplified by the two-out-of-five test (Billson 2017). The number of samples can extend to 12, as observed in the 6-out-of-12 test, or can be adapted to the practical limits of an assessor's capability within a single test. The distribution of samples within each test can be asymmetrical, such as in the AAA-B configuration of the one-out-of-four test (also known as 4-alternative forced choice or 4-AFC), or symmetrical, such as in the AA-BB arrangement of the tetrad test (Sanderson 2017) or dual-standard test (Vietoris 2017) (Category 3).

One of the key distinctions among sensory discrimination tests lies in whether the attribute of difference is explicitly specified (Category 4). For instance, in the 3-AFC (3-alternative forced choice) test, assessors may be required to identify the sample that exhibits the most pronounced bitterness, with bitterness serving as the specified attribute. On the other hand, the triangle test entails assessors identifying the odd one out of three samples, without any specific attribute being provided. Consequently, the triangle test represents the unspecified version of the 3-AFC test.

Certain discrimination tests incorporate a reference within the test itself (Category 5). Consider the duo-trio test (Purcell 2017), where assessors are presented with three samples, two of which are identified to the assessor with 3-digit codes, while one sample is explicitly labelled as the control or reference. One of the two coded samples is identical to the control, or reference, sample. The third sample represents the product being assessed, such as the reduced-fat version of the reference product.

The task assigned to assessors (Category 6) can vary across different discrimination tests. For example, in the same-different test, assessors provide a binary response of yes or no to indicate whether they perceive a difference between samples. In the duo-trio test, assessors match the samples to a reference or control. In the triangle test, assessors select the sample that differs from the others or appears as the odd one out. In the 3-AFC test, assessors identify the sample with the most intense attribute. While in the tetrad test, assessors categorize samples into groups or pairs.

Response bias (Category 7) is important when planning discrimination testing approaches (O'Mahony 1992). Response bias relates to the decision-making process employed by assessors when determining whether a product exhibits a difference or not. Response bias is not directly linked to the assessor's sensitivity; rather, it relates to their thought processes and individual personality. Different assessors may adopt varying approaches, with some being more cautious and stringent in their decision-making, ensuring they can confidently detect differences before reporting them, while others may have a more relaxed or less cautious approach, potentially reporting differences that the more cautious assessors might not. This decision approach can fluctuate from day to day, depending, for example, on the assessor's mood. Consequently, it becomes crucial to design tests in a manner that minimizes the influence of response bias. This is a key reason for the implementation of forced choice in the majority of discrimination tests. For instance, in the triangle test, assessors are explicitly informed that two of the samples are identical, while one sample differs. By presenting the test in this manner, response bias is controlled, as it removes the assessor's judgment regarding when to identify a sample as different or not. If the panellists were told that one of the samples *might* be different, then response bias would not be controlled, as it would again be left to the panellist's judgment about when to say a sample was different or not.

This discussion leads to the final category (8): the presence of a rating scale within the test. The "R-index" is one such rating scale where assessors are asked to rate their confidence level in their response. It assists sensory scientists in accounting for the effects of response bias. In tests like the same-different test, where assessors are presented with two samples and asked to determine if they are the same or different, they are also provided with additional options to indicate their level of confidence. Typically, a four- or six-point scale is used, using responses such as "the two samples are different, and I am sure," "the two samples are different, and I am not sure," "the two samples are the same, and I am not sure," and "the two samples are the same, and I am sure." Another rating scale is employed in the difference from control (DFC) test, where assessors rate the degree of difference between a sample and a labelled control sample. This test is also referred to as the degree of difference (DOD) test.

## 13.3 Current/Recent Developments

### 13.3.1 Which Test?

It would be beneficial for test users to explicitly describe their testing procedures, allowing them to align with established standards or, if variations exist, to appropriately rename the test. An illustrative example is found in a recent interesting study by He et al. (2023), where the authors mentioned using the 3-AFC and 4-AFC methods. However, their actual approach involved asking panellists to identify the odd sample, aligning more closely with the triangle and one-out-of-four tests. Had the authors based the calculations of the number of participants required on the 3-AFC sample size tables, their study might have been significantly underpowered due to the inherent differences in the test structure. A clear definition of testing processes not only ensures accuracy in reporting but also guards against potential underestimation of statistical power in research studies.

### 13.3.2 Analysis of Discrimination Tests

The use of critical value tables gives the sensory scientist a quick and easy way to check for statistical significance. However, the use of sensory software can give so much more information than the simple "yes there is a statistically significant difference." First, the software can help design the study, helping the scientist determine the number of participants needed to meet the objective, but it can also provide the actual calculated $p$-value which can give the scientist more information than the simple cut-off achieved from tables. Additional information such as confidence intervals and $d'$ ("dee prime") can help the sensory scientist interpret the data in relation to the action standards and make the correct decisions based on the data collected. A package built for the open source software R (R Core Team 2020) allows anyone to analyze the outputs of their discrimination tests as quickly and easily as looking up the results in a critical values table. For a very useful introduction to using R for discrimination tests, please refer to Brockhoff and Linander (2017) and Chapter 8 of this book. However, a few examples are shown below which were carried out in R Studio (https://www.rstudio.com/). The first example shows how the critical value for a triangle or tetrad can be found. To find the critical value for other tests, simply change the p0 value accordingly.

```
library(sensR)
findcr(78, alpha = 0.05, p0 = 1/3)
[1] 34
```

The second example gives the *p*-value (0.1992) and additional information such as *d′* (0.76524):

```
discrim(30, 78, method = "triangle")
Estimates for the triangle discrimination protocol with
30 correct
answers in 78 trials. One-sided p-value and 95 % two-sided
confidence intervals are based on the 'exact' binomial test.

        Estimate Std. Error  Lower  Upper
pc       0.38462    0.05509 0.3333 0.5017
pd       0.07692    0.08263 0.0000 0.2525
d-prime  0.76524    0.43145 0.0000 1.4750

Result of difference test:
'exact' binomial test:  p-value = 0.1992
Alternative hypothesis: d-prime is greater than 0
```

If using the test to determine similarity using a proportion of discriminators approach (pd) at 30% (i.e. the two products are similar if no more than 30% can detect a difference in the population), we get a critical value of 33.

```
findcr(78, alpha = 0.05, p0 = 1/3, pd0 = 0.30, test =
"similarity")
[1] 33
```

### 13.3.3 Difference or Similarity Testing

Often, sensory discrimination tests are conducted to determine if a difference exists between two products (Kemp et al. 2009). For example, a company may be attempting to improve the creaminess of a low-fat yoghurt, the absorption of a skincare cream, or the effectiveness of a worktop cleaner. In each of these cases, the discrimination test requester is hoping that the null hypothesis for the test can be rejected and that the samples will be different: the yoghurt is perceptibly creamier, the cream absorbs more easily, or the worktop cleaner is actually better at cleaning. In the ASTM survey referred to earlier, 56% of the respondents stated that their main testing objective was testing for difference, while 23% stated it was for similarity (see Figure 13.2). Eighteen percent of respondents said that the reasons for testing were equally split between difference and similarity testing.

There are several situations in which businesses aim to maintain consistent sensory characteristics in their products and Chapter 5 discusses the practical approaches for using

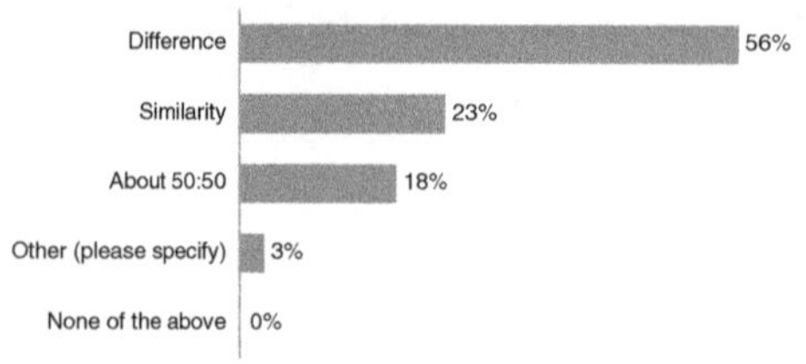


**Figure 13.2** ASTM survey Q5. Are your discrimination tests mainly for difference or similarity?

discrimination tests to test for similarity (or equivalence). One important scenario is quality control, where daily production must deliver a consistent sensory experience despite process and production variations (Munoz 2013). Another use of similarity testing is the stability of sensory characteristics throughout a product's shelf life and discrimination tests are a vital part of the sensory scientist's toolbox for this application (Kilcast 2011). Businesses may also make changes in product formulations, such as using new suppliers, ingredients, or recipes and, again, they have to ensure these changes do not negatively impact the sensory experience for consumers. The business is really looking for information that the reformulated product is similar enough to be used interchangeably with the current.

Environmental consciousness drives companies to make eco-friendly products, processes, and packaging changes, which can impact sensory characteristics. Additionally, sensory similarity is sought after for innovation and marketing, allowing claims like "as good as brand X" or "less sugar, same great taste" (ISO 2021). Chapter 5 includes information to help the sensory scientist approach this important topic, such as test methods, panel size and type, and the use of historical data, especially linking discrimination testing data to other datasets. The chapter includes a discussion about the various practical and statistical approaches to designing and analyzing the results from similarity tests and finishes with a case study to show one of the approaches in action.

Similarity testing can be more complex than testing for a difference, because when we reject the null hypothesis in a difference test we know that it is likely that there is a difference, but when we are unable to reject the null hypothesis, all we know is that we were unable to reject the null hypothesis, not that the products in the test are "similar" (Bi 2005). In addition, on many occasions, the question to be answered is really "what is the size of the sensory difference and will consumers notice any difference?"

The ASTM have a very useful approach to difference and similarity testing which is detailed in ASTM E3009 (ASTM 2023) for the tetrad test, summarized nicely in the scope: "This test method covers a procedure for determining whether a perceptible sensory difference exists between samples of two products or to estimate the magnitude of the perceptible difference." The number of assessors for the test is chosen based on the level of sensitivity desired for the test, for example based on the Type I ($\alpha$ risk ) and Type II ($\beta$ risk) errors, as well as the Thurstonian measure of sensory difference ($\delta$). The standard also includes a useful classification of $\alpha$ and $\beta$ values and their relationship to the "level" of difference. For example an $\alpha$ of 0.10–0.05 "indicates slight evidence that a difference was apparent," while an $\alpha$ of 0.01–0.001 "indicates strong evidence that a difference was apparent." The same approach for the values for $\beta$ is also given. For example a $\beta$ of 0.10–0.05 "indicates slight evidence that a difference was *not* apparent," while an $\alpha$ of 0.01–0.001 "indicates strong evidence that a difference was *not* apparent." For $\delta$, there are three ranges: small (less than 0.5), medium (0.5–1.0), and large (greater than 1). A process change cost savings case study is included with the standard which shares a very useful approach for balancing the statistical risks, and tables are provided for choosing the relevant number of assessors. The case study uses $\alpha$ at 0.10, $\beta$ at 0.10 and the company's $\delta$ is set at 1.25. The ASTM publishes several discrimination testing standards (a full list can be found here: https://www.astm.org/get-involved/technical-committees/committee-e18/subcommittee-e18) and a new ASTM guide for selecting sensory discrimination methods is currently under development (https://www.astm.org/workitem-wk78520).

### 13.3.4 Which Assessor?

Determining the most suitable assessors for sensory discrimination tests is a critical consideration in the field of sensory science. The choice often boils down to whether to use consumers or opt for screened and trained panellists. In sensory science, we use consumers for many different hedonic objectives such as preference mapping, focus groups, and concept testing to name a few. However, analytical sensory tests, those generally with objective sensory aims, have traditionally been performed with screened, selected, and subsequently trained panellists. It is obvious that a small group of screened and trained panellists is unable to give the scientist reliable information in hedonic tests, but can consumers take part in analytical sensory tests? For an interesting debate about this area, see Ares and Varela 2017, as well as the comments and replies in Food Quality and Preference.

According to Ares and Varela (2017), the debate about who to use for analytical sensory tests was started by Moskowitz in 1996 when he argued that consumers were able to produce similar results to a trained panel in intensity rating of sensory attributes. The authors considered consumers for analytical sensory tests due to the development of various alternative and flexible methods for sensory characterization, such as the methods shared by Valentin et al. (2012). These "alternative" methods can be adapted to panels with different degrees of training, fuelling the debate about the use of untrained assessors for analytical tasks in sensory science. Additionally, the ecological validity implications around specific research questions and the need to account for individual differences in perception and decision making have also contributed to the consideration of consumers for analytical sensory tests.

There are several advantages and disadvantages of using consumers for analytical tests. When deciding between consumers and trained panellists (who have been screened for their sensory acuity and selected based on their abilities), several factors must be considered. Consumer panels, composed of individuals from the target market, offer insights into how the general public perceives the products' differences or similarities. Consumers are the ultimate judges of product acceptability. So, using them for analytical sensory tests can provide more real-world relevance and practical insights into product performance. This can be particularly relevant for market-driven industries seeking feedback aligned with consumer preferences. Consumer panels can provide a diverse range of sensory experiences and preferences, which can be useful for understanding the variability in consumer responses to products. Consumer panels may also be more cost effective for certain projects, as the methods they can take part in do not require them to have extensive training. With the fairly recent increase in remote testing, as well as technological advances, we might well be able to conduct discrimination testing in a more ecological manner in the home and with consumers.

On the other hand, screened and trained panellists bring a level of expertise and consistency, having undergone screening to check their sensory acuity and training to evaluate specific sensory attributes accurately. Consumers may lack the sensitivity and consistency achieved by trained panellists, which can lead to lower sensitivity and accuracy in detecting subtle differences among samples. Consumers may have individual preferences and biases that can influence their sensory evaluations, leading to greater variability in the results. This, in turn, may require greater numbers of participants for the test which can have hidden costs, such as the increase in the amount of sample required, the time for preparation and distribution, as well as the time involved in management of the study. In addition, and this is more applicable to the descriptive analytical sensory tests, consumers may have limited vocabulary to describe sensory characteristics, which can make it difficult to obtain detailed and precise sensory information. Consumer panels may not be suitable for all types of products, particularly those with complex sensory characteristics or specific off-flavors that consumers may not be able to detect.

The cost-effectiveness of each approach, the desired level of expertise, the product type, and the nature of the sensory attributes being evaluated all play a role in determining the most suitable assessor type for a given sensory discrimination test. Striking the right balance between these two assessor types depends on the objectives of the study and the specific requirements of the industry in question (Meiselman et al. 2022). One useful approach is to use both trained panellists and consumers, linking their outputs through the use of a standardized measure of sensory difference ($d'$) that determines the threshold at which a sensory difference becomes significant enough to generate a meaningful consumer response (Rousseau 2015).

## 13.4 Future

### 13.4.1 Ethical Considerations in Sensory Discrimination Testing

Ethical considerations in sensory discrimination testing emphasize the importance of obtaining informed consent and respecting participant rights. Prior to conducting any sensory tests, researchers should provide clear and comprehensive information to participants about the purpose, procedures, potential risks and benefits, and their rights regarding participation. However, researchers should strive to minimize biases in the information provided, study design, data collection, and analysis. This includes employing randomization techniques, blinding procedures, and balanced sample presentation to mitigate any systematic effects.

Informed consent ensures that participants have a full understanding of the study and voluntarily agree to take part. Researchers should also inform participants of their right to withdraw from the study at any time without penalty or consequences. Confidentiality and anonymity are vital ethical considerations in sensory discrimination testing to protect participants' privacy and maintain data confidentiality. Researchers must implement measures to safeguard participants' personal information and ensure that their identities remain confidential throughout the study. This includes storing and handling data securely, using coded or anonymous participant identifiers, and avoiding the disclosure of sensitive information to unauthorized individuals. Additionally, when reporting study findings, researchers should ensure that individual participant data cannot be identified and take necessary precautions to maintain anonymity.

### 13.4.2 Experimental Design of Discrimination Tests

Experimental design plays a crucial role in sensory discrimination testing, ensuring reliable and meaningful results. By carefully considering factors such as sample selection, randomization, and balancing techniques, researchers can minimize biases and confounding variables that may influence sensory discrimination outcomes. Proper experimental design helps maximize the statistical power of the tests and allows for the accurate interpretation of the results. Chapter 3 includes an overview of a wide range of sensory discrimination tests and how each test should be conducted and analyzed. The references within Chapter 3 give detailed information about the wording for each discrimination test, as the wording of the discrimination test can impact the manner in which the participant makes their decision and thus the end results (Castura et al. 2018). It will be interesting to see further research on the precise wording and instructions given for all discrimination tests to enable standard developers to construct relevant and reliable guidelines for the sensory scientist to follow.

There are several discrimination tests that require product familiarization prior to, or during, the test, such as the A-Not A with reminder. As the authors of Chapter 9 suggest, more research is required into the protocols of these reminder-type tests, especially when the research is interested in understanding both preference and discrimination ability.

### 13.4.3 Replicated Discrimination Testing

Discrimination tests are often duplicated or even triplicated to make the best use of the resources available, and this area is likely to grow due to budgetary and sustainability issues (Meyners et al. 2023). The number of panellists required to meet the test objectives, particularly in similarity testing, can seem quite daunting, and therefore repeating the test with the number of panellists available seems like a good solution. However, the scientist needs to take into account how they will analyze the data, as simply converting, say 30 panellists conducting a test twice into 60 panellists and using critical value tables, can lead to erroneous conclusions. Chapter 7 includes incredibly useful approaches for the design and analysis of replicated discrimination tests as well as an excellent glossary of statistical terms and notations.

In a related area, the use of historical discrimination testing data alongside machine learning techniques could allow sensory scientists to compare ingredients, recipes, and formulas globally, and possibly predict the outcomes for new products and reformulations. Using these types of techniques could also offer the sensory scientist detailed information about human sensitivities to specific product components. More information on this topic can be found in Chapters 7 and 10.

### 13.4.4 New Discrimination Tests

The tetrad test is seen as a "recent" development in sensory science, however, the test has been around for many years as discussed in Chapter 1 (Wood 1949). There are more than 40 discrimination tests currently available with around 10 in common use (see Table 13.2).

**Table 13.2** Some example discrimination test methods.

| Discrimination tests (bold if in common use in sensory science) | |
|---|---|
| **Paired comparison** | **Same-different** |
| **2-Alternative Forced Choice (AFC)** | 2-AFC with Reminder (2-AFCR) |
| 2-AC | 2AFC (matching task) |
| Self-defined 2-AFC | 3-AC |
| 3-AFC | 4-AC |
| 4-AFC | 5-AC |
| 5-AFC | 6-AC (and so on through x-AC) |
| 6-AFC (and so on through x-AFC) | **A-Not A** |
| One-out-of-four | One-out-of-five (and so on) |
| **Triangle** | A-Not A with Reminder (A-Not AR) |
| **Duo-trio constant reference** | Ratings of A-Not A |
| **Duo-trio balanced reference** | Specified tetrad |
| Dual reference duo trio | **Unspecified tetrad** |
| Dual-pair | ABX discrimination task |
| Two-out-of-five | **Difference from control/degree of difference** |
| **Ranking** | Multiple standards |
| Dual-standard | Multiple pairs |
| Three-out-of-six/hexagon | Dual-pair/A interval AX |
| Multiple comparison | Degree of satisfaction-difference |

It is interesting to consider the potential for new tests to be developed or even adjustments to current tests such as warm-up procedures and matching tasks as opposed to selection tasks (as used in the 2AFC in Chapter 6). The increased use of attribute-specific tests, especially the option of exploiting multisample designs, is an interesting build on current protocols, and with the benefit of new analysis methods using the information from Chapter 12, can only improve the sensory scientist's toolbox.

Other options might include using the panellists' learnings from a test to build a subsequent test, such as a triangle followed by a 3-AFC once the results of the triangle have been analyzed. As the author of Chapter 5 concludes, new discrimination methods are needed for heterogeneous products such as snacks and meals, and products that change over consumption such as breakfast cereals and chewing gums.

### 13.4.5 Participants in Discrimination Tests

Chapter 10 of this book and Gómez-Corona (2020) introduced the idea of using blind and visually impaired panellists, persons with disabilities, as well as panellists from vulnerable social conditions for sensory discrimination tests. This is an interesting and socially responsible idea, and it will be interesting to see new publications investigating this area further. The further development of conducting discrimination testing with children, as outlined in Chapter 11, will benefit markets that produce products for the younger age groups. This idea can easily be extended to products developed for the older generation, therefore, we will hopefully see more research in this area. Using discrimination testing for cross-cultural research, as suggested by the authors of Chapter 7, by applying R-index procedures, could help with the understanding of global sensitivities – very useful for products that are available across the world. Developing this further, using discrimination testing to monitor sensitivities and preferences over time, will further aid companies that manufacture products across different countries.

## 13.5 Conclusions

Sensory discrimination tests have been in use since the beginning of the last century and are likely to stay as an important part of the sensory scientist's toolbox. Their use spans so many applications – from product design and development through to manufacturing and shelf life – it is obvious from the development of this book that discrimination tests are here to stay. Sensory discrimination testing is a fundamental tool in sensory science, enabling researchers and industry professionals to make informed decisions regarding assessor selection, product development, quality control, and consumer preferences, as well as investigating sensitivity and sensory acuity. By employing appropriate sensory discrimination tests and robust statistical analysis, valuable insights can be gained to meet the demands and expectations of consumers and to further our knowledge of human sensitivity, eating behaviors, and health.

## References

Ares, G. and Varela, P. (2017). Trained vs. consumer panels for analytical testing: fuelling a long lasting debate in the field. *Food Qual. Prefer.* 61: 79–86.

ASTM International (2023). *Standard Test Method for Sensory Analysis – Tetrad Test (E3009).* ASTM International. https://www.astm.org/e3009-15e01.html.

Bi, J. (2005). Similarity testing in sensory and consumer research. *Food Qual. Prefer.* 16 (2): 139–149.

Billson, S. (2017). Chapter 8 – Two-out-of-five test. In: *Discrimination Testing in Sensory Science* (ed. L. Rogers), 171–181. Woodhead Publishing. https://doi.org/10.1016/B978-0-08-101009-9.00008-3.

Brockhoff, P.B. and Linander, C.B. (2017). Chapter 15 – Analysis of the data using the R package sensR. In: *Discrimination Testing in Sensory Science* (ed. L. Rogers), 303–344. Woodhead Publishing. https://doi.org/10.1016/B978-0-08-101009-9.00015-0.

Castura, J.C., King, S.K., and Phipps, K. (2018). How task instructions affect performance on the unspecified tetrad test. *J. Sens. Stud.* 33 (3): e12329.

Garcia-Burgos, D., Andres, F., Trier, S. et al. (2022). Discrimination of sweet-fat ingredients in people with weight-and eating-related problems using a signal detection theory. *J. Sens. Stud. 37* (6): e12783.

Gómez-Corona, C. (2020). Sensory and consumer research for good: a review on social responsibility. *Curr. Opin. Food Sci. 33*: 115–123.

Greenaway, R. (2017). Chapter 13 – ABX task. In: *Discrimination Testing in Sensory Science* (ed. L. Rogers), 267–288. Woodhead Publishing. https://doi.org/10.1016/B978-0-08-101009-9.00013-7.

He, Y., Chen, J., Shi, W. et al. (2023). Can non-volatile tastants be smelled during food oral processing? *Chem. Senses 48*: bjad028.

Hunter, S.R., Hannum, M.E., Pellegrino, R. et al. (2023). Proof-of-concept: SCENTinel 1.1 rapidly discriminates COVID-19-related olfactory disorders. *Chem. Senses* 48: bjad002.

ISO 20884 (2021). *Sensory Analysis – Guidance on Substantiation for Sensory and Consumer Product Claims*. Geneva: International Organisation for Standardisation.

ISO 8586 (2023). *Sensory Analysis— Selection and Training of Sensory Assessors*. Geneva: International Organisation for Standardisation.

Kemp, S.E., Hollowood, T., and Hort, J. (2009). *Sensory Evaluation, A Practical Handbook*. Wiley Blackwell.

Kilcast, D. (2011). Sensory evaluation methods for food shelf life assessment. *Food Beverage Stab. Shelf Life* https://doi.org/10.1533/9780857092540.2.350.

Lawless, H.T. and Heymann, H. (2010). *Sensory Evaluation of Food. Principles and Practices. Sensory Evaluation of Food – Principles and Practices*. New York: Springer.

Meiselman, H.L., Jaeger, S.R., Carr, B.T., and Churchill, A. (2022). Approaching 100 years of sensory and consumer science: developments and ongoing issues. *Food Qual. Prefer. 100*: 104614.

Meyners, M., Carr, B.T., and Kunert, J. (2023). Equivalence and non-inferiority tests using replicated discrimination and preference data. *J. Sens. Stud.* e12882. https://onlinelibrary.wiley.com/doi/abs/10.1111/joss.12882.

Moskowitz, H.R. (1996). Experts versus consumers: a comparison. *J. Sens. Stud.* 11: 19–37.

Munoz, A.M. (ed.) (2013). *Sensory Evaluation in Quality Control*. Springer Science & Business Media.

Norton, V., Lignou, S., Faka, M., and Methven, L. (2022). Individual variation in mouthfeel sensitivity: investigating influences of whey protein content, consumer age, food format and fat addition. *Food Qual. Prefer.* 101: 104638.

O'Mahony, M. (1992). Understanding discrimination tests: a user-friendly treatment of response bias, rating and ranking R-index tests and their relationship to signal detection. *J. Sens. Stud.* 7: 1–47. https://doi.org/10.1111/j.1745-459X.1992.tb00519.x.

Purcell, S. (2017). Chapter 10 – Duo-trio. In: *Discrimination Testing in Sensory Science* (ed. L. Rogers), 197–207. Woodhead Publishing. https://doi.org/10.1016/B978-0-08-101009-9.00010-1.

R Core Team (2020). *R: A Language and Environment for Statistical Computing*. Vienna: R Foundation for Statistical Computing. https://www.R-project.org/.

Rousseau, B. (2015). Sensory discrimination testing and consumer relevance. *Food Qual. Prefer.* 43: 122–125.

Sabiniewicz, A., Wrage, F., and Hummel, T. (2021). The impact of cognitive and noncognitive factors on odor discrimination performance. *J. Sens. Stud.* 36 (5): e12683.

Sanderson, T. (2017). Chapter 9 – Tetrad test. In: *Discrimination Testing in Sensory Science* (ed. L. Rogers), 183–195. Woodhead Publishing. https://doi.org/10.1016/B978-0-08-101009-9.00009-5.

Valentin, D., Chollet, S., Lelièvre, M., and Abdi, H. (2012). Quick and dirty but still pretty good: a review of new descriptive methods in food science. *Int. J. Food Sci. Technol.* 47: 1563–1578.

Vietoris, V. (2017). Chapter 14 – Dual-standard test. In: *Discrimination Testing in Sensory Science* (ed. L. Rogers), 289–301. Woodhead Publishing. https://doi.org/10.1016/B978-0-08-101009-9.00014-9.

Wood, E.C. (1949). Organoleptic tests in the food industry. III.-Some statistical considerations in organoleptic tests. *J. Chem. Technol. Biotechnol.* 68: 128–131. https://doi.org/10.1002/jctb.5000680411.

Yang, Q. and Ng, M.L. (2017). Chapter 5 – Paired comparison/directional difference test/2-alternative forced choice (2-AFC) test, simple difference test/same-different test. In: *Discrimination Testing in Sensory Science* (ed. L. Rogers), 109–134. Woodhead Publishing. https://doi.org/10.1016/B978-0-08-101009-9.00005-8.

# Index

Note: *Italic* page numbers refer to *figure* and **Bold** page numbers reference to **tables**.



*Discrimination Testing in Sensory Evaluation*, First Edition. Edited by Lauren Rogers, Joanne Hort, Sarah E. Kemp and Tracey Hollowood.

## g



## h



## i



## k

## o



## p